Recent Advances In Indian Entomology

Recent Advances In Indian Entomology

Edited by
O.P. Lal
Principal Scientist
Division of Entomology
Indian Agricultural Research Institute
New Delhi-110012

2016
Associated Publishing Comapny
A Division of
Astral International Pvt. Ltd.
New Delhi - 110 002

First Published, 1996
Reprinted, 2016

ISBN: 978-81-86580-17-2 (Hardbound)

Publisher's note:

Every possible effort has been made to ensure that the information contained in this book is accurate at the time of going to press, and the publisher and author cannot accept responsibility for any errors or omissions, however caused. No responsibility for loss or damage occasioned to any person acting, or refraining from action, as a result of the material in this publication can be accepted by the editor, the publisher or the author. The Publisher is not associated with any product or vendor mentioned in the book. The contents of this work are intended to further general scientific research, understanding and discussion only. Readers should consult with a specialist where appropriate.

Every effort has been made to trace the owners of copyright material used in this book, if any. The author and the publisher will be grateful for any omission brought to their notice for acknowledgement in the future editions of the book.

Published by : **Associated Publishing Company**
A Division of
Astral International Pvt. Ltd.
– ISO 9001:2008 Certified Company –
4760-61/23, Ansari Road, Darya Ganj
New Delhi-110 002
Ph. 011-43549197, 23278134
E-mail: info@astralint.com
Website: www.astralint.com

Lasertypeset : **Apna Computer,** New Delhi-110 005

Digitally Printed at : **Replika Press Pvt. Ltd.**

PRINTED IN INDIA

Foreword

Agriculture is the back bone of India for providing food to the growing population and raw materials to numerous industries. There are, however, many insect pests that cause tremendous losses in fields and stores, and transmit dangerous plant and human diseases. No doubt, pesticides have been playing a vital role in minimising the losses, but, irrational and overuse of pesticides has led to many problems like insecticide resistance, adverse effects on non-target organisms, resurgence of insect pests, health hazards and environmental pollution. Also, the changing cropping pattern with intensive farming and high inputs have changed the pest scenario in the country. In recent years, the use of synthetic pyrethroids had caused serious outbreaks of white fly on cotton in Andhra Pradesh, Gujarat and Tamil Nadu, and *Helicoverpa armigera* on cotton in Guntur and Prakasham districts of Andhra Pradesh. Twenty years ago, an estimated annual crop loss in India was about Rs 6000 crores which has now crossed Rs 20,000 crores. Therefore, not only for preventing the losses, but also for avoiding pollution, it is important to find effective control measures for a sustainable agriculture. There should be conscious efforts for Integrated Pest Management using all the available tactics of pest control. Prophylactic measures should be replaced by need based application of pesticides with proper surveillance and monitoring of insect pests.

Significant changes have taken place in the last 2 to 3 decades in entomological techniques to control various insect pests. It is in this context that this book on ***Recent Advances in Indian Entomology*** edited by Dr. O.P. Lal is a timely one. It has contributions from eminent scientists in their respective fields and covers wide areas like taxonomy, physiology, toxicology, ecology, biological control, varietal resistance etc. The recent developments made in the field of genetic engineering, remote sensing, semiochemicals and post harvest grain technology for their significant role in pest management have been included in this book. Besides, it has many other chapters rendering the publication broad based.

I am confident, the book will prove quite useful to all those engaged in the field of agriculture, extension, plant protection, environment and public health. It will help evolve ecologically compatible and economically sound pest management programmes for a sustainable agriculture.

The book is primarily based on the lectures delivered at the Summer Institute on Recent Advances in Entomology held from May 23 to June 11, 1994 at the Indian Agricultural Research Institute, New Delhi sponsored by the Indian Council of Agricultural Research. Dr. O.P. Lal, Principal Scientist, Division of Entomology and Director of the Summer Institute, deserves great appreciation and congratulations for having brought out this useful publication.

Prof. R.B. Singh
Director
Indian Agricultural Research Institute
New Delhi-110 012

New Delhi
October 10, 1995

Preface

After independence in 1947, the first problem before India was to provide food to its vast population as the food production was not sufficient at that time. But, thanks to the policy makers, administrators and scientists that agriculture was given the top priority in India and with the use of high yielding varieties, fertilizers, irrigation facilities, plant protection measures and good agronomic practices, India was able to achieve the Green Revolution. The total production of food grains which was only 50 million tonnes in the year 1950-51 has gone upto 186 million tonnes in the year 1994-95. This is a matter of pride for Indian agriculture. We have now not only enough food production but are also in a position to export it to other countries.

One can not deny the fact that heavy application of pesticides was made to save the crops from ravages of pests, diseases and weeds. But, the indiscriminate use of pesticides has polluted the environment and affected our health adversely besides deleterious effects on non-target organisms. More and more insects are being reported resistant to different insecticides year after year posing a serious threat to agriculture. However, there has been also work on alternate control measures like biopesticides, varietal resistance, transgenic plants, remote sensing etc. for a better pest management strategy.

The present book ***Recent Advances in Indian Entomology*** gives a vast knowledge about the progress made in India on different entomological fronts. It contains vital information on changing scenario of Indian agriculture, plant protection set up in India, registration and regulation of pesticides, surveillance and monitoring of insect pests, post harvest grain technology, management of vectors of viral diseases, apiculture, vermiculture, vertebrate pests and management of insects of public health importance. It has many chapters on taxonomy, physiology, biological control, toxicology and IPM in different crops. The book is mainly based on lectures given by eminent scientists during a Summer Institute held from May 23 to June 11, 1994 at Indian Agricultural Research Institute, New Delhi, financed by Indian Council of Agricultural Research. It has covered almost all the important areas related to insect pest control.

The book will be of great value to researchers, teachers and students. All those engaged in the field of plant protection, transfer of technology, extension work, agriculture, horticulture, environment, bio-control and natural farming will also be immensely benefited.

Since it is the first edition, it may have many mistakes and shortcomings for which valuable suggestions from readers will be highly appreciated.

O.P. Lal

Acknowledgements

I express my sincere thanks to the Indian Council of Agricultural Research, New Delhi, for providing funds to conduct the Summer Institute. I would like to place on record my deep gratitude to all the faculty members and guest speakers for providing and sharing valuable information.

I am extremely grateful to Prof. R.B. Singh, Director, for kindly writing the foreword, and Prof. S.K. Sinha, Ex-Director, Dr. Anupam Varma, Dean and Joint Director (Education), Dr. N.N. Goswami, Ex-Dean and Joint Director (Education), Dr. Panjab Singh, Joint Director (Research), Dr. K.M. Singh, Head, Division of Entomology, IARI and Dr. S.L. Mehta, Deputy Director General (Education), ICAR, New Delhi for their constant help and guidance during the Summer Institute and also at various stages of the publication of this book.

My special thanks are also due to all my colleagues in the Division of Entomology, IARI, for their active involvement and whole hearted support.

O.P. Lal

Acknowledgements

List of Contributors

Dr. N.P. Agnihotri, Principal Scientist, Division of Agricultural Chemicals, IARI, New Delhi

Dr. Mrs. Chandra Mohini Anand, Senior Scientist, Division of Entomology, IARI, New Delhi

Dr. R.K. Anand, Principal Scientist, Division of Entomology, IARI, New Delhi

Dr. Atma Ram, Principal Scientist, Division of Entomology, IARI, New Delhi

Dr. R.K. Bhatnagar, Principal Scientist, Division of Entomology, IARI, New Delhi

Dr. Dinesh Chandra, Senior Scientist, Division of Entomology, IARI, New Delhi

Dr. S. Chaudhary, Senior Scientist, Division of Entomology, IARI, New Delhi

Dr. Mrs. Swaran Dhingra, Senior Scientist, Division of Entomology, IARI, New Delhi

Dr. S.I. Farooqi, Principal Scientist, Division of Entomology, IARI, New Delhi

Dr. R.D. Gautam, Senior Scientist, Division of Entomology, IARI, New Delhi

Dr. G.K. Girish, Joint Commissioner, Storage and Research, Ministry of Food, Krishi Bhawan, New Delhi

Dr. M. Gopalan, Director, Centre for Plant Protection Studies, Tamil Nadu Agricultural University, Coimbatore, Tamil Nadu

Dr. G.T. Gujar, Senior Scientist, Division of Entomology, IARI, New Delhi

Dr. S.L. Gupta, Senior Scientist, Division of Entomology, IARI, New Delhi

Dr. Prem Kishore, Senior Scientist, Division of Entomology, IARI, New Delhi

Dr. Rajesh Kumar, Technical Officer, Division of Entomology, IARI, New Delhi

Dr. O.P. Lal, Principal Scientist & Ex-Head, Division of Entomology, IARI, New Delhi

Dr. Sone Lal, Director, Indian Grain Storage Institute, Hapur, Uttar Pradesh

Dr. Y. Mahto, Senior Scientist, Division of Entomology, IARI, New Delhi

Dr. K.K. Marwaha, Principal Scientist, Division of Entomology, IARI, New Delhi

Dr. K.N. Mehrotra, Emeritus Scientist, Division of Entomology, IARI, New Delhi

Dr. M. Naim, Senior Scientist, Division of Entomology, IARI, New Delhi

Dr. B.S. Parmar, Division of Agricultural Chemicals, IARI, New Delhi

Dr. A.V.N. Paul, Senior Scientist, Division of Entomology, IARI, New Delhi

Dr. V.P.S. Panwar, Professor, Division of Entomology, IARI, New Delhi

Dr. Miss Amrit Phokela, Senior Scientist, Division of Entomology, IARI, New Delhi

Dr. R. Prasad, Former Head, Division of Agronomy, IARI, New Delhi

Dr. S.K. Prasad, Senior Scientist, Division of Entomology, IARI, New Delhi

Dr. V. Ragunathan, Plant Protection Advisor to the Government of India, Ministry of Agriculture, New Delhi

Dr. A.K. Raheja, Assistant Director General (Entomology and Biological Control),

ICAR, New Delhi

Dr. V.V. Ramamurthy, Senior Scientist, Division of Entomology, IARI, New Delhi

Dr. N.S. Randhawa, Fromer Director General, ICAR and Secretary, Department of Agricultural Research and Education, New Delhi

Dr. Y.N. Rao, Head, Division of Agricultural Physics, IARI, New Delhi

Dr. N.K. Roy, Former Head, Division of Agricultural Chemicals, IARI, New Delhi

Dr. V.S. Saxena, Senior Scientist, Division of Entomology, IARI, New Delhi

Dr. R.P. Sharma, National Research Centre on Plant Biotechnology, IARI, New Delhi

Dr. V.K. Sharma, Professor & Head, Department of Entomology, G.B. Pant University of Agriculture and Technology, Pantnagar, Nainital, Uttar Pradesh

Dr. K.H. Siddiqui, Senior Scientist, Division of Entomology, IARI, New Delhi

Dr. D.S. Singh, Principal Scientist, Division of Entomology, IARI, New Delhi

Dr. P.B. Singh, Scientist, S-I, Division of Entomology, IARI, New Delhi

Dr. R.P. Singh, Senior Scientist, Division of Entomology, IARI, New Delhi

Dr. Shivendra Singh, Research Associate, Division of Entomoloy, IARI, New Delhi

Dr. Vijai S. Singh, Principal Scientist, Division of Entomology, IARI, New Delhi

Dr. P. Sircar, Principal Scientist, Division of Entomology IARI, New Delhi

Dr. B.G. Srivastava, Senior Scientist, Division of Entomology, IARI, New Delhi

Dr. Mrs. Chitra Srivastava, Scientist, S-I, Division of Entomology, IARI, New Delhi

Dr. K.L. Srivastava, Senior Scientist, Division of Entomology, IARI, New Delhi

Dr. Y.N. Srivastava, Senior Scientist, Division of Entomology, IARI, New Delhi

Dr. B. Subrahmanyam, Senior Scientist, Division of Entomology, IARI, New Delhi

Dr. S. Suresh, Assistant Professor, Agricultural Entomology, Tamil Nadu Agricultural University, Coimbatore, Tamil Nadu

Dr. Gopal Swarup, Former Head, Division of Nematology, IARI, New Delhi

Dr. T.P.S. Teotia, Chief Entomologist, Dhamuka Agricultural Research Centre, (R & D Unit of M/s Northern Minerals Ltd.), Daulatabad Road, Gurgaon, Haryana

Dr. Mrs. Shashi Verma, Principal Scientist, Division of Entomology, IARI, New Delhi

Dr. T.D. Yadav, Principal Scientist, Division of Entomology, IARI, New Delhi

Dr. Anupam Varma, Dean & Joint Director (Education), IARI, New Delhi

Dr. K.P. Srivastava, Principal Scientist, Division of Mycology and Plant Pathology, IARI, New Delhi

Dr. V.P. Sharma, Director, Malaria Research Centre, Delhi

Dr. G. Ravi, Assistant Professor, Tamil Nadu Agricultural University, Coimbatore, Tamil Nadu

Contents

Recent Advances in Indian Entomology
Editor: O.P. Lal
APC Publications Pvt. Ltd., New Delhi, 1996, pp. 1-8

Recent Advances in Entomological Researches : An Overview and Technical Options

O. P. LAL and PREM KISHORE

The use of synthetic insecticides has been the principal method of insect control practiced by the farmers or growers. The demand of chemical treatment has been enormous accounting almost 30 % of production costs. The ability of modern pesticides to control pests is undeniable, but irrational and overuse of these pest control chemicals has led to several pesticide-related problems such as resistance development, hazards to non-target organisms, environmental pollution, poisoning, and residues in crops. Use of synthetic pyrethroids in recent years caused serious outbreaks of whitefly in cotton in Andhra Pradesh, Gujarat and Tamil Nadu and *Heliothis armigera* in cotton in Guntur and Prakasham districts of Andhra Pradesh. The persistent residues of DDT and BHC have been found in milk, butter, eggs, meat, oil and vegetables.

This situation has created a crisis and there arises an urgent need to develop more effective and efficient methods for the management of insect pests based on sound ecological principles. Advances in biological means to combat insect pests have furthur opened up new vistas in insect control. The current methods of insect control may show a great variety but have been developed to reflect compromise involving technological capability, cost effectiveness, durability and environmental friendliness.

Sampling for determination of pest populations and pest monitoring

For the effective pest management programmes, it is essential to know the dynamics of pests and feel the pulse of the surrounding environment which ultimately determines the status of the pest. The collection of representative samples is an essential step in monitoring of crop pests. The sampling methods should be simple, rapid and adquate in their reliability. At the time of sampling prior consideration of the distribution and behavior of pest is essential. The mobile pests like white flies and midges can only be sampled by sweeping techniques. They may occur in larger numbers on the edges than in the interior parts of crops. These should be sampled earlier. Certain insects cause severe injury on individual plants or in small loci long before minimal number of pests can be found on neighbouring plants. Detection of such early infestation is of great importance and requires sampling on a wide range to cover the field.

There are two kinds of monitoring programmes :

i) monitoring for research purposes and,

ii) monitoring for use by pest managers.

The purpose of first type is to establish economic thresholds for evaluating the influence of abiotic and biotic factos or to determine the most damaging stage of the pests. This is of exploratory nature and requires careful and extremely sensitive watch on the components of ecosystem. In the second type, monitoring systems must be quick, inexpensive and simple in practice.

There are four ways to sample insect pests viz. random sampling, point sampling, trap sampling and sequential sampling. Sampling procedures designed for research purposes are reviewed by Southwood (1966), Anonymous (1969), Ruesink and Kogan (1975).

The sequential sampling is used for classifying populations in which the sample size is not fixed but variable. The advantage of this sampling method over other sampling methods is its efficiency in terms of number of observations required to reach a decision e. g. very high or very low infestation levels can be established by comparitively few observations while intermediate infestation levels will generally require more observations.

Monitoring is an important tool in pest management and helps in pin pointing early outbreaks and the progress of infestation. This method also enables dicisions to be made on control strategy and the efficiency of the treatment. The application of suitable control strategy at the right time results in its efficent use and reduction in the frequency of treatment. The pest surveillance should be used as an integral component of IPM programme for early detection of the pest (Mathur and Prem Kishore 1987; Price and Waldbauer, 1982).

Economic thresholds

It is not known in most of the crops that what pest population density will cause economic damage. Serious efforts should be made to work out economic thresholds i. e. the population level at which economic loss is inflicted. The cost of protection at this population level is equal to cost of produce obtained. If the threshold levels can be predicted then timely control action can be taken to cut crop losses. Data should be generated through survey and surveillance programmes for developing economic thresholds. To achieve this goal effective sampling techniques are required. The economic damage threshold is complex and requires experience. It involves prognosis of certain weather factors and of crop and pest development.The various aspects to be considered here includes:

i) Dynamics of pest multiplication

ii) Susceptibility and growth vigour of the crop

iii) Efficiency of monitoring services

iv) Insecticide related aspects specially efficacy and effect of pesticides.

Semio-chemicals and insect growth regulators

These chemicals are able to modiy the behaviour of a perceiving organism at sub-micro/nanogram levels. Semiochemicals are categorized as naturally secreted, botanical constituents or synthetic. These behaviour modifying chemicals (BMC) include pheromones, kairomones, allomones, attractants, deterrents and other parapheromones. A single chemical secreted may have both intra and inter-species-activities or only one activity. These chemicals are becoming increasingly valuable in insect pest management as they offer enormous potential for controlling many devastating insect pests. Many of the chemicals involved in insect communication may be exploited as pest detection agents.

Pheromones are generaly species specific and are widely used as traps for assessing the population dynamics of the pest. The main application of the pheromone is to determine the seasonal abundance of pest, making possible the judicious application of insecticide or other control measures by careful interpretation of trap catch. Mating disruption is also facilitated by pheromones. It envolves release of large amounts of synthetic pheromone which results in disruption of communication between two sexes thereby preventing mating. Pheromones can also be used for mass trapping.

The synthetic pheromone gossyplure is widely used for the monitoring of male pink bollworm of cotton while potato tuber moth pheromone PTM-1 and PTM-2 reduce the larval infestation in tubers significantly. The population of codling moth, *Cydia pomonella* can be suppressed by Delta traps baited with "codling moth lure H. F." suspended from trees only at 1.8 m. height from ground. The sex pheromone has potential for suppressing the sweet patoto weevil, *Cylas formicarius,* by mass trapping of males and by disrupting mating.

The behaviour modifying chemicals for fruit flies offer greater potential as these are technologically simpler to produce. Methyl eugenol is naturally found in several plants and can be used in a wider range situations for more than one species of the fruit fly.

Kairomones are natural chemical substances present in many plants or animal hosts that direct the insect pests towards feeding sites. An innovative approach of exploiting BMC's is the use of kairomones for modyfying behaviour and stimultating searching capacity of natural enemies.

The insect growth regulators teflubenzuron at 50g/ha and Flucyclozuron at 70g/ha. sprayed at 15 days interval gave control of seveal insects including borer pests.

Parasites and predators

Biocontrol agents, unlike pesticides, multiply, migrate and compete with other organisms. Therefore, their potential tend to even out overtime and as such there is need to release them often to achieve effective pest control just as pesticide is sprayed more than once. Biocontrol takes little more time to keep the pest under check hence, is to be followed in ritualistic manner. Several predators, parasites and pathogens thriving on insect pests are available in nature in abundance and can be successfully tapped for pest control (DeBach 1966, DeBach and Bartlet 1964, Askew 1971, Huffaker and Messenger 1976, Ridgeway and Vinson 1977).

Egg parasitoid, *Trichogramma* is important parasitoid of several pests and is widely augmented entomophagous arthropod. The other important parasitoids attacking insect pests and successfully employed in pest control are *Goniozus, Elasmus, Eriborus, Bracon, Trichospilus, Tetrastichus* and *Chelonus. The* inundative release at weekly intervals of *Trichogramma* at 50,000/ ha has been found very effective in controlling internode borer of sugarcane in Tamil Nadu and Bihar. Similarly the release of important parasitoid of *Pyrilla, Epiricrania melanoleuca* at 4,000-5,000 cocoons and 4-5 lakh eggs/ha in U. P. , Bihar, Punjab and Rajasthan, has resulted in effective control of *Pyrilla.* The parasitoid, *Chelonus blackburni* has been found attacking bollworms of cotton in large numbers in Karnataka, TamilNadu, Maharashtra, Gujarat. A Braconid parasitoid, *Bracon kirkpatricki* has also shown promise in controlling bollworms of cotton. Three important parasites *viz. Bracon brevicornis, Goniozus nephantidis* and *Trichospilus pupivora* have been successfully utilized for the control of pests of oil seeds. *Microplitis maculipenis* and *Diaeretiella rapae* are effective against castor semilooper and aphids respectively in oil seeds.

Trichogramma spp, *Campoletis* spp., *Bracon hebetor* and *Telenomus* sp. have shown potential for controlling serious pests of pulses including pod borer, *Heliothis armigera. The egg parasitoid, Telenomus manolus* has been found effective against red hairy caterpillar. The releases of *T. politor* against castor semilooper and *T. remus* against *Spodoptera litura* were found very effective. The tomato fruit borer was successfully controlled by the releases of *Trichogramma brasiliensis* at 2,50,000 adults/ha. The parasitoids, *Apanteles plutellae* and *Tetrastichus sokolowskii* gave effective control of diamond back moth and aphids on cabbage and cauliflower. The cabbage butterfly is well parasitized by *Apanteles glomeratus* at larval stage and *Pteromalus puparum* at pupal stage (Lal and Chandra 1976). The chalcid parasitoid, *Aphelinus mali* has been successfully utilized against wooly aphid

Amongst predators, the coccinellids are important group of predators which feed on number of insect pests. Coccinellids like *Pharoscymnus, Cryptolaemus, Scymnus, Chilocorus* and *Menochilus* feed on mealy bugs, coccids, scales and mites of citrus, grapevines and guava. The green lace wing, *Chrysopa,* feeds on large number of important pests like aphids, mealy bugs and number of young caterpillars. The mirid predator, *Cyrtorhinus lividipennis* attacks brown plant hopper and green leaf-hopper of rice. The large populations of sugarcane scale were suppressed in Andhra Pradesh by the release of two indigenous predators, *Pharoscymnus horni* and *Chilocorus nigritus.* The predator *Cryptolaemus montrouzieri* is being mass multiplied for the control of mealy bugs of citrus, grape, coffee etc. Two predatory mites *viz. Amblyseius tetranychivorus* and *Phytoseillus parasimilis* have been reported effective in checking phytophagous mites of brinjal and ladyfinger.

Entomopathogens

Of late insect pathogens have been successfully utilized in controlling important crop pests. They are non-toxic to many forms of life and have mainly reported from insects belonging to orders Lepidoptera, Hymenoptera and Diptera. The

Baculoviruses including Nuclear polyhedrosis viruses (NPV) and Granulosis viruses (GV) are the most important group of viruses and successfully tested against cotton bollworm, *Heliothis* spp. which also attacks chickpea and millets; red hairy caterpillar of groundnut; and tobacco caterpillar, *Spodoptera litura* which attacks 17 important economic crop plants (Allen et. al., 1978; Heimpel and Angus, 1959).

Fungal pathogens of insects are also equally important. Over 500 species of fungal pathogens have been reported from different insects. The important insect pathogens belonging to genera, *Entomophthora*, *Massospora, Beauveria, Metarhizium, Verticillium, Spicaria, Aspergillus, Isaria* and *Hirsutella* provided effective pest suppression (Ignoffo and Hosteller, 1977).

Some of the spore forming bacteria are important insect pathogens. *Bacillus thuringiensis*(B. t.) has been reported to infect as many as 525 insect species of various insect orders. Lepidopteran group is highly susceptible to *B. thuringiensis.* A number of B.t. formulation which is available as dust and spray formulations and are available in other countries. Thuricide is one important B.t. formulations and gave effective control of diamond-back moth and pierid butterfly larvae (Steinhaus 1954, Heimpel and Angus 1959, Van der Lan 1967, Jayraj 1986). In India red flour beetle has been controlled by utilizing protozoan, *Farinocysitis triboli.* The protozoan pathogen also infect insects of the orders like Lepidoptera, Coleoptera, Orthoptera, Hemiptera and Diptera.

Many of these insect pathogens are compatible with insecticides and can be easily incorporated in insect pest management system. Their large scale culture and application are easy and inexpensive in most of the cases.

Resistant crop varieties

The best plant protection for future should be one based on host plant resistance. This method is particularly relevant to subsistance farming systems of the semi-arid tropics. Breding for pest resistance has received little attention in past, but now the importance of breeding not only for yield, but for quality, adaptability and pest resistance is being increasingly recognised. Plant breeders in general lack an understanding of insects and their hosts and tend to regard the insect populations as fixed environmental parameter with all the consequent implications. The release of superior but otherwise susceptible crop varieties will not achieve real yield potential. However, the cultivars that help increase crop yields (eg CSH-1) but change the pest complex should be watched carefully and newer pest problems countered effectively until cultivars with adquate levels of resistance are developed (Prem Kishore 1992 a, b, Prem Kishore and Gahukar 1992).

Some of the important crop varieties developed with agronomic and other desirable traits are :- IET 7918, IET 6178, IET 3127, IET 2812 for stem borer, Bhadra, IET 7575, IET7943 for brown plant hopper, IET 7301, IET 7302 for green plant hopper in paddy; Ganga 5, Him 123, Deccan 101 and 103 for stem borer and pink borer in maize; P-311, SPV 1015 (PGS-1) (Prem Kishore 1987, 1992b) for shootfly and stem borer in sorghum and ICP-909, ICC-506, ICP 10466 for pod borer in pulses. Most of the agronomically good varieties of cabbage like 'Golden Acre', 'Pusa Mukta' are susceptible to cabbage aphid and cabbage caterpillar while 'Red pickling', 'Red Rock Mammoth' and 'Red Drum Head' are comparatively resistant to cabbage caterpilar (Lal 1991a,b). In brinjal 'Sm 202', Pant Samrat', Sm 17-4' and 'Pusa Purple Cluster' are comparatively resistant to shoot and fruit borer *(Leucinodes orbonalis)* (Lal 1991c).

The availability of new sources of resistance to certain specific pests has now broadened the base for developing varieties to suit endemic areas, hot spots or regional/location specific pest problems. The multiple pest resistant varieties will be suitable for inter-regional cropping or at the national level.

Development of transgenic plants

The protein produced by *Bacillus thuringiensis* (B. t.) is lethal to the caterpillars of lepidopterous insects. The B. t. gene is responsible for production of toxin when transferred to the crop plants through genetic engineering, the plants produce their own biocides.

The B. T. protein gene has been transferred into tomato (Fischhoff et al, 1987; Delanny et al. 1989), tobacco (Vaeck et al., 1987) and several other crops. Delanny et. al. (1989) evaluated transgenic tomato plants under field conditions against tobacco hornworm, tomato fruitworm and tomato pin worm and found that transgenic plants suffered less feeding as compared to heavy feeding on non-trasgenic plants.

Botanicals as future pesticide

More and more emphasis is being given now on developing pesticides from naturally occuring plants. Different pesticidal formulations prepared from Neem and other botanicals like Pongamia, Datura, Thevatia etc. have proved promising in checking pest populations. Various brands of such botanicals are already in market. Use of botanicals does not have harmful residues. The tetranortriterpenoids, meliantrial, salannin, and azadirachtion, occuring mainly in seeds of neem, act as antifeedants or disruptants of insect growth and development or toxicants, making neem suitable for practical use as pesticide.

Remote sensing

It uses the properties of electromagnetic spectrum to survey and diagnose various phenomena in the envornment. Remote sensing is based on the principle that each body reflects or absorbs radiant energy as electromagnetic waves with specific properties. There are two types of remote sensing. They are passive sensing and active sensing. Passive sensing employs the sun as source of energy as in photography and active sensing transmits energy, as in the use of radar The passive sensing is employed in agriculture with wavelengths of 0.3 to 15 mm. This range is easy to distinguish between reflected and emitted energy. The objective of remote sensing is to make monitoring more effective and economical not to replace the monitoring of crops from the ground. As regards the early detection of pest it must be taken into account that this referes primarily to damage caused in the upper parts of crops. Infestations restricted in their early stages to lower parts of the plants may be masked and undetectable from above (Blazquez and Horn, 1980).

Pest avoidance tactics

Agronomic adjustments also help in reducing pest populations. Early sowings have been found to reduce populations. Sowing of sorghum during *Kharif* immediately on the onset of monsoon considerably reduces the incidence of shootfly. Deep ploughing exposes the immature stages of number of serious pests to the activity of predators mainly birds and sun. Burning of trashes, threshed left overs, stubbles and consumption of harvested stalks before the next crop season help in reducing the incidence of stem borers of different crops and pink bollworm. Clean cultivation proper drainage, timely cultural operations and balanced application of fertilizers and irrigation also help in reducing pest populations.

Number of pest problems can be minimised by using high quality pest free seeds. sowing of healthy planting material also reduces pest incidence, specially in case of sugarcane borers.

These technical options have great promise for developing future pest control strategies and their adoption in right earnest will prove ideal alternative to pesticides and will improve the quality of the environment.

References

Anonymous 1969. Insect pest management and control. Principles of plant and animal pest control. *Vol 3. Nat. Acad. Sci. Publ.* 1969, 508 pp.

Allen, E. , Ignoffo CM, and Jaques, R. P., eds. 1978. Micorbial control of insect pests. Furture strategies in pest management systems. N. S. F.-USDA, University of Florida Workshop, Gainesville, Florida; 290 pp.

Askew, R. R. 1971. Parasitic insects. American Elsevier, New York, 316pp.

Blazquez, C. H. and Horn, F. W. Jr. 1980. Infrared photography : application in citriculture. NASA Ref. Publ. 1067, 82 pp

DeBach, P. 1966. The competive displacement and coexistence principles. *Annu. Rev. Entomol.,* **11** : 183-212.

DeBach, P. and Bartlet, B. R. 1964. Methods of colonization, recovery and evaluation. Pages 402-428 In P. DeBach., ed., Biological control of insect pests and weeds, Reinhold, New York. 844 pp.

Delanny, X., LaVallee B. J., Proksch, R. K., Fuchs, R. L., Sims, S. R., Greenplate, J. T., Marrone P. G., Dodson, R. B., Augustine, J. T., Layton, J. C. and Fischhoff, D. D. 1989. Field performance of transgenic tomato plants expressing the *Bacillus thuringiensis* var. kurstaki insect control protein. *Bio. Technology* **7** : 1265-1269.

Fischhoff, D. A., Bowdish, K. S., Perlak, F. J., Marrone, P. G., McCormick, S. M., Niedermeyer, J. G., Dean, D. A. Kusano-Kretzmer, K., Mayer, E. J., Rochester, D. E., Rogers S. G. and Fraley, R. T. 1987. Insect tolerant transgenci tomato plants. *Bio. Technology* **5** : 807-813.

Heimpel, A. M. and Angus T. A. 1959. The site of action of crystalliferous bacteria in lepidopterous larvae. *J. Invertebr. Pathol.* **1** : 152-170.

Huffaker, C. B. and Messenger, P. S., eds, 1976. Theory and practice of biological control. Academic Press, New York 788 pp.

Ignoffo, C. M. and Hostetter, D. L. 1977. Summary of the Ignoffo, C. M. and Hostetter, D. L. 1977. Summary of the environmental stability of microbial insecticides. *Misc. Publ. Entomol. Soc. Am.* **10** : 117-119.

Jayaraj, S. 1986. Role of insect pathogen in plant protection. *Proceedings of Indian National Science Academy*, **52** (1) : 91-107.

Lal O. P. 1991a. Varietal resistance in cabbage against the cabbage aphid, *Brevicoryne brassicae* L. (Homoptera : Aphididae) in Kulu Valley, India. *Journal* of *Plant Disease* and *Protection,* **98** (1) : 84-91.

Lal, O. P. 1991b. Studies on field resistance of different cultivars and crosses of cabbage against cabbage caterpillar, *Pieris brassicae* L. (Lepidoptera : Pieridae). *Journal* of *Plant Disease* and *Protection,* **98** (3) : 293-300.

Lal, O. P. 1991c. Varietal resistance in the egg plant, *Solanum melongena* against the shoot and fruit borer, *Leucidondes orbonalis* Guen. (Lepidoptera : Pyralidae). *Journal* of *Plant Disease* and *Protection,* **98** (4) : 405-10.

Lal, O. P. and Chandra, J. 1976. Some parasites of cabbage worm, *Pieris brassicae* Linn. (Lepidoptera : Pieridae) from Kulu valley, Himachal Pradesh *Current Science,* **45** (21) : 766-767.

Mathur, Y. K. and Prem Kishore 1987. Recent concepts of integrated management of key pests of agricultural crops. Pages I-X, In Y. K. Mathur, A. K. Bhattacharya, N. D. pandey, K. D. Upadhyay and J. P. Srivastava eds., *Recent Advances in Entomology*, Gopala Prakashan, Parade, Kanpur (India).

Price, P. W. and Waldbauer, G. P. 1982. Ecological aspects of pest management. pages 33-68, In R. L. Metcalf and W. H. Luckmann, eds., *Introduction to insect pest mangement, John* Wiley and Sons, New York, 577 pp.

Prem Kishore 1987. P-311. A new sorghum variety resistant to stem borer, *Chilo partellus* (Swinhoe), *J. ent. Res.* **11** (1) : 115-118.

Prem Kishore, 1991. Insect plant relationship studies in sorghum and pearl millet. *Ann. Pl. Resist., Insects Newslet,* **17** : 64-64.

Prem Kishore and Gahukar, R. T. 1992. Operational integrated pest management strategy involving cultural-cum-host plant resistance in sorghum. *J. ent. Res.* **16** (2) : 155-162.

Prem Kishore 1992a. Evaluation of twin resistance sources amongst advance generation derivatives in sorghum to shootfly, *Atherigona soccata* Randani and stem borer, *Chilo partellus* (Swinhoe). *J. ent. Res.* **16** (3) : 236-241.

Prem Kishore 1992 b. SPV1015 (PGS-1) - A new sorghum variety endowed with multiple resitance to shootfly, *Atherigona soccata* Rondani and stem borer, *Chilo partellus* (Swinhoe) *J. ent. Res.* **16** (4) : 319-321.

Ridgeway, R. L., and Vinson, S. B., eds. 1977. Biological control by augmentation of natural enemies. Plenum press. New York. 480 pp.

Ruesink. W. G. and Kogan, M. 1975. The quantitative basis of pest management : sampling and measuring, *Introduction to Insect Pest Management. In* R. L. Metcalf and W. Luckman (eds). John Wiley & Sons, New York, pp. 309-352.

Southwood, T. R. E. 1966. Ecological methods. Methuen, London, U. K. 391 pp.

Steinhaus. E. A. 1954. The effect of disease on insect populations. *Hilgardia* **23** : 197-281

Vaeck. M., Raynaerts, A., Hofte. H. Jansens, S, De Beuckeleer, M, Dean, Cm Zabeau, M., Van Montagu, M, and Leemans, J. 1987. Transgenic plants protected from insect attack. *Nature* **328** : 33-37.

Van der Lan, P. A., ed. 1967. *Insect Pathology and microbial control,* North Holland. Amsterdam. 262 pp.

Recent Advances in Indian Entomology
Editor: O.P. Lal
APC Publications Pvt. Ltd., New Delhi, 1996, pp. 9-13

Changing Scenario in Indian Agriculture

RAJENDRA PRASAD

When the freedom dawned on India on 15 August, 1947 the first problem before the Indian Government was the provision of adequate food to all its citizens. Even today after 47 years of freedom we still have one or the other component of food problem unresolved. Nevertheless, we have made some success of which we can be proud of. In crop year 1950-51, the year when we started the planned development of this country, the total production of food grains was only about 50 million tonnes. In 1992-93, we produced 180million tonnes of food grains. Much of this increase in food grains was made possible by the development of high yielding varieties of cereals. This programme began in mid 50's with hybrid maize and in early 60's with rice but the real Gold was struck with the large scale import of seeds of two dwarf wheats: (i) Lerma rojo and (ii) Sonora-64. The increase in total cereal production from 66 million tonnes to 83 million tonnes in 1967-68 was so great that we celebrated the Green Revolution in 1968 when the Prime Minister of India late Smt. Indira Gandhi released a special postage stamp on Wheat production in India. The people all over the world specially those in developmed world could not believe that the Indians could do so. In cereal production we have never looked back since then.

In the finer analysis, these new dwarf wheats did not show up their yield potential with the traditional agronomic package. The Indian Agronomists took only 2-3 years to show that dwarf mexican wheats required the following 4 major changes in their agronomic package.

1. Shallower seeding.
2. Seeding about a month later than the traditional mid-October sowing.
3. Advancing the first irrigation by about a week from that practised in the traditional wheat cultivation. This is generally known as giving first irrigation at crown root initiation (CRI) stage.
4. Adequate balanced fertilization.

Since yield advantage with dwarf wheat varieties was tremendous, the Indian farmers immediately took to these new recommendations and wheat production has since then been moving up and up; it increased from mere 6.46 million tonnes in 1950-51 to 56.8 million tonnes in 1992-93, a nearly 10-fold increase (Table 1). This increase was possible both by an increase in wheat area from 9.7 million hectares in 1950-51 to 24.2 million hectares in 1992-93 (Table 2) as well as in the productivity from 663 kg/ha in 1950-51 to 2348 kg/ha in 1992-93 (Table 3). The increase in acreage under rice was also much less; 30.8 million hectares in 1950-51 to 41.2 million hectares in 1992-93 and so was the increase in rice production from 20.6 million tonnes in 1950-51 to 71.9 million tonnes in 1992-93. We can thus regard the increase in wheat production in the country as the first major phase in the changing scenario of Indian Agriculture.

Table 1. Production of selected crops - All India (1950-51 to 1992-93)

'000 tonnes

Year	Rice	Wheat	Total cereals	Total Pulses	Total foodgrains
1950-51	20576	6462	42414	8411	50825
1954-55	25219	9043	57085	10950	68035
1959-60	31676	10324	64872	11799	76672
1964-65	39308	12257	76939	12417	89356
1969-70	40430	20093	87810	11691	99501
1974-75	39579	24104	89812	10104	99826
1979-80	42330	31830	101129	8572	109701
1984-85	58336	44069	133576	11963	145539
1989-90	73573	49850	158179	12858	171036
1992-93	71910	56855	165560	14733	180293

Table 2. Gross area under selected crops - All India (1950-51 to 1992-93)

'000 hectares

Year	Rice	Wheat	Total cereals	Total Pulses	Total foodgrains
1950-51	30810	9746	78230	19091	97321
1954-55	30764	11259	85944	21914	107856
1959-60	33820	13380	90990	24833	115818
1964-65	34462	13422	94237	23875	118112
1969-70	37680	16626	101547	22023	123570
1974-75	37889	18010	99051	22024	121075
1979-80	39414	22172	102947	22259	125206
1984-85	41159	23565	103358	23415	126773
1989-90	42167	23502	103358	23415	126773
1992-93	41206	24213	100795	23858	124653

Table 3. Average yield per hectare of selected crops - All India (1950-51 to 1992-93)

'000 tonnes

Year	Rice	Wheat	Total cereals	Gram	Total Pulses (including gram)	Total foodgrains
1950-51	668	663	542	482	441	522
1954-55	820	803	664	608	500	631
1959-60	937	772	713	544	475	662
1964-65	1078	913	816	651	520	757
1969-70	1073	1202	865	715	531	805
1974-75	1045	1388	907	570	455	824
1979-80	1074	1436	982	481	385	876
1984-85	1417	1870	1285	661	526	1149
1989-90	1745	2121	1530	652	549	1349
1992-93	1745	2348	1643	752	618	1446

The second phase in the changing scenario of Indian agriculture started from the discovery that dwarf wheats needed seeding by mid November as compared to mid-October in the case of tall Indian wheats. This permitted the cultivation of rice in (singh and Pande, the non traditional north western India and *rice wheat rotation* became the most important crop rotation in this part of the country (Singh & Paroda, 1994). Again the success with rice in this non-traditional area was due to less humidity and rainfall associated with lesser incidence of diseases and pests.

Also it was discovered that dwarf wheats could be planted as late as the middle of January and yet they gave 2-3 t/ha grain. Therefore, a number of new multiple cropping systems were developed namely cotton-wheat, maize-potato-wheat etc.

Another change that needs mention is the development of *short duration arhar* varieties which reduced the crop duration from 200 days to 140-150 days. This also permitted developemtn of pigeonpea-wheat cropping system. Thus the development of these *multiple cropping systems* (Palaniappan, 1985) could be considered as a major change in the scenario of Indian agriculture, which has contributed towards the increased food production. These multiple cropping systems have brought food security in the country and despite severe and moderate droughts (Table 4) we have never encountered a serious foodgrain shortage. However, all our problems of food grains have not been solved. For example, although per capita availability of foodgrains/day has increased from 395 g/day to 511 g/day, the proportion of cereals to pulses has gone down. The aviability of pulses decreased from 60.7 g/

person/day to only 40.3 g/person/day; about 50 per cent decrease in the availability. This would explain the rising prices of the pulses in the Indian market. The area under pulses has remained at 22± 1 million hectares from 1954-55 onwards and the production also remained almost stagnant at 11 ± 1 million tonnes/year except in 1992-93 when record production of 14.7 million tonnes was obtained in pulses. The main reason is that pulses are grown mostly under the dryland conditions and marginal and submarginal lands and therefore, despite the production potential of 2-2.5 tonnes/ha of available varieties of most pulses crops the national average has changed only from 608 kg in 1954-55 to 752 kg/ha in 1992-93. This calls for serious action programme in the pulse production. One of the major causes for low yields is the attack of pod borer and other insect pests and the entomologists have a very important role to play in increasing pulse production in the country.

Table 4. Years of drought in India

Year	Area affected (Millions sq. km)	% area of the country affected	Category
1951	1.04	33.2	Moderate
1952	0.81	25.8	Slight
1965	1.35	42.9	Moderate
1966	1.01	32.3	Moderate
1968	0.45	20.6	Slight
1969	0.62	19.9	Slight
1971	0.42	13.3	Slight
1972	1.39	44.4	Severe
1974	0.92	29.3	Moderate
1979	1.24	39.4	Moderate
1982	1.04	33.1	Moderate
1985	0.95	30.1	Moderate
1986	0.60	19.0	Slight
1987	1.55	49.2	Severe

Similarly we also ran in the shortage of edible oil during the last decade or so and a large sum of money had to be invested by the Government of India in importing it from other countries. This problem is nearly solved, but with the ever increasing population in India we have always to keep our fingers crossed.

During the couple of years we have also landed into the serious problem of rising fertilizer costs and alternative sources and ways of supplying plant nutrients are being investigated upon. These include bio-fertilizers, changes in fertilizer application practices, better understanding of residual effects of fertilizer applied to the preceding crop on the succeeding crop etc. (Tandon, 1992). Also there is search for more efficient fertilizer materials (Prasad and Goswami, 1992).

Since the agriculture makes up the major proportion of our gross national product (GNP), proper planning and carefull implementation can also lead it towards making a major contribution in country's export earning. The current thrust in the agricultural development is therefore shifting from agricultural to horticultural crops, such as apples, grapes etc. Needless to add that for producing good quality fruits, and their good preservation, the entomologists have a major contribution to make. Unless adequate plant protection is provided successful horticultural production is not possible.

Thus from the date of entring of dwarf wheats in Indian agriculture there has been a major transformation in cropping systems and crop production practices. However, for the real success and prosperity in the farming comminity, adequate marketing facilities including good stroage houses, good roads and credit need to be provided (Gangadharan, 1994). These are important future needs of Indian Agriculture.

References

Gangadharan, T.P. 1994. Rural credit needs - perceptives. p. 107-118. *In*. Agricultural Credit System in India. Mruthyunjaya, V.C. Mathur and Praduman Kumar (eds.). Agricultural Economics Research Association (India), New Delhi.

Palaniappan, S.P. 1985. Cropping systems in Tropics - Principles and Management. pp. 215. Wiley Eastern Ltd., New Delhi.

Prasad, Rajendra, 1992. Research and development of new fertilizer materials in India. *In*. Soil Fertility and Fertilizer use. vol. 5. V. Kumar, U.N. Mishra, B.P. Govil, G.C. Shrotriya and S.V. Kaore (eds.), IFFCO, New Delhi.

Prasad, Rajendra and Goswami, N.N. 1992. Soil fertility restoration and management for sustainable agriculture in South Asia. *Adv. Soil Sci.* **17** : 37-77.

Singh, R.B. and R.S. Paroda. 1994. Sustainability and productivity of rice-wheat systems in Asia-pacific region : Research and technology development needs. p. 1-35. *In*. Sustainability of Rice-Wheat Production System in Asia, Regional Office for Asia and the Pacific (RAPA), FAO, Bangkok, 1994.

Tandon, H.L.S. 1992. Fertilizers, Organic Manures, Recyclable Wastes and Biofertilizers. Fertilizer Development and Consultation Organisation, New Delhi, pp. 147.

Recent Advances in Indian Entomology
Editor: O.P. Lal
APC Publications Pvt. Ltd., New Delhi, 1996, pp. 14-21

Plant Protection Set Up in India

V. RAGUNATHAN

In the food and fibre policies of our country, Plant Protection has a pivotal role to play. To protect is to conserve and to conserve crop production is to sustain human and animal life.

One of the main conclusion of the Indian Famine Enquiry Commission in 1945 was that "if full benefits of irrigation, manuring and improved varieties are to be assured, effective action must be taken to deal with diseases, pests, vermins and weeds, etc. Crop protection is an important factor in increasing crop production". The Commission, therefore, recommended that crop protection being an important factor in increasing crop production separate plant protection staff should be employed. The Government of India, therefore, established on permanent basis in 1946 the Central Plant Protection Organisation (now called the Directorate of Plant Protection, Quarantine & Storage, Faridabad) under the Minister of Agriculture with its headquarters at New Delhi. It was the beginning of an organised scientific plant protection set up in the country Prior to this, there was no institutional set up for plant protection at Central level to monitor pest and disease situation in the field except for locusts in "scheduled Desert Area" of India.

Agriculture and Plant Protection being in the concurrent list of the Constitution, the full plant protection activities in the States and Union Territories are their responsibilities. However, the Central Government is directly responsible for foreign plant quarantine and locust control in the Scheduled Desert Area. The Government of India through the Directorate of Plant Protection, Quarantine & Storage initiates, plans, supports and encourages problems and activities related to pest and disease control, pesticides, development of plant protection equipment, Integrated Pest Management and enforcement of rules under Destructive Insects and Pests Act, 1914 and Insecticides Act, 1968.

In order to meet the local needs, the States and Union Territories had established plant protection services in their respective Departments of Agriculture and now almost all the States/UTs have an exclusive Plant Protection Wing. However, additional assistance and supports are provided by the Directorate of Plant Protection, Quarantine & Storage and related Organisations of the Government of India whenever required.

The global concern on adverse effects of pesticides in the environment led India to rethink and reorient its policy on Plant Protection. The minimum need-based application of chemical pesticides on the basis of pest surveillance and monitoring became the guiding principles in pest control operations. The concept of Plant Protection technology has been further modified to conserve the natural enemies and eco-system by propagating Integrated Pest Management (IPM) approach. This approach has now become the cardinal principle and the major plank of our policy on Plant Protection in agricultural production during the recent years.

The IPM approach aims at maximising the natural control resources and to utilise every pest control methods and pesticides only on need basis with minimum environmental disturbances. In view of the inherent risks and ill effects of chemical pesticides on eco-system and environment, it has become necessary to implement IPM approach effectively over large cropped areas. Further, by their very toxic nature, pesticides are required to be used judiciously with greater care to safeguard wild life, birds, environment besides the users themselves. On this background, the Government of India has enacted an Act called the Insecticides Act, 1968, which regulates the import, manufacture, sale, transport, distribution and use of pesticides in the country. With the enforcement of the Act, considerable awareness has been created to ensure supply of quality and safer pesticides to the users. To facilitate the enforcement of the said Act, a network of pesticides testing laboratories has been created in the country.

As the first line of defence, plant quarantine plays a major role in plant protection as well as it is an international obligation as per International Plant Protection Convention (IPPC) of FAO, 1951.

Vigil against the entry of exotic pests and diseases through the international trade is being provided in order to implement the statutory provisions as laid down under Destructive Insects and Pests Act, 1914" and "The Plant, Fruits and Seeds (Regulation of Import into India) Order, 1989" issued thereunder.

The Plant Protection responsibility is being discharged through the following major activities.

(1) Enforcement of Insecticides Act, 1968
(2) Implementation of Destructive Insects and Pests Act, 1914
(3) Integrated Pest Management (IPM)
(4) Training in Plant Protection and
(5) Locust Control

The plant protection coverage since the beginning of the planning is given below:-

Five Year Plans	**Plant Protection Coverage** (In million ha.)
1st Plan	2.40
2nd Plan	6.40
3rd Plan	17.20
4th Plan	60.50
5th Plan	70.00
6th Plan	108.00
7th Plan	137.00
8th Plan	202.00 (target)

The plant protection during each Plan period have been fairly satisfactory and the country has witnessed phenomenal increase in plant protection coverage. The physical coverage under Plant Protection has gone upto 137 million ha. by the end of Seventh Plan from a level of 2.4 million ha. at the betinning of the Planning era.

Essentially the coverage is by using chemical pesticides. However, the need for caution for judicious and need based application of pesticides has been emphasized particularly during the recent years.

1. Enforcement of Insecticides Act, 1968

The production, import, export, registration, quality and the use of pesticides are regulated under the Insecticides Act, 1968.

Registration of Pesticides

The Registration Committee (RC) constituted under Section 5 of the Insecticides Act, 1968 registers pesticides after verifying their efficacy and safety to human beings and animals. Any person desiring import and manufacture of any pesticide can apply for registration of such pesticides under Section 9 of the Act. The Committee is ordinarily expected to take a final view in the matter within a period of 12 months. The registrations are granted under three different sub-sections which are as under:

(i) Provisional registration under Section 9 (3b):

Provisional registration is granted to a pesticide which is introduced for the first time in India for a period of two years for data generation under Indian conditions.

(ii) Regular registration under Section 9 (3):

Once the applicant submits the data as per the requirements to the satisfaction of the Registration Committee, registration is granted on regular basis under Section 9 (3).

(iii) Repeat registration under Section 9 (4):

Once registration is granted under Section 9 (3) in favour of an applicant, any other person can apply for registration under Section 9 (4) by putting an application in the prescribed proforma alongwith the required fee and registration is granted on the same condition on which the pesticide was originally registered.

Quality control of pesticides

Quality control of pesticides has been accorded the highest priority in order to ensure the chemical pesticides used at requisite degree of efficacy and safety. In this context, there is a network of Government, pesticides testing laboratories which comprise 39 State Pesticides Testing Laboratories with the analysis capacity of 45,556 samples per annum, in addition to Central Insecticides Laboratory at Faridabad and 2 Regional Pesticides Testing Laboratories at Kanpur and Chandigarh. Pesticides samples are tested at Central Insecticides Laboratory, Faridabad for:

(i) Bioefficacy, phytotoxicity, compatibility and resistance,

(ii) Acute oral/dermal LD 50 estimation,

(iii) Analysing samples and their formulations and

(iv) Packaging, labelling and shelf life.

Monitoring of the availability of pesticides

The Union Department of Agriculture and Cooperation reviews the production and availability position of major and commonly used pesticides with the nodal department of the concerned State Governments wherever required and ensures the supply of adequate quantity of pesticides. The consumption rate is experiencing a downward trend; reduced to 70,000 M.T. from 75,000 M.T. (Tech. Grade), due to introduction of IPM technology.

II. Plant quarantine (Implementation of Destructive Insects and Pests Act, 1914)

Import

The first recorded plant quarantine measures in India dates back to 1906 when perceiving the danger of introducing the Mexican boll weevil, the Government of India directed all the cotton seeds imported from the new world should be imported into India after mandatory fumigation with carbon-di-sulphide at the port of entry. However, it soon became apparent that cotton is not the only source for bringing exotic pests and pathogens into the country. The danger through plant imports was equally serious and consequently a committee was constituted in 1911 to go into this entire question. The committee recommended among other things that:

(i) Plant imports likely to introduce pests should be fumigated.

(ii) Importantion of plants from foreign countries should be permitted through principal ports only and,

(iii) All living plants, excepting vegetables or fruits in India for consumption and seeds and few other specified exceptions, should be fumigated with HCN gas at the port of entry.

Consequently based on the recommendations of the committee, the DIP Act was passed on 3rd February, 1914 to regulate the introduction into India and transport from one State to another in India of any insect, fungus or other pest which is/may be destructive to crops. The DIP Act included provisions under Section 4 empowering the officers of the Customs Department to implement the rules framed under the Act. These officers continued to discharge this function until the Directorate of Plant Protection, Quarantine and Stroage was establidhed in 1946.

The first Plant Quarantine and Fumigation Station was formally inaugurated in December, 1951 at Bombay. During the successive Plan periods, plant quarantine and fumigation stations were established at all major sea ports, international airports and land frontiers. At present, there are 26 plant Quarantine and Fumigation Stations located at 10 international airports, 9 sea ports and 7 land frontiers. In order to enforce the provisions of the DIP Act, two glass/screen houses had been commissioned at Faridabad with the assistance of the Government of Australia with a view to augmenting post entry quarantine facilities.

With a view to providing best quality seeds/planting materials to the Indian farmers for increasing the productivity, the farming income and earning, the Government of India has liberalised the Import Policy by introducing New Policy on

Seed Development (NPSD) with effect from 1st October, 1988 without compromising plant quarantine requirements.

Export

In order to comply with the provisions of the International Plant Protection Convention (IPPC), 1951 sponsored by FAO to which India is a signatory, identified officers have been authorised by the Government of India for issue of phytosanitary certificates in respect of agricultural commodities.

Besides, the Government of India have also approved three national agencies as recommended by the National High Level Scientific Committee for Plant and Animal Introduction and Quarantine, for the introduction into India of planting materials of various crops for experimental purposes. These agencies are:

(a) National Bureau of Plant Genetic Resources, New Delhi for agricultural and horticultural plants,

(b) The Forest Research Institute, Dehradun for forest plants and

(c) Botanical Survey of India, Calcutta for ornamental and other plants.

Domestic Quarantine

There are certain exotic diseases which have alredy been introduced and are causing extensive damages to the crops in the country which are required to be contained/controlled through the domestic plant quarantine measures. A few of such examples are as under:

a. Banana bunchy top
b. Banana mosaic
c. Apple scab
d. Potato wart
e. Potato golden nematode
f. Coffee berry borer
g. Apple woolly aphis

III. Integrated pest management

For sustaining crop production potential from the ravages of pests and diseases, in the wake of introduction of high yielding varieties during the late 60s, chemical pesticides became the major tools to the farmers to tackle the problem, resulting in their indiscriminate and excessive use of pesticides. The indicriminate use of pesticides resulted in the adverse effects like pest resurgence, pesticide residue, poisoning, health hazards, destruction of natural biocontrol fauna, ecological imbalances and environmental pollution. In view of the safety to the eco-system and preservation of environment, the Government of India has reoriented its policy on plant protection and the Integrated Pest management approach has been recognised as a major thrust for sustaining agricultural production and improving the quality of agricultural produces. In this context, 26 Central IPM Centres have been established by merging erstwhile 43 stations (11 Central Biological Control Stations, 19 Central Surveillance Stations and 13 Central Plant Protection Stations) which act as model stations for disseminating IPM technology among the State extension functionaries and farmers.

A modest beginning to transfer the IPM technology through IPM demonstration-cum-training to the State extension functionaries has been made by these Central IPM Centres in rice and cotton crops in collobaration with the State Departments of Agriculture. In addition, these Stations are also undertaking pest surveillance, monitoring and mass rearing of biocontrol agents. In order to disseminate IPM technology upto grassroot level workers including farmers, a massive approach towards IPM training-cum-demonstration is being undertaken considering the training and demonstration as the key areas for popularising the IPM approach. During the VIII Plan period, it is proposed to train about 35,000 State extension functionaries in IPM technology. The components of IPM are adopted and popularised while executing the IPM programmes in case of rice, sugarcane and potato and demonstrated at farming level. Some of the components of IPM on pulses, oilseeds, vegetable crops have also been evolved and further researches are in progress to develop IPM packages for these crops.

The Indian Council of Agricultural Research and the State Agricultural Universities are also engaged in research for developing and updating IPM packages in rice, cotton, sugarcane, vegetables, oilseeds, pulses and potato, etc. Also there are specific research programmes for IPM in apple, coconut, weed control, biocontrol, white grub control, etc.

The FAO Intercountry Programme in rice is in operation in India since 1981 under the aegis of FAO-IPC programme. Ten intensive field training programmes of 8-15 days duration each have been organised and 359 SMS of States have been trained in IPM skills and technologies who are serving as master trainers. The FAO Intercountry Programme in rice has immensely helped India to create a solid core rice IPM trained master trainers. The second phase of the programme has also been approved. In addition, a UNDP programme for strengthening IPM activities in India has also been finalised.

In view of the ill-effects of chemical pesticides, emphasis has been laid upon plant origin pesticides which are comparatively cheaper and at the same time safer for use by the users than chemical pesticides. In this context, the Registration Committee has decided to promote the use of bio-pesticides namely Bacillus thuringiensis (B.t.) and Bacillus sphaericus (B.s.) for use in agriculture and public health and allowed commercialisation in order to promote IPM programme. Besides, plant origin pesticides, neem based pesticides have also been registered for use in agriculture. Simultaneously, the data requirements for registration of these pesticides have been simplified to the bearest minumum in order to boost their production and export.

IV. Training in plant protection

To cope with the increasing demand for plant protection specialists and to enhance human resource development with the ultimate objective of providing at least one plant protection worker for each block and to augment the progress of State Governments and other agencies, the Central Plant Protection Training

Institute (CPPTI) under the Directorate of Plant Protection, Quarantine & Storage was established at Hyderabad in 1966. The Institute was established especially to provide appropriate training to the extension workers dealing in plant protection with emphasis on Integrated Pest Management. Elaborate facilities for conducting regular tests and trials with various pesticides and plant protection equipments by adaptive research have also been provided.

The Institute has been expanded and strengthened since 1974 under FAO and UNDP assistance. Besides, the Institute has also been recognised as an advanced training centre in plant protection under NAEP III of the Directorate of Extension, Ministry of Agriculture. The Institute conducts regular training courses in various disciplines on plant protection and pesticide formulations and residue analysis besides arranging specially designed short duration courses of pest and disease surveillance, plant quarantine, weed control, efficacy of plant protection practices by adopting IPM approach, demonstration of improved methods and techniques of plant protection by organising seminars, symposium and workshops.

A ten months Post Graduate Diploma course in plant protection has been introduced from 1977. Apart from imparting training to the in-service personnel, special provision has been made in the course for training unemployed agricultural graduates for better utilisation of their potential in pest management on modern scientific lines.

A new ten months Post Graduate Diploma Course in Pesticide Chemistry and two months Orientation Course in Plant Protection (twice in a year) have been introduced since 1993 for inservice personnel to enhance their technical skill in their respective field.

V. Locust warning and control

The Indian Council of Agricultural Research sponsored a scheme or research on the desert locust in 1931 following the desert locust plague of 1925-31. After the termination of this scheme in 1939, the Government of India established a Locust Warning Organisation (LWO) on permanent basis with a nucleus staff under the supervision of the then Imperial Entomologist to the Government of India. In October, 1946 with the establishment of the Directorate of Plant Protection, Quarantine and Storage under the Ministry of Agriculture, Government of India, the Locust Warning Organisation was merged with the Directorate and further strengthened.

With a view to improving the control techniques, a field research unit was set up at Bikaner in 1957 for conducting studies on various aspects of locust biology and ecology in the field. At present, there is a network of 4 circle offices, one field station for investigation of locust and 24 outposts for keeping constant vigil on locust activities in the Scheduled Desert Area of Rajasthan, Gujarat and Harayana.

In order to study the locust migration from one country to another and for effective control, international cooperation is essential which is being coordinated by FAO since 1951. India actively participates in international control for locust and also is a member of FAO Desert Locust Control Committee and the Regional Commission for Control of Desert Locust in South East Asia.

Under a FAO sponsored project, a remote sensing laboratory has been set up to detect the potential locust breeding ground at the field headquarters of the Locust Warning Organisation, Jodhpur to analyse imageries received from the satellites which are coordinated by the National Remote Sensing Agency, Secunderabad (Andhra Pradesh).

Thrust areas and future strategies

Efforts to overcome pesticide hazards:

To overcome the serious hazards due to pesticides as well as keeping in view the growing concern both at the national and international level, concerted efforts have been initiated by Central and State Governments for safer and judicious use of pesticides.

(i) Regulatory measures to ensures safety while manufacturing/formulating, sale, transport, storage and use of pesticides.

Different provisions have been laid down under the Insecticides Act, 1968 and the Rules framed thereunder to ensure safety of insecticides to human beings and animals. The Registration Committee constituted under the Insecticides Act registers an insecticide only after satisfying itself on various safety parameters. On the basis of toxicity, the Committee has banned about 12 pesticides for use in the country and restricted the use of 13 pesticides. The endeavours are still on to register safer pesticides which may be lethal to target species but comparatively harmless to the nontarget species including human being.

(ii) Minimising the dependence on pesticides by promoting IPM.

As a result of implementing Integrated Pest Management through pest surveillance, conservation and augmentation of beneficial species, promotion of tolerant and resistant varieties and IPM demonstration cum training, there has been some reduction in the pest outbreak as well as in the use of pesticides. There are many success stories in the States of Tamil Nadu, Kerala, Andhra Pradesh and Orissa where IPM has gained acceptance among the paddy farmers and among the cotton growers in Andhra Pradesh, Tamil Nadu, Punjab, Haryana and Gujarat.

(iii) Education through training, demonstration and publicity on the safe and judicious use of pesticides.

Both at Central and State level, periodical training are organised to educate the users of pesticides especially the farmers, the workers in public health programmes, extension field functionaries and pesticide dealers. However, keeping in view the vastness of the country and the large number of pesticide users, systematic efforts are needed for publicity through various media such as radio, TV, bulletins, pamphlets, posters, etc.

Recent Advances in Indian Entomology
Editor: O.P. Lal
APC Publications Pvt. Ltd., New Delhi, 1996, pp. 22-27

Genetic Engineering of Plants for Insect Pest Management

R. P. SHARMA

The insect pest management is one of the most important yield destabilizing factors in almost all the crops. While for some insects, genetic sources of resistance are available in the respective plant germplasm, for many others germplasm is almost devoid of effective resistance genes. It is because of this factor that conventional breeding strategies have not been successful in breeding insect pest resistant varieties. The only means available at present for managing the insects is extensive application of insecticides and pesticides. Excessive usage of the pesticides causes, on one hand, environmental pollution and on the other, it forces the insects to develop resistance against these chemicals. Application of the pesticides also imposes additional economic burden on the farmers.

In recent years, the area of genetic engineering has provided an altogether new approach for circumventing the use of chemical insecticides for controlling the insects and pests. This approach employs the techniques of molecular biology whereby genes coding for insecticidal proteins can be identified, isolated, mobilized an expressed in crop plants. Such transgenic plants thus acquire the ability, which is heritable, to protect themselves from insect pest attack.

The most common gene candidates being used for engineering plants resistant to insects pests are:

a. Delta endotoxin coding gene from *Bacillus thuringiensis* strains.
b. Protease inhitors genes from cowpea, soyabean etc.
c. Amylase inhibitors from cereal and legumes
d. Lectins, primarily from wheat germ, snowdrop plant etc.

Bacillus thuringiensis crystal proteins :

Bacillus thuringiensis, a gram positive soil bacterium is omnipresent. It produces intracellular crystal proteins or delta endotoxin during its sporulation (Barton et. al., 1987). These proteins often appear mocroscopically as distinctively shaped crystals and constitute about 20-30% of the dry weight of sporulated cultures (Boutler et. al., 1990). The crystal proteins generally exhibit a molecular weight in the range of 70 kD to 130 kD. These proteins are proteolytically cleaved and activated to a core fragment of about 60 kD inside the larval midgut. The toxic fragment of the 130 kD proteins is localized in the N-terminal helf and the highly conserved C-terminal half is not essential for toxicity (Fietelson et. al., 1992).

The crystal proteins charaterized by their insecticidal activity and spectrum are grouped in four classes. Lepidoptera-specific (Cry I), Lepidoptera and Diptera-specific (Cry II), Coleoptera-specific (Cry III) and Diptera-specific (Cry IV).

Today more than 45 nucleotide sequences of crystal protein genes have been determined. Although they constitute a family of related genes, there are 17 distictly different crystal protein genes. Feitelson *ei.a1.*(1992) have expanded this range to 29 distinct crystal proteins based on their toxin domain structures. Identification of new strains of B. t. exhibiting toxicity to mosquitoes and beetles has stimulated extensive screening programes for novel variants of crystal proteins.

In terms of toxicity, the target organ for Cry toxin is the larval midgut. Upon ingestion, the insoluble crystalline protoxins are solubilized in the gut which has a high pH of 12. The protoxins of 130 kD are activated by gut protease and the resulting toxic fragments bind to specific receptors in the target tissue. After binding to a specific receptor, located on the brush border membrane of the columnar cells, the toxins insert irreversibly into the plasma membrane of the cells which renders the cell's volume regulation mechanisms ineffective. Accordingly, the cell swells and ultimately bursts resulting in the death of the insect from starvation of septicaemia.

The delta endotoxin gene/s, cloned in plant transformation vectors like *Agrobacterium tumifaciens* binary vectors, have been successfully introduced into a number of plants. The first reports of insect resistant transgenic plants appeared in 1987., (Gill et al., 1992; Gould, 1991; and Higuchi et. al., 1984). Tobacco and tomato plants were stably transformed by using *Agrobacterium tumefaciens*. carrying vectors comprising different cry gene constructs. The transgenic plants differed in their resistance to tobacco hornworm (*Manduca sexta)*, tobacco budworm/ cotton boll worm (*Heliothis zea*) and beet army worm (*Spodoptera exigua*). depending on the type of gene construct used. The effectiveness of the transgenic plants in insect pest management, however, turned out to be very limited, primarily because of the sub-optimal expression of the B.t. crystal protein gene in plant cells. This became an impediment in the commercial exploitation of transgenic plants. However, in the last six years, spectacular progress has been made in enhancing the cry gene expression in the transgenic plants which include important monocot crops like rice and maize (Table 1).

The expression of delta endotoxin gene in the transgenic plants was shown to be suboptimal when the full length gene was introduced. It has now been shown that this impediment can be circumvented by either truncating the gene i.e. removing the c-terminal half coding part of the gene which is not necessary for insect toxicity or by modifying bacterial biased codons towards plant codon usage. Synthetic genes with plant biased condons and having m-RNA stabilizing features ensuring maximum levels of expression in transgenic plants have been constructed and are being used in generating genetically engineered plants for insect resistance.

Since the first reports in 1987, tremendous progress has been made in engineering *cry* gene into important crop plants like potato, tomato, maize and cotton. These transg*ᵉ* plants are at advanced stages of testing and awaiting

Table 1. Crop plants transformed for cry gene expression

Crop	Target insect	cry gene	Major groups involved
Tabacco	*Manduca sexta, Spodoptera sps.,*	*cryIA (b)*	Monsanto, calgene, Rohm and Has
	Heliothis sps.,	*cryIA (c)*	Sandoz, CIBA-Geigy, Plant Genetic System
Tomato	*Heliothis sps., Keiferia sps.,*	*cryIA(b)* *cryIA(c)*	Agrigenetics, Campbell, Monsato, Rogers NK, Plant Genetic systems
Potato	*Phthorimaea operculella,*	*cryIA (b)*	Monsanto, Calgene
	Leptenotarsa decemlineata	*cryIIIA*	Frito-Lay
Cotton	*Heliothis sps.,*	*cryIA (b)*	Monsanto, Calgene,
		cryIA (c)	Agracetus, Northrup King
Corn	*Ostrinia nubilalis*	*cryIA (b)*	CIBA-Geigy Calgene, Monsanto
	Chilo suppressalis,	*cryIA (b)*	Plantech Research
	Cnaphalocrocis medinalis		
Broccoli	*Plutella xylostella*	*cryIA (c)*	Cornell University
White spruce	*Chorisoneura fumiferana*	*cryIA (a)*	Agracetus

Other crop species include rapeseed, apple and walnut.

clearances from licensing authorities for commercial cultivation. It is speculated that the first commercial products will be marketed by 1995-96. In order to ensure prolonged effectivity and to widen the application to a broad spectrum of crops and a wider range of insects, there is need for intensive research in different areas.

Designing synthetic cry genes for specific crops, keeping in view the plant's codon usage, will further improve the protein expression and insect toxicity. In addition, indentification of novel and more toxic crystal protein genes using PCR technology wherein primers corresponding to the toxic domain of the gene can be used to amplify the DNA from naturally occurring bacterial strains is expected to be quite useful in such an endeavour. Strategies to introduce truncated cry genes into chloroplast genome will circumvent the codon usage problems because of the prokaryotic nature of the chloroplast genome and transcriptional and translational apparatus. Such an approach will be more useful for leafy vegetables and crops in which green tissues are predated upon by insects. Chloroplast transformation, though difficult, has been achieved recently (Hilder et al., 1987). What is needed is construction of chloroplast vectors carrying a truncated version of native cry gene and its expression under a suitable chloroplast promoter. It will be of interest to investigate the levels of cry gene expression in the chloroplasts and its effectiveness.

In recent years, concern has been raised about the development of resistance in insects to B.t. crystal proteins of genetically engineered plants (Ishimoto and Kitamura 1988). The fact that different crystal proteins can bind to different receptors in the insect midgut suggests that a combination or alteration of different proteins could help retard the development of resistance (Johnson et al., 1989)

Construction of vectors carrying two different cry genes or vector cassettes containing a cry gene and another insect resistance gene like proteinase or amylase inhibitor or lectins will help to avoid the possibility of insects developing resistance. Expression of proteins in specific tissues (using tissue specific promoters) during certain stages of crop growth and development viz. seed development will reduce the selection pressure on insects allowing limited damage to less important plant parts. Agronomic practices like raising a crop of transgenic plants interspersed with normal plants is yet another approach which could be investigated.

Protease inhibitors

Other potential candidates in engineering insect resistance in crop plants include protease inhibitors, amylase inhibitors and lectins. Many protease inhibitors have been identified from many plant species, which are active against certain insect species both *in vitro* assays against insect gut proteases and in *in vitro* aritifical diet bioassays. The mechanism of antimetaboilic action of protease inhibitors is not yet fully elucidated. Direct inhibition of digestive enzymes is unlikely to be the only effect and possibly the proteases vital in insect development, such as those involved in moulting are also affected.

Several transgenic plants have been developed for enhanced insect resistance using protease inhibitor genes. The first gene to be successfully transferred was that isolated from cowpea encoding a trypsin inhibitor (Peferoen, 1992). This protease inhibitor has been shown to be an effective antimetabolite against a range of field and storage pests including members of the Lepidoptera, Coleoptera and Orthoptera. A full length cDNA clone encoding a trypsin inhibitor from cowpea was placed under the control of a CaMV 35S promoter and tobacco plants were transformed via *Agrobacterium tumefaciens*. The cpT1 - expressing plants showed only minor damage compared with the control plants against first instal larvae of the tobacco budworm (*Heliothis virescens).* This protection has been subsequently demonstrated for other lepidopteran pests, including *H. zea. Spodoptera littoralis* and *Manduca sexta.*

Not only the genes encoding protease inhibitors isolated from cowpea have conferred resistance when expressed in tobacco but the tomato inhibitor II gene when expressed in the same model plants has also been shown to confer insect resistance (Schnepf and Whiteley 1981).

α - amylase inhibitors

α-amylase inhibitors present in cereal and legume seeds have been shown to be inhibitory to insect gut α-amylases. Ishimoto and Kitamura (1988) have claimed that an α-amylase inhibitor was responsible for the protection of *Phaseolus vulgaris* seeds against attack by *Callosobruchus chinesis.* a storage pest of certain legumes. the use of α-amylase inhibitor genes to confer insect resistance in transgnic plants is yet to be demonstrated.

Lectins

Lectins are carbohydrate binding proteins found in many plant tissues and are abundant in the seeds and storage tissues in some plants. the seeds of certain legumes are particularly rich sources of lectins. For example, in *Phaseolus vulgaris* lectins constitute upto 3% of the seeds protein. Most of the lectins are toxic to mammals, birds and insects. The mechanism of lectin toxicity is analogous to that believed to occur in rat, namely that the ingested lectins causes disruption of midgut epithelial cells. This leads to a breakdown of nutrient transport and facilitates the absorption of potentially harmful substances.

Certain lectins like bean lectins and winged bean lectins have been identified as insecticidal. Soybean lectin has been shown to be toxic to the larvae of the lepidopteran, *Manduca sexta.* Similarly, wheat germ agglutinin has been shown to have an ihibitory effect on European Corn borer *(Ostrinia nubilalis)* and corn root worm *(Diabrotica undecimpunctella).* The disadvantage, however, is using the above mentioned lectins is that they have also bean shown to exhibit mamalian toxicity (Vaeck et al., 1987).

The pea lectin, on the other hand has been shown to be innocuous to mammals but insecticidal only in the purified form. In addition, letins purified from snowdrop and garlic are insecticidal and non-toxic to mammals. Snowdrop lectin is toxic to brown planthopper (*Nilaparvata lugens),* a pest of rice.

Genes encoding pea lectin have been expressed at high levels in transgenic tobacco plants and the transgenic plants had enhanced levels of resistance to *Heliothis virescens*. The gene encoding snowdrop lectin has also been successfully transferred to tobacco, where the results indicate that it too has conferred a measure of resistance.

Conclusion

It should be borne in mind that engineering plants for insect pest resistance offer only one of the routes towards insect pest management. Insects themselves are dynamic living systems and are expected to face the adverse situation challenging their existence. Thus, continuous exposure to transgenic plants might force the insect to develop genetic resistance in due course of time. As outlined earlier in this article, deployment of gene cassettes comprising of various Bt. genes, protease inhibitors and lectins can be an answer to reduce the chances of insects developing resistance. Alternatively, and which seems to be a preferred solution, is the development of integrated pest management practices suited to crop-insect-location requirement. It is, therefore, extremely important that concerted efforts should be directed wherein biopesticides, transgenics technology, biological control measures and chemical insecticides are integrated in a well defined fashion for effective and durable insect pest management.

References

Barton, K.A., Whiteley, H.R. and Yang, N.S. L *Bacillus thuringensis* delta endotoxin in transgenic *Nicotiana tobacum* provides resistance to Lepidopteran insects. *Plant. Physiol* **85** : 1103-1109.

Boulter, D., Edwards, G.A., Gatehouse, A.M.R., Gatehouse, J.A. and Hilder, V.A. 1990. Additive protection effects of incorporating two different higher plant derived insect resistance genes in transgenic tobacco plants. *Crop Protection;* **9** : 351-354.

Feitelson, J.S., Rayne, J. and Kim, L. 1992 *Bacillus thuringiensis* : Insect and beyond., *Bio/Tech* **10** : 271-275.

Fischoff, D.A., Bowdish, K.S.; Perlak, F.J., Marrone, P. Gr, Me Cormick, S.M.; Niedermeyer, J.G.; Dean, D.A.; Kuşano-Kretzmer, K.; Mayer, E.J.; Rochester, D.E. : Rogers, S.G. and Fraley, R.T. 1987. Insect tolerant transgenic tomato plants. *Bio/Tech,* **5** : 807-813.

Gill, S.S., Cowles, E.A. and Pietrantonio, P.V. 1992. The mode of action of *Bacillus thuringienis* endotoxins. *Ann. Rev. Entomol.* **37** : 615-636.

Gould, F. 1991. The evolutionary Potential of crop pests. *Amer. Sci. 79* : 496-507.

Higuchi, M., Tsuchiga, I. and Iwai, K. 1984. Transgenic plants protected from insects attack. *Agric. Biol. Chem.* **48** : 695.

Hilder, V.A., Gatehouse, A.M.R., Scheerman, S.E., Barker, R.F. and Boulter, D. 1987. A novel mechanicm of insect resistance engineered into tobaco. *Nature; 330* : 160-163.

Ishimoto, M. and Kitamura, K. 1988. Identification of the growth inhibitor on azuki. bean weevil in Kideny bean *(Phaseolus vulgaris* L.). *Jap. J. Breed.* **38** : 367-370.

Johnson, R., Narvaez, J., An, G. and Ryan, C.A. 1989. Expression of protlease inhibitor I & II in transgenic tobacco plants : Effect on natural defense against *Manduce sexta* larvae *Proc. Natl. Acad. Sci. USA* **86** : 9871-9875.

Meeusen R.L. and Warren G. 1989. Insect control with genetically engineered crops. *Ann. Rev. Entomol.* **34** : 373-381.

Peferoen, M. 1992. Engineering of insect resistant plants with *bacillus thuringiensis* crystel protein genes. In Plant Genetic manipulation for crop protection (Edits A.M.R. Gratehouse, V. A. Hilder, D. Boulter) CAB. Intl. pp. 135-153.

Schnepf, H.E., and whiteley H.R. 1981. Cloning and expression of the *Bacillus thuringiensis* crystal protein gene in *E. Coli. Proc. Natl. Acad. Sci. USA.,* **78** : 2893-2897.

Svab, Z. and Maliga, P. 1993. High frequency plastid transformation in tobacco by selection for a chimeric *aad A* Gene. *Proc. Natl. Acad. Sci. U.S.A.,* **90** : 913-917.

Vaeck, M., Reynaerts, A., Hofte, H., Jansens, S. , De Beuckeleer, M., Dean, C., Zaleau, M., Van Montagu, M. and Leemans, J. 1987. Transgenic Plants protected from insect attack. *Nature* : **328.** 33-37.

Recent Advances in Indian Entomology
Editor: O.P. Lal
APC Publications Pvt. Ltd., New Delhi, 1996, pp. 28-37

Application of Remote Sensing in Entomology

Y. N. RAO

What is remote sensing

Everything in nature reflects, absorbs or emits some type of electromagnetic radiation. 'Remote Multispectral Sensing' is the detection and recording, from a distance or remote location, of reflected or emitted electromagnetic radiation in many discrete, relatively narrow spectral bands between 0.3 micron and 14 micron wavelengths and radar and Spassive microwave bands between 1 mm and 1 m wavelenghts (fig. 1).

The response of earth surface materials to incident radiation (i.e., their reflectance) and/or the energy emitted by all objects as a function of their temperature and structure (i.e., emittance) essentially determines the signatures (fig. 2).

The generalised processes and elements involved in electromagnetic remote sensing of earth resources are data acquisition and data analysis. In short, we use sensors and record variation in the way earth surface features reflect and emit electromagnetic energy. The data analysis process involves examining the data using various viewing and interpretation devices to analyze pictorial data, and/or a computer to analyze digital sensor data (fig. 3).

Spectral characteristics of vegetation

Spectral characteristics of vegetation describe the nature of energy-matter interaction that takes place when the energy in the (a) visible, (b) near-infrared and (c) middle-infrared wavelength bands strikes a leaf. The spectral reflectance of green vegetation is distinctive and quite variable with wavelenght (fig. 4).

Plant leaves reflect, absorb and transmit radiation. The reflected energy is equal to the incident energy reduced by the energy, which is either abosrbed or transmitted.

The wavelength region 0.5-0.75μm is the visible light absorptance region, dominated by pigments, primarily chlorophylls a and b, carotenes and xanthophylls.

Chlorophyll absorbs 70-90% of light in the blue (~0.45μm).and red part (~0.68μm). Absorptance is smallest in the green wavelength region (~0.55μm), where a reflectance peak of < 20% occurs from the upper leaf sufaces. Low pigment content results often in higher reflectance (fig. 5).

The wavelenth region 0.75-1.35μm is the near infrared region, and a region of high reflectance and low absorptance, affected considerably by internal leaf structure (fig. 6).

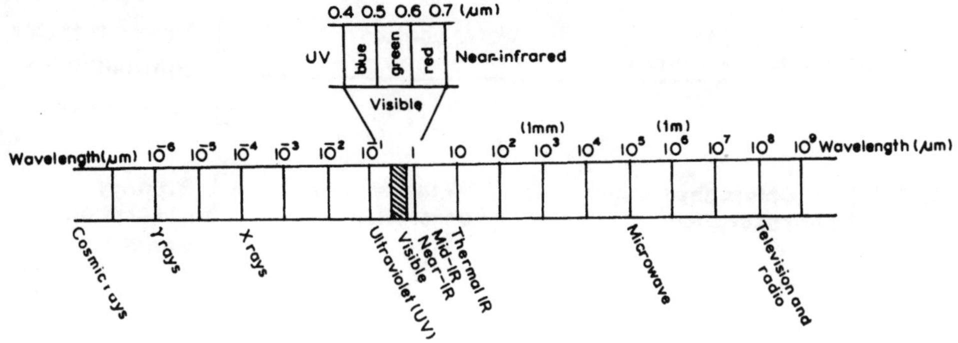

Fig. 1. The electronmagnetic spectrum (Lillesand and Kiefer 1987).

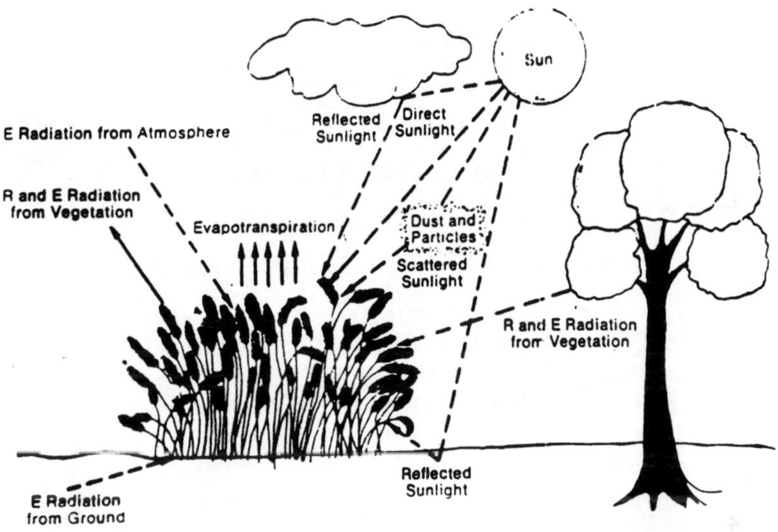

Fig. 2. Interaction of radiations with the terrestrial vegetation. Measurement of radiations reflected (R) or emitted (E) by objects as a result of this interaction is the basis of remote sensing techniques (Courtesy : LARS).

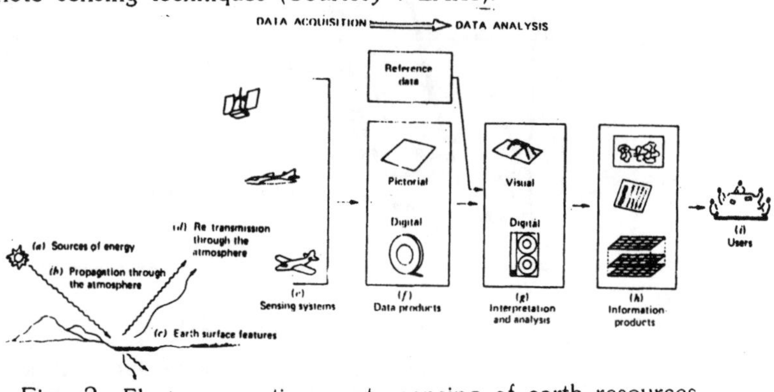

Fig. 3. Electromagnetic remote sensing of earth resources.

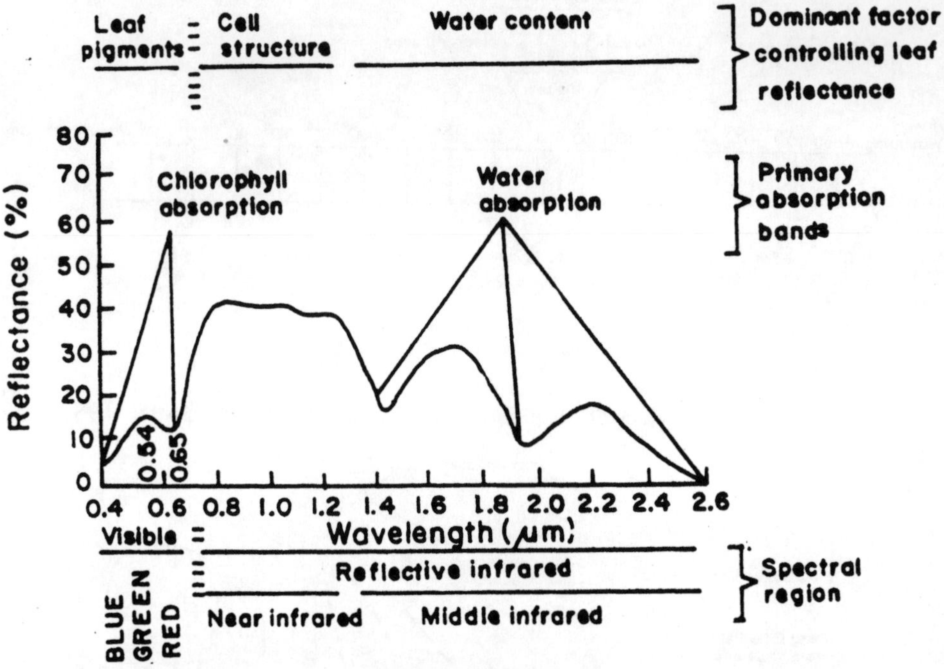

Fig. 4. Significant pectral response characteristics of green vagetation. (Courtesy : LARS).

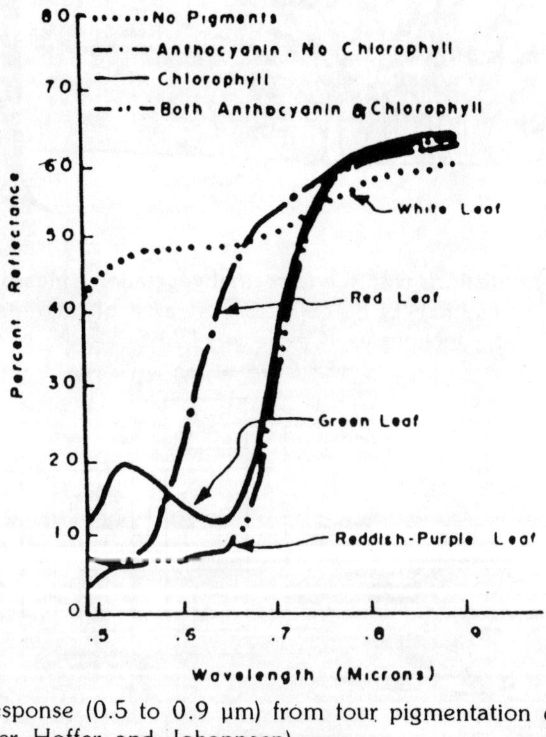

Fig. 5. Spectral response (0.5 to 0.9 μm) from four pigmentation conditions of *Coleus* leaves (After Hoffer and Johannsen).

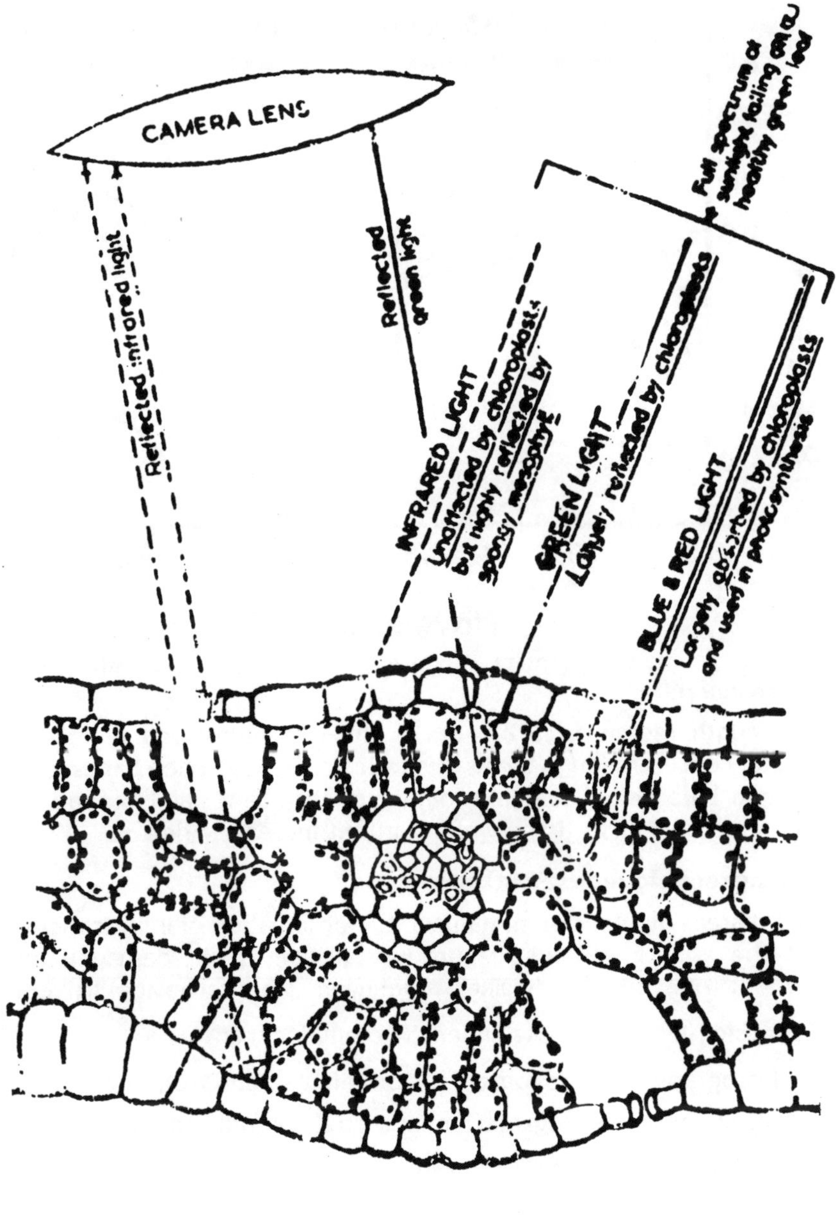

Fig. 6. Structure of a plant leaf and its reflectance characteristics (source : Colwell, 1969).

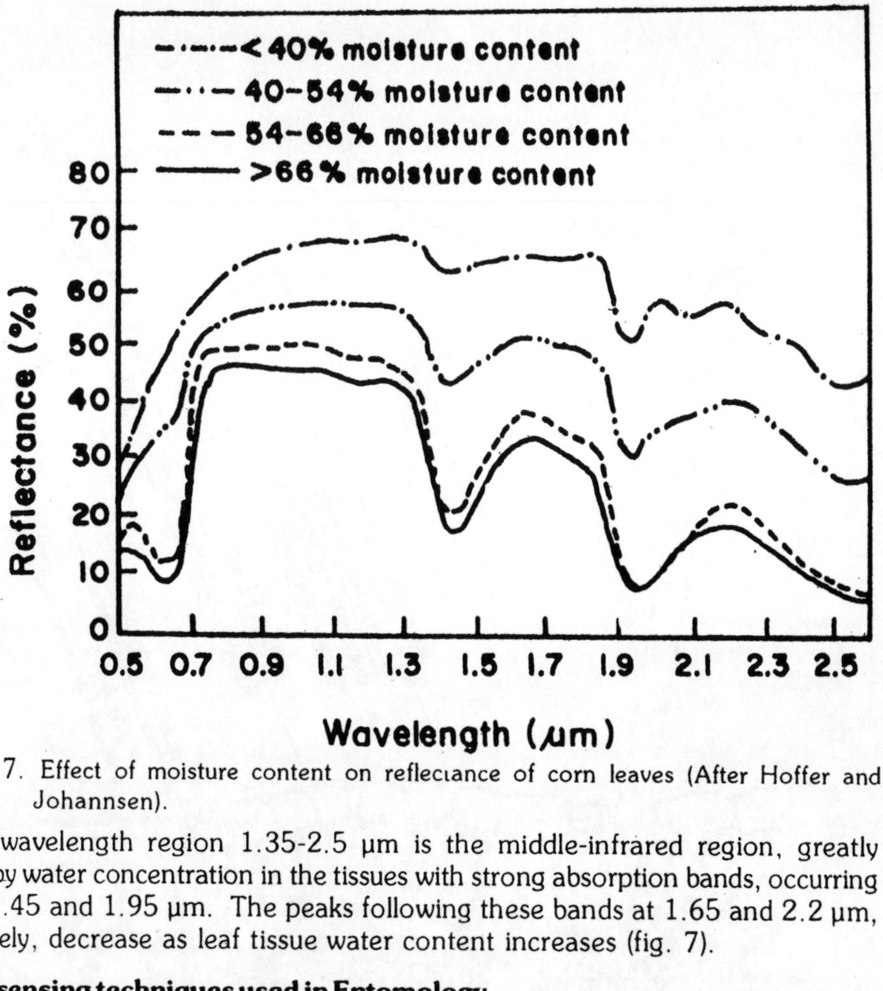

Fig. 7. Effect of moisture content on reflecıance of corn leaves (After Hoffer and Johannsen).

The wavelength region 1.35-2.5 μm is the middle-infrared region, greatly affected by water concentration in the tissues with strong absorption bands, occurring at both 1.45 and 1.95 μm. The peaks following these bands at 1.65 and 2.2 μm, respectively, decrease as leaf tissue water content increases (fig. 7).

Remote sensing techniques used in Entomology

There are 3 areas of application, namely, the observation of insects themselves, detection of the effects that insects produce (usually plant damage), and the monitoring of environmental factors likely to influence insect behaviour (Riley 1989).

Remote sensing techniques used in entomology include :

a. Photography and videography from aircraft and from ground

b. Satellite-borne photography, multisectral scanning and thermal imaging.

c. Ground-based and airborne radar

All entomological applications of remote sensing depend on the propagation of electromagnetic waves between the target being observed and a sensing device.

Aerial photography

Insect identification is a major problem in almost all entomological remote sensing experiments. In aerial photography, particular attention is paid to methods of detecting the activity of insects from the changes they produce in the appearance of plant foliage. Foliage may be changed if insects leave deposits on leaves or if they induce alterations of leaf color, shape, or density as a direct or indirect result of feeding. Photographic method generally relies on the detection of plant stress or damage, which in many cases could have a variety of causes quite unrelated to insect activity. In these instances identification is dependent on circumstantial evidence such as the previous history of insect activity in the area and the timing and pattern of the symptoms. Finally, identification depends on on-the-spot visual inspection by entomologists. However, some insect pests produce honeydew that serves as host for a sooty mold fungus, which balckens foliage and is readily detectable by aerial photography.

Aerial photography was used to investigate the distribution of host plants of tropical fruitflies. Aerial photography is now routinely used in annual surveys to monitor the post harvest plant removal of cotton to prevent the outbreak of boll weevils, *Anthonomus grondis* in Texas.

Multispectral sensing (MSS)

Multispectral scanning can form images over a much wider range of electromagnetic wavelengths (0.4-14 µm) than photographic method. Instead of generating an instantaneous image of a whole scene, multispectral scanners use sensors with very narrow fields of view to scan an area of interest systematically, and an image is built up as the scan progresses. The sensors are responsive to different segments of the electromagnetic spectrum, so each forms an image that represents the reflectance of the scene in its particular wave band. The widely used multispectral scanners in the Landsat series of satellites provide images in four bands with a ground resolution of 57 x 79m : MSS 4 : 0.5-0.6 µm, MSS 5 : 0.6-0.7 µm, MSS 6 : 0.7-0.8 µm, MSS 7: 0.8-1.1 µm.

Landsat 4 and 5 carried in addition a more powerful scanner called a thematic mapper (TM). This device also formed images in the thermal infrared region (10.2 - 12.2 µm), i.e., it responded to infrared radiation emitted by the scene, and thus provide a measure of temperature.

The Indian Remote Sensing Satellite System, i.e., IRS-1A and IRS-1B provide images in the following bands, with 36.5 m and 73.0 m ground resolution :

1. 0.45-0.52 µm
2. 0.52-0.59 µm
3. 0.62-0.68 µm
4. 0.70-0.86 µm

MSS images can be stored on magnetic recording media and subsequently corrected for geometric distortion prior to digital processing, display on an image tube, and transfer to film as color composites.

Satellite based remote sensing has been concentrated in two principal areas, naemly, the detection of changes in vegetation and the measurement of meteorological data.

The spectral response data has been used to derive different vegetation indices. A parameter called the ratio vegetation index, formed by dividing the near infrared by the red respnse (MSS band 4/MSS band 2), was found to be a good indicator to detect insect-induced defoliation of forest areas and to delineate vegetation that might support locust breeding in normally desiccated habitats. Because reflectance in the red region is inversely proportional to chlorphyll density, and reflectance in the near-infrared region is proportional to leaf density, a large index is indicative of a high green-leaf biomas (Jackson et al. 1983). The normalized vegetation index (band 4 - band 2)/(band 4 + band 2), is currently considered somewhat better indicator of biomass than the simple band 4/band 2 ratio used in earlier studies (fig. 8).

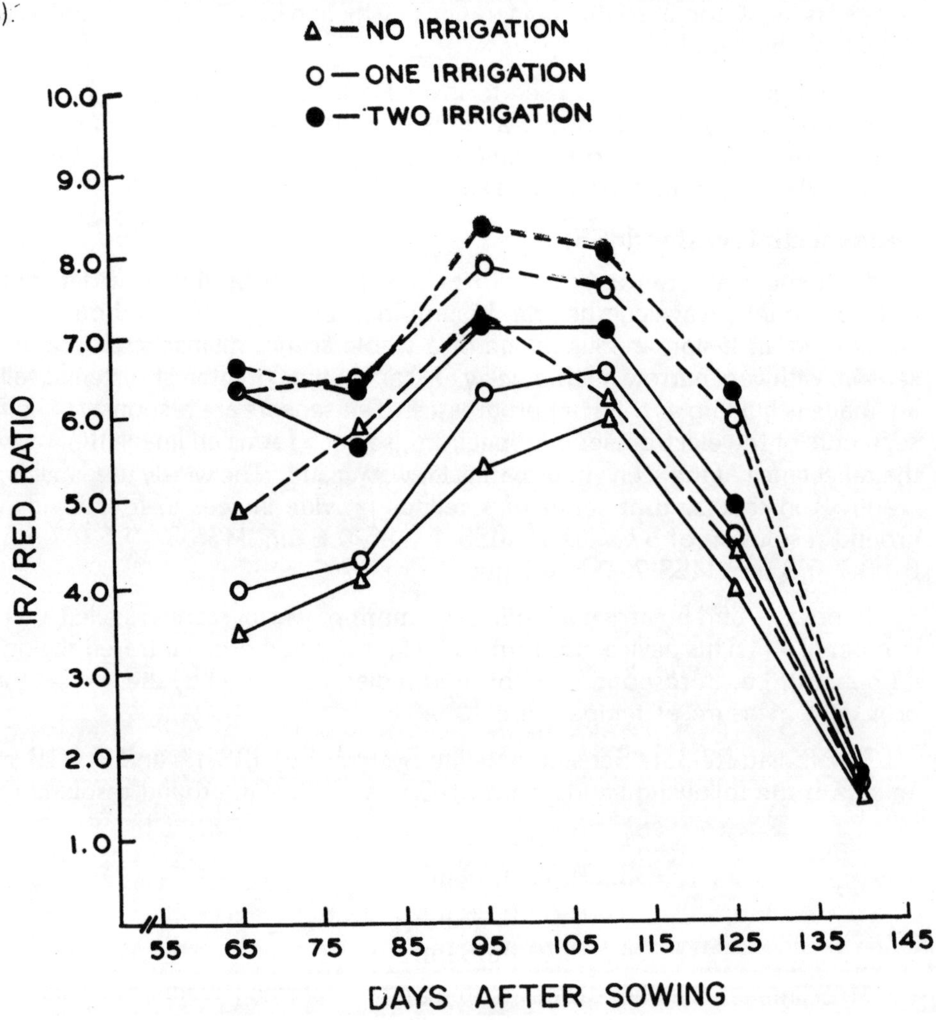

Fig. 8. Variation in IR/Red Ratio during crop Growth : Two Cultivars of Chickpea (After Nagarajarao and Ramana). Δ - NO IRRIGATION O - ONE IRRIGATION ● - TWO IRRIGATIONS

Satellite remote sensing is an ideal tool to investigate the environmental factors controlling the pest development, especially in monitoring rainfall and air temperature. Radiation from the scene is measured (usually at several wavelenghts within the 15-μm band), and air temperature profiles are computed using radiative transfer equations. Rainfall is estimated indirectly from the extent, appearance, and life history of clouds imaged in the visible and infrared regions.

Flying moths of East African armyworm, *Spodoptera exempta,* are concentrated by convergent winds associated with rainstronms and subsequently lead to dense outbreak of caterpillars. Meteosat data has been used to monitor these rainstorms and for rapidly locating potential outbreaks. Landsat data has been used to identify areas of ephemeral vegetation capable of supporting desert locusts (*Schistocerca gregaria*). US/FAO researchers have found the vegetation index maps and rainfall estimates made from AVHRR of NOAA meteorological statellite, a valuable contribution to the forecasting of desert locust.

Radar

Radar has been successfully used as a tool in the study of long distance migration and flight behaviour of windborne / air borne insects. These raders use a wavelength of 3 to 10 cm. In case of small insects, millimeter radar is used. Insect migration can be deduced from the orientation and directive properties of the radar antenna, and target range from the time elapsing between transmission of the illuminating pulse and the reception of the echo. Typical maximum detection range would be 1.5 to 2.8 km for individual insects and up to several tens of kilometers for dense swarms. Flight behaviour of locusts, grasshoppers and moths of various species of *Heliothis, Spodoptera* etc. have been studied.

Application of satellite remote sensing for pest management in India

Vegetation indices, when plotted over the cropgrowing period are highly related to growth variables. Stressed crops have a lower vegetation index than healthy crop. However, the vegetation indices developed so far cannot be used to distinguish different classes of stress and early warning of stress, before its effects in leaf area and light interception are noticed. (Nageswararao et al., 1991).

Desert locusts

The desert locust is endemic over Rajasthan, Gujarat and Haryana and is a part of the world recession area. Successful breeding, triggered by suitable ecological conditions (soil moisture after widespred rainfall, shade and green vegetation development) in the locust recession area, result in a large number of highly mobile and devastating swarms. Destroying locusts is a long and ardous affair, particularly if early action is not taken.

Vegetation index maps and the rainfall estimates generated from NOAA/AVHRR data offer unique capabilities for maintaining a close watch on the desert locust population development. Temporal and spatial distributions of desert vegetation biomass estimates based on AVHRR vegetation indices were closely correlated with the locust population and rainfall data. The establishment of quantitative relationship between a satellite-derived potential breeding activity factor and the observed desert locust populations was attempted by Hielkema et al. (1986). These authors have also suggested that the operational use of satellite data would narrow down the actual area for taking pest control measures to a mere fraction.

Many researchers have reported that the path of the desert locust journey is conditioned by large scale weather systems like the position of the Intertropical Convergence Zone (ITCZ) and low level wind circulation converging air current leading to vertical upward motion. The position of the ITCZ can be ascertained from the cloud signatures observed by INSAT/VHRR and areas likely to be affected by the locust can be visualized in advance. This preliminary forecast can be used to gear up actions to countrer the locust menace.

Brown plant hopper on rice crop

Long distance migration of brown plant hopper (*Nilaparvatha lugens)* is found to be associated with synotptic weather conditions (depressions). The continuous cultivation of paddy, season after season, without crop rotation is commonly considered as being responsible for the endemic BPH areas. Staggering of sowing and harvesting dates of rice over a region is also one of the major causes of BPH damage to rice. Hence, a monitoring system to secure the implementation of crop rules, would help in taking control measures against BPH. In this context, the satellite remote sensing technology can be used as a "watch dog" for regulating the cropping and sowing/harvesting patterns of rice in the BPH prone areas. Optimal period of data acquisiton has enabled the detection of BPH affected rice areas in Bantumalli area of Kristna Dt. (A. P.) during October 1981. The conditions favourable for BPH attacks on rice were monitored for Mandya District of Karnataka.

Whitefly on cotton crop

The cotton crop in Guntur District of A. P. was severely hit by white fly (*Bemisia tabaci*) during 1984-85 and 1985-86 kharif seasons. Adults and nymphs of whitefly feed on the leaves and they suck sap from plants. The larval feeding causes yellow chlorotic spots and mottling, which gradually leads to leaf shedding and reduction in fruit formation. Sooty mold fungi (*Cladosporium sp.*) developing on the honey dew excreta of the insects discolor the leaf and reduce photosynthesis and quality of lint. The cotton crop spatial delineation using Landsat MSS data of 1973 and 1981 confirms the development of the monocultured cropping pattern of cotton in some parts of Guntur/Prakasham districts. In addition, the irrigated crop land adjoining the cotton belt where many vegetable crops are grown in summer, provided alternate hosts for most of the cotton pests. Landsat 5 (MSS and TM) false colour composites were used for the assessment of the condition of cotton crop. Detection of changes in the red colour on Landsat FCC on 28.11.1985 in comparison to previous 'normal' year, supported by ground truth data (collected on 09-12-1985), enabled delineation of areas affected by the whitefly. Areas of moderate (50% crop loss) and severed (80% crop loss) damage by the whitefly were easily identified. The damaged crop area estimated from the Landsat false colour composites on scale 1:2, 50,000 (app) were found to be within plus or minus 12% of the area reported by the Directorate of Agriculture.

Remote sensing techniques could be used :

i. to implement the legislative enactments to stop continuous cotton growing in a specified area,

ii. to demarcate areas requiring crop rotation,

iii. to monitoring the time of harvest of cotton so as to break the life cycle of the pest, and

iv. to estimate the acreage under alternate hosts (tomato, chilli, tobacco etc.) for assessing the carry over potential of the pest in the off season.

Future perspectives for remote sensing in Entomolgy

Support-research in the application of remote sensing in plant protection is necessary. Case studies carried-out in the country need validation and large scale extension. This requires support from researcher and users in this technology. It is also necessary to investigate the economics of using remote sensing techniques in combination with the conventional method of farmers or extension workers inspecting the fields at frequent intervals. Cost-benefit analysis and standardization of methodology have to be carried out for each group of stress-causing and yield reducing factors.

References

Curran, P. J. 1985. Principles of Remote Sensing Longman, London.

Hielkema, J. U., Roffey, J, and Tucker, C. J. 1986. Assessment of ecological conditions associated with the 1980-81 desert locust plague upsurge in West Africa using environmental statellite data. *Int. J. Remote Sensing* **7** : 1609-1622.

Jackson, R. D., Slater, P. N. and Pinter, P. J. Jr. 1983. Discrimination of growth and water stress in wheat by various vegetation indices through clear and turbid atmospheres. *Remote Sens. Environ.* **13** : 187-208.

Lillesand, T. M. Kiefer, R. W. 1987. Remote Sensing and Image Interpretation. John Wiley and Sons, New York.

Nageswararao, P. P., Jayaraman, V. and Chandrasekhar, M. G. 1991. Applications of Remote Sensing in Plant Protection. Technical Report ISRO - NNRMS-TR-89-91.

Riley, J. R. 1989. Remote Sensing in Entomology . *Ann. Rev. Entomol.* **34** : 247-271.

Swain, P. H. and S. M. Davis (eds) 1978. Remote Sensing : The Quantitative Approach. Mcgraw-Hill, New York.

Recent Advances in Indian Entomology
Editor: O.P. Lal
APC Publications Pvt. Ltd., New Delhi, 1996, pp. 38-42

Ecofriendly Pesticides: Its Impact in Agriculture

N. K. ROY

The country's requirement of foodgrains in 2000 AD is estimated around 250 million tonnes. But the current production is around 178 million tonnes. Since the cultivable land is limited, the additional requirement can be met by increasing the productivity. The enhanced productivity may be achieved by timely supply of adequate inputs, such as seeds, irrigation, fertilizers and pesticides. The role of Pesticides is of special significance to provide protection against insects and pests.

The pesticide consumption inIndia is around 70,000 metric tonnes. More than 50% of our pesticides is of chlorinated hydrocarbons viz. DDT, BHC, endosulfan etc. followed by organophosphates, carbamates and pyrethroids. After Bhopal tragedy the use of carbamate pesticides is very much restricted. It is now being imported to some extent for use in cotton.

Although a large number of pesticides are used in agriculture, contamination is mostly from DDT and BHC. The Government of India has banned the use of DDT is agriculture and use of BHC is restricted on few crops and a committee has been constituted to review its future use in agriculture.

What is ecofriendly pesticides

The pesticides which are less harmful to environment (including human health) or have benign impact on the environment through the various stages of development, manufacture, use, packing, distribution, consumption and disposal and recycling may be called ecofriendly product (EEP) and will be labelled as Eco. Mark.

A large number of safer biodegradable chemicals are now available in the world which are ecfriendly. A short list of ecofriendly pesticide is given below.

Insecticides

A large number of synthetic pyrethroids viz. cypermethrin, decamenthrin; fenvalerate, fluvalinate etc. are used in agriculture. These compounds have many advantages as they are biodegradable and do not leave toxic residues in the environment.

It is now reported that these compounds tend to concentrate on leaf waxes and are extremely toxic to fish and are inactive towards ticks or mites. A large number of oxime ether pyrethroids have been discovered whichare found effective against ticks or mites. The fish toxicity was lower than the earlier group of pyrethroids.

They show quick toxic action and low mammalian toxicity. The most important advantages is the photostability which is greatly increased. The most potent commercial insecticide is 4-chlorophenyl ketoxime (3-phenoxy) benzyl ether (1). Thus a series of oxime ether ecofriendly pyrethroids are now available as insecticides for use in agriculture.

A series of novel dithiacyclic carbomoyl oximes which contain a readily oxidizable sulfide linkage and a thiohydroximate group in the same molecule has been synthesised, and its insecticidal and miticidal properties was determined. One of the most active compound in this series is 4-(methyl carbamonyl) oximino-1, 3-dithiolanes with one or two methyl substitutents at the C-5 position (2). These are found biodegradable and ecofriendly.

Propesticides

The concept of propesticides is completely new and gaining importance currently. These compounds are metabolised into active pesticides within the pests or plant biological system e.g. carbosulfan (3) and acephate (4). These compounds are very safe towards mammals and also environment friendly as these are easily converted into carbofuran (5) and methamidophos (6) respectively in the biological system.

Now targetted pesticides

The insecticides so far discussed all act on central nervous system common to both insects and mammals. A few effective compounds which are available in market are chlordimeform (7) and dimilin (8). These are called as ecofriendly as they do not leave any toxic residues in the environment.

Plant origin pesticides

Pesticides of botanical origin have numerous advantages in comparison to synthetic pesticides because of their quick biodegradability, safety to environment and hence ecofriendly. The neem tree (Azadirachta indica A. Juss : Meliaceae) contains a number of compounds having pesticidal properties. Salannin (9) is an *effective* housefly deterrent. Meliantriol (10) and Azadirachtin (11) are effective antifeedant for locusts. Azadirachtin is also used as repellant. It also disrupts the metamorphosis of several insect larvae.

Similarly, Ryanodine (12) isolated from *Ryania speciosa vahl* (Farn. Flacowtiaceae) has good insecticidal property against *Tribolium casteaenum* larvae and ecofriendly too.

A number of isobutylamides of unsaturated C_{10}-C_{18} acids such as pellitorine (13) and its analogues have been isolated having insecticidal activity. Several potent synthetic analogues like synthetic pyrethroids have been developed. N-isobutyl 12-(3-trifluoromethyl phenoxy-(2E, 4E)-2, 4 clodecodienamide (14) was the most active compound.

Fungicides

A new class of systemic fungicides belonging to azole group viz. etaconazole (15) and tridemenol (16) are also now available. these are required at low dosages and do not leave any toxic residues in the environment.

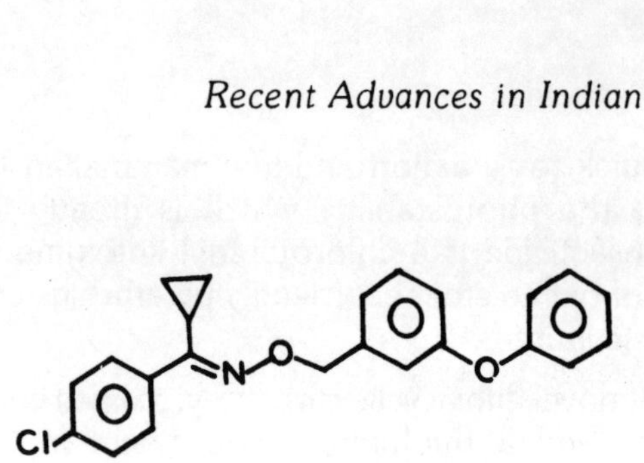

(1)

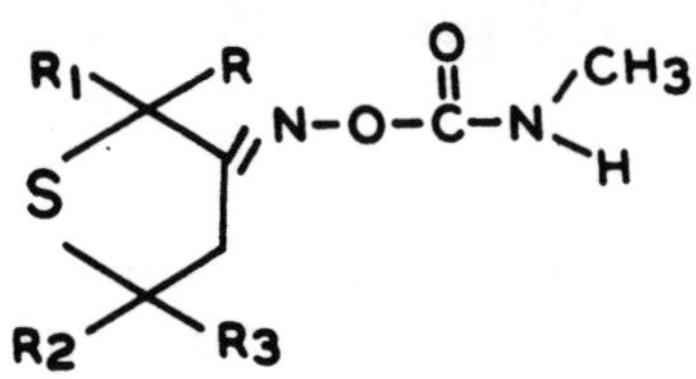

R	R1	R2	R3
CH_3	H	H	H
CH_3	CH_3	H	H

(2)

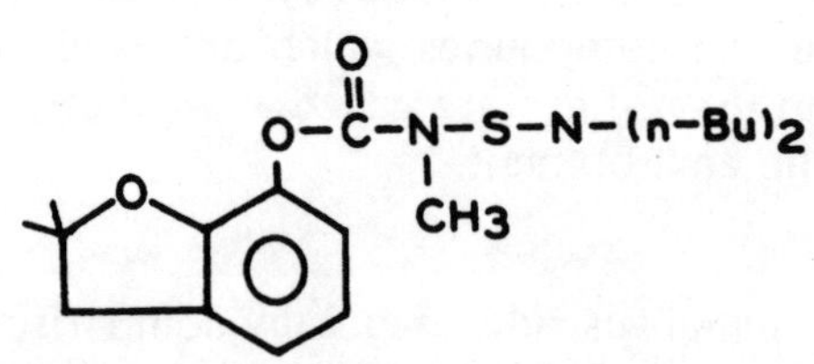

(3)

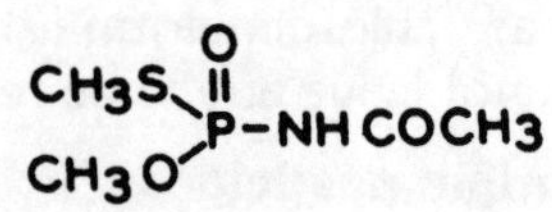

(4)

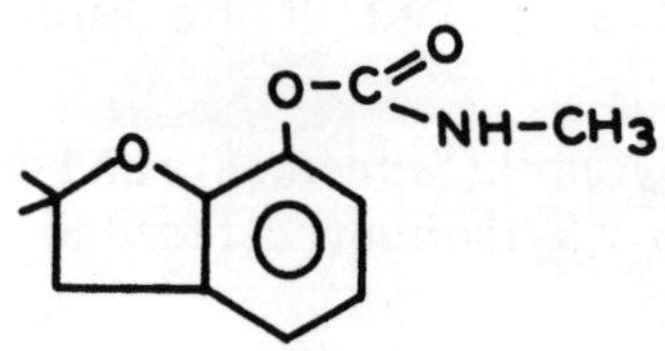

(5)

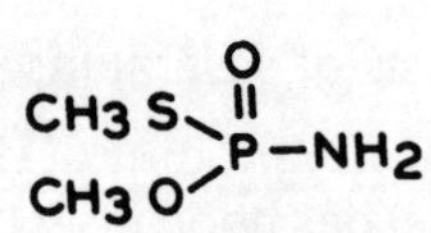

(6)

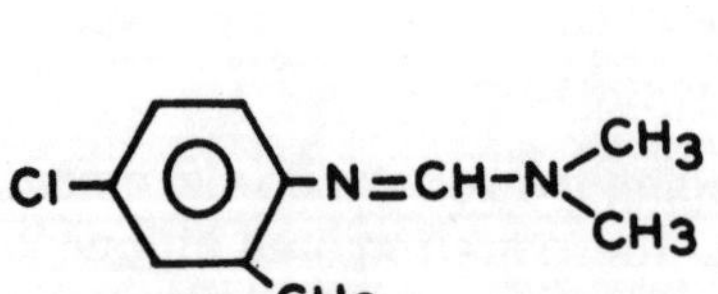

(7)

(8)

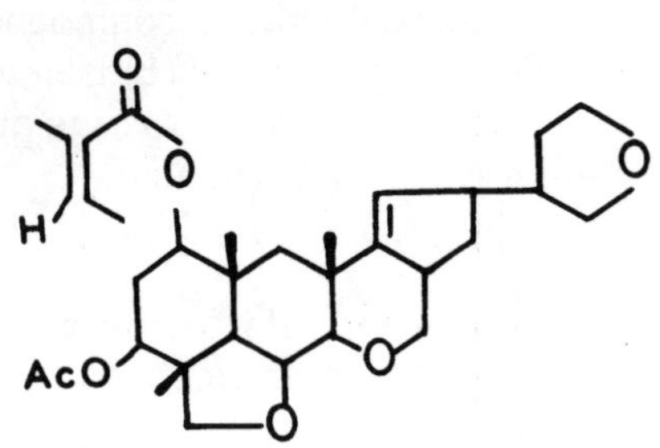

(9)

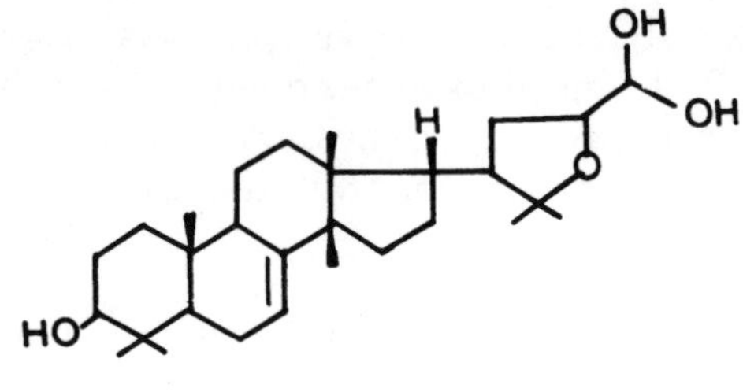

(10)

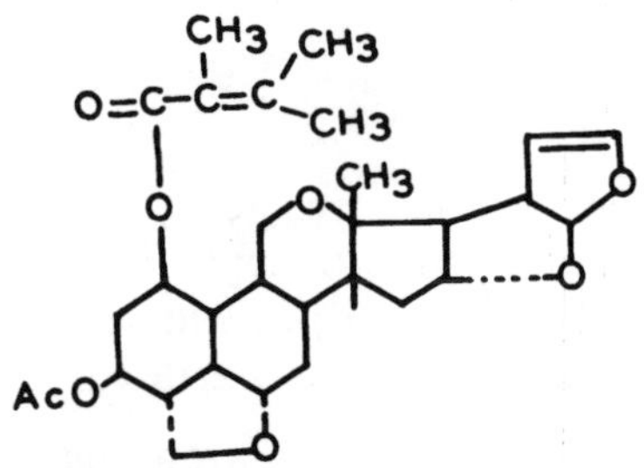

(11)

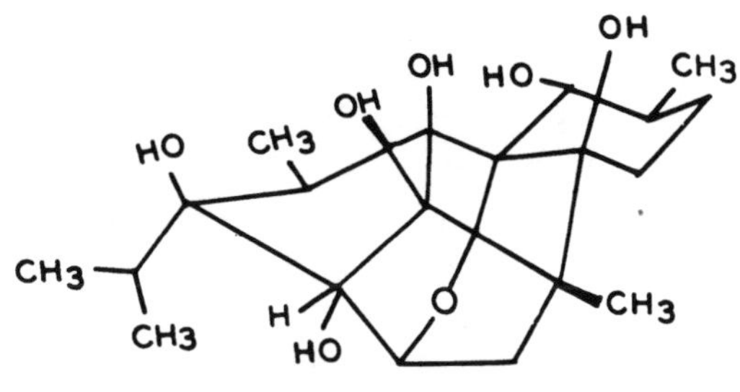

(12)

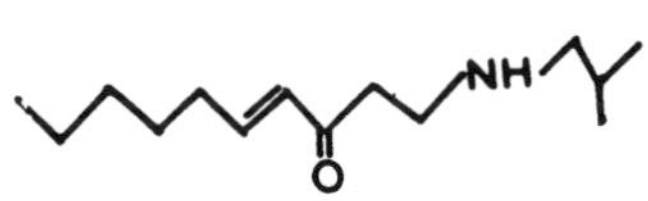

(13)

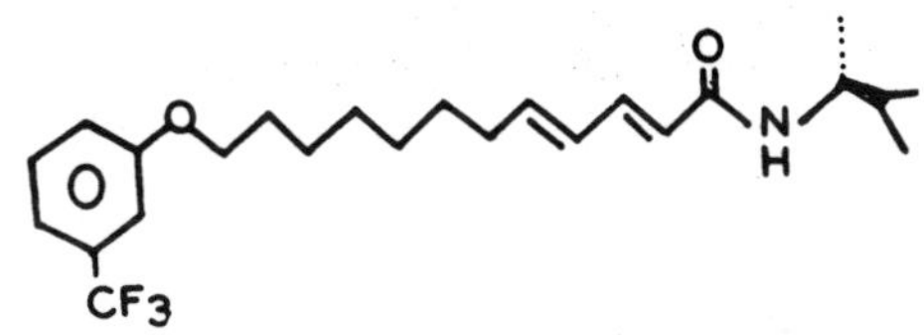

(14)

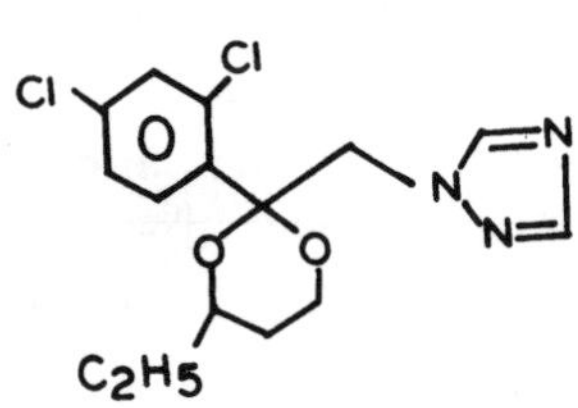

(15)

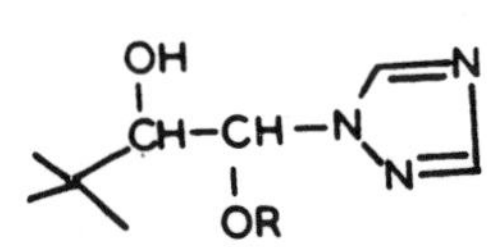

(16)

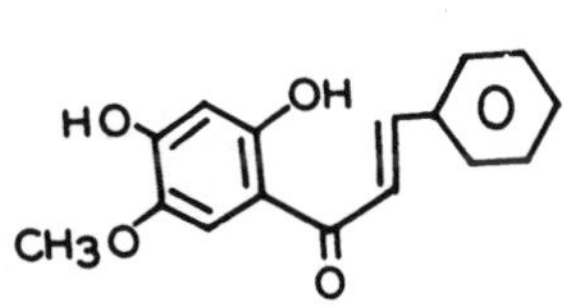

(17)

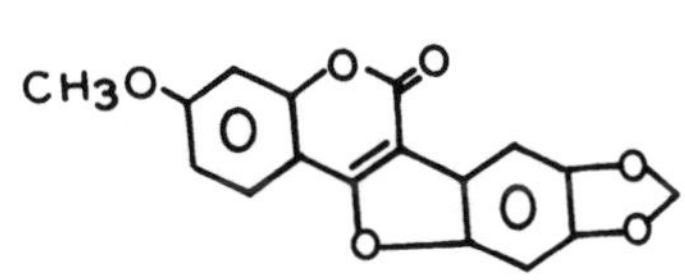

(18)

Similarly many plant products have exhibited good fungicidal activity. compounds 2, 4 - dihydroxy-chalcone (17) flemichapparin and flemichapparin-C (18) isolated from *F. chapper* have sown 50% growth inhibitory activity against *H. oryzae* and *a. solani* at a concentration of 70 ppm. These are environment friendly.

Herbicides

In the area of herbicides, new high potency herbicides viz. chlorosulfuron and sulfometuron methyl are now available in the market. These are effective at very low dosages and do not leave any toxic residues in the environment.

Natural occuring compound Bialaphos, a tripeptide metabolite of *streptomyces hydrogcopicus* is a potent herbicide which controla number of weeds (monocot and dicot).

Biopesticides

The exploitation of microorganisms in pest control has been receiving greater attention in the recent years. The most successful commercial product is *Bacillus thuringensis* (Bt). This product is effective to control number of pests in various crops and there will be no residual toxicity in the environment, hence it is ecofriendly.

Therefore the pesticides which are target selective and less hazardous to theenvironment should be encouraged for future use in agriculture in future.

To quote the words of Buchal 1983 "The challanges are enormous. the more we understand about iteraction between chemical compounds and the physiological and biochemical reactions or organism the more we well be able to design ecofriendly compounds with predictable behaviour. In this regard pesticide science is still at an infant stage.

References

Aditychaudhury, N., Bhattacharya, A., Chowdhury and Pal, S. 1985, Chemical constituents of plants exhibiting insecticidal, antifeeding and insect growth regulating activities. J. Sci. Ind. Res. **44** : 85-101.

Baker, R.D. (1987). Synthesis and chemistry of Agrochemicals (Eds D.R. Baker, J.G. Fenyes, W.K. Mobery and B. Cross) A.C.S. Series 355 Washington, D.C.

Crosby, D.G. 1971, Minor insecticides of plant origin. In Naturally Occuring insecticides. p. 177-239 (Eds. M. Jacobson and D.G. Crosby) Marcel Dekker, New York.

Parmar, B. S. 1993. Botanical and Biopesticides, (Eds. B.S. Parmar and C. Devakumar) pp.-1-73. Westvill publishing House, N. Delhi - 12.

Recent Advances in Indian Entomology
Editor: O.P. Lal
APC Publications Pvt. Ltd., New Delhi, 1996, pp. 43-62

Advances in the Management of Insect Vectors of Viral Diseases of Plants

ANUPAM VARMA and K. P. SRIVASTAVA

Economic losses due to plant viruses are estimated to be arounds $ 15 billion per annum. Reduction in yield is more in plants infected early and even greater when infected with more than one virus. Cowpea banding and cowpea cholorotic spot viruses reduce yields of cowpea by 41 and 56% respectively (Sharma and Varma, 1975). Virus infection not only reduces yield but also impairs quality of the product, and in perennial plants causes gradual and steady decline (Varma, 1983).

Apart from the viral dieases, economic losses are also caused by the diseases due to Mollicutes. These diseases are commonly known as yellows type of diseases (YTDs). More than 300 YTDs are known to affect plants. Some of which result in upto 95% loss in yield (Varma and Rajamannar, 1989)

The casual pathogens of almost all the viral and yellows type of diseases are transmitted in nature by vectors like insects, mites, nematodes and fungi. Of these,

Table 1. Important group of insect vectors of plant viruses.

Toxonomic group	Number vector spp.	No. of viruses transmitted	Virus-vector relationship
Homoptera			
Aphididae	200	160	N S P
Psyllidae	3	(B)	P
Coccidae	15	2	S
Aleyrodidae	3	25	(N) S P
Cicadellidae			
Auchenorryncha	130	35	S P
Coleoptera	30	20	S P
Thysanoptera	6	3	P

N : non-persistent; (N) : non-persistent transmission rare; P : persistent; S : semipersistent; (B) known to transmit only a BLO insects are the most important vectors of plant viruses, because of their ecological and feeding behaviour. More than 70% of all insect vectors of plant viruses belong to the Order Homoptera (Table 1). Orders Coleoptera and Thysanoptera also contain some important vectors.

Mode of transmission of plant viruses by insect vectors

Insect-transmitted plant viruses are divided into three basic types, namely non-persistent, semi-persistent and persistent (Watson and Roberts, 1939; Sylvester 1969; Garrett, 1973). The non-persistently transmitted viruses are also referred as stylet-borne viruses (Kennedy *et al.*, 1962; Pirone, 1969). Viruses transmitted in a persistent manner are called circulative. The persistent (circulative) viruses can be further divided into propagative and non-propagative, depending on whether they multiply in the vector or not (Harris, 1981 a).

Properties of viruses transmitted by insect vectors

The properties of viruses transmitted by insects in different manner differ (Table 2). Non-persistent type of transmission is only found for the viruses transmitted by aphids (Table 1), except for cowpea mild mottle virus which is transmitted by the whiteflies in a non-persistent manner. The viruses belonging to alfalfa mosaic, cucumo and potyvirus groups are transmitted by aphids in a non-persistent manner. The vectors can acquire these viruses in a very short acquisition feeding time and transmit without any latent period (Table 2).

Table 2. Properties of viruses with different virus-vector relationships.

	Non Circulative		Circulative	
Property	Non-persistent	Semi-persistent	Non-propogative	Propagative
Pre-acquistion fasting	required	not required	not required	not required
Acquistion time	< 1 min	> 30 min	hrs to days	hrs to days
Latent period	nil	nil	hrs to days	days
Persistence in the vector	1hr	<100hrs	Several days	All life
Retention after moult	No	No	Yes	Yes
Transovarial transmission	No	No	No	Yes
Mechanical transmission	++	+	+	-

Semi persistently transmitted viruses namely beet yellows and parsnip yellow fleck, have transmission characteristics in between non-persistent and persistent type of viruses. The vectors retain the ability to transmit these viruses for as long as 3-4 days. The virus may be acquired by the vector in as little as 30 min acquisition feeding and preacquisition fasting has no effect on transmission efficiency.

Persistently transmitted viruses are acquired by the vectors through long acquisition feeding periods (hrs to day). These viruses accumulate within insect vectors, pass through the gut wall, enter the haemolymph and salivary glands, and are then transmitted to plants through saliva during feeding. Hence these viruses are called circulative viruses. Some of these viruses also multiply in their vectors. These are, therefore called circulative propogative viruses. The circulative viruses require a distinct latent period in their vectors before transmission as they have to circulate in the vector and the propogative-circulative viruses also have to multiply to acquire enough concentration to cause infection. The transmitting ability of the vectors lasts for many days to life as these viruses are not lost during moulting. The salient features of the circulative viruses are: long acquisition feeding time, a distinct latent period in the vector, persistence in the vector for several days to throughout the life of the vector, retention of the virus even after moulting, and in some cases transovarial transmission (Table 2).

Persistently transmitted viruses usually show a high level of specificity in their relationship with the vector, whereas non-persistently transmitted viruses are less specific. For example *Myzus persicae* is known to transmit more than 100 viruses in a non-persistent manner and onion yellow dwarf virus (a non-persistently transmitted virus) can be transmitted by more than 100 aphid species.

Dependent transmission

.Some plant viruses are transmitted only in the presence of other viruses which are known as helper viruses. Such dependent transmission has been found for all the types of insect vector transmissions. Potato aucuba mosaic (PAMV), potato virus C (PVC) or tobacco etch (NAT strain) potyviruses require the presence of potato virus Y (PVY) for transmission by aphid vector (Fig. 1). This has been demonstrated (Thornbury *et al.*, 1985) to be due to coding of a 53 to 58 KD helper protein, required for vector transmission, by potato virus Y but not by viruses like PAMV and PVC.

Table 3. Dependent transmission of carrot mottle virus (CMoV).

Helper virus present in CMoV source plants	Transmission by *C.aegopodi*		*M. persicae*	
	CMoV	helper	CMoV	helper
Nil	0	-	0	-
Carrot red leaf virus	+++	+++	0	0
beet western yellow virus	-	-	++	+++
potato leaf roll	-	-	+	+++

Dependent transmission is also known for the viruses transmitted in a semipersistent or persistent manner. Heracleum virus 6 (HV6; 1200 nm flexuous filament) and *Anthriscus* yellows virus (AYV; isometric 28 nm) are efficiently transmitted by the vector aphids in a semipersistent manner but Heracleum latent virus (HLV 730 nm flexuous filament) and parsnip yellow fleck virus (PYFV; 28 nm isometric) are not transmitted on their own. In mixed infections, however, HV6 helps in the transmission of HLV through end to end aggregation (Fig. 2) and AVY

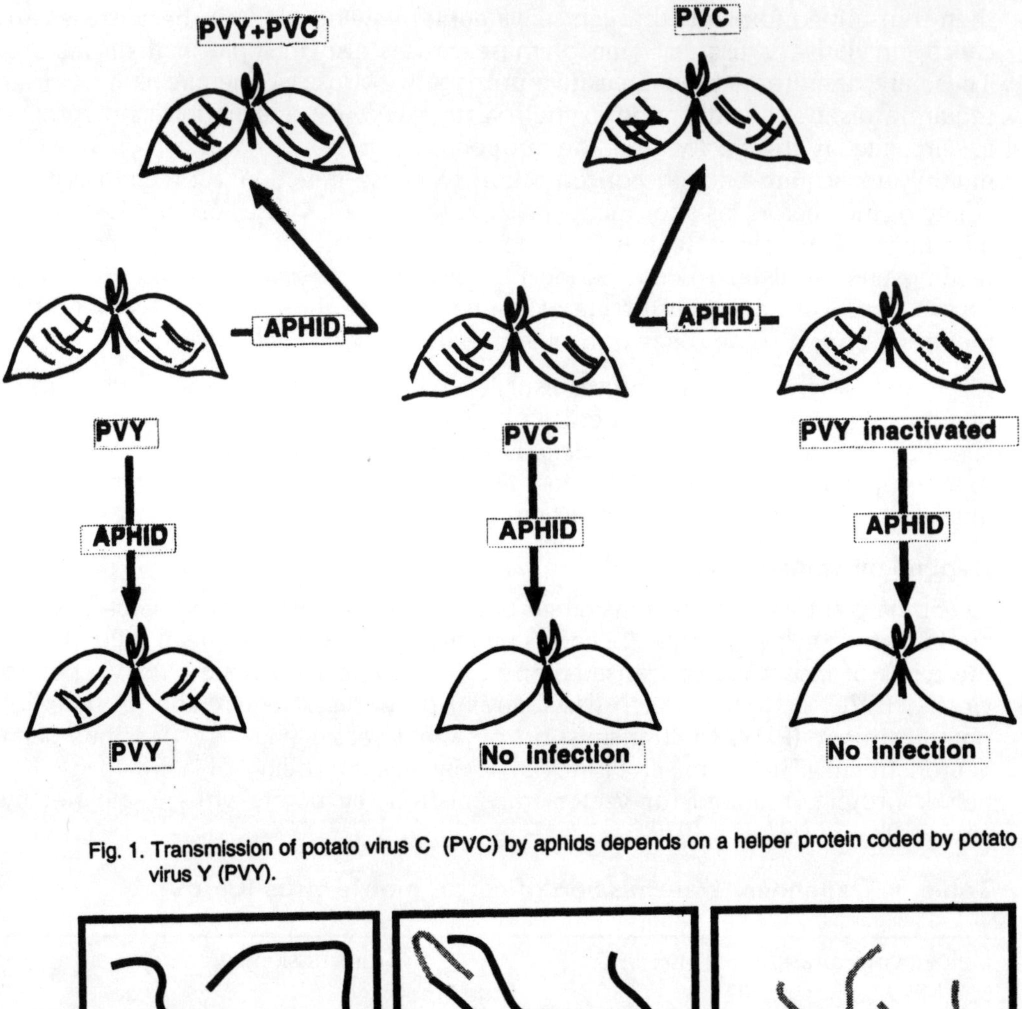

Fig. 1. Transmission of potato virus C (PVC) by aphids depends on a helper protein coded by potato virus Y (PVY).

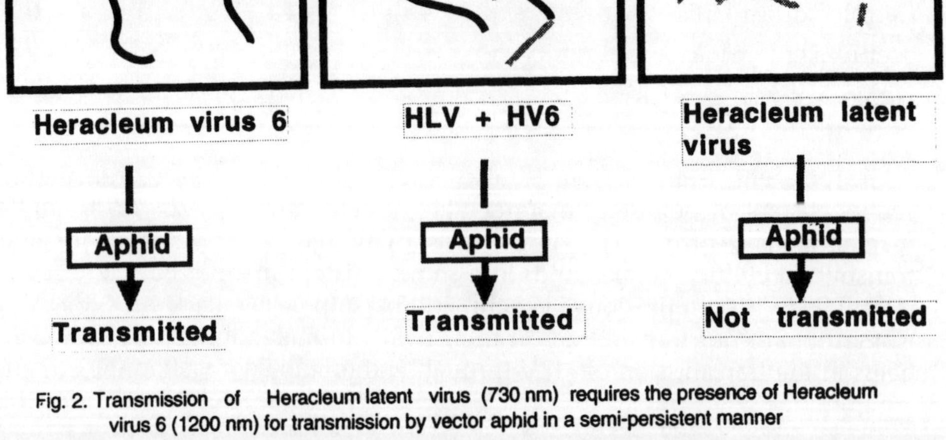

Fig. 2. Transmission of Heracleum latent virus (730 nm) requires the presence of Heracleum virus 6 (1200 nm) for transmission by vector aphid in a semi-persistent manner

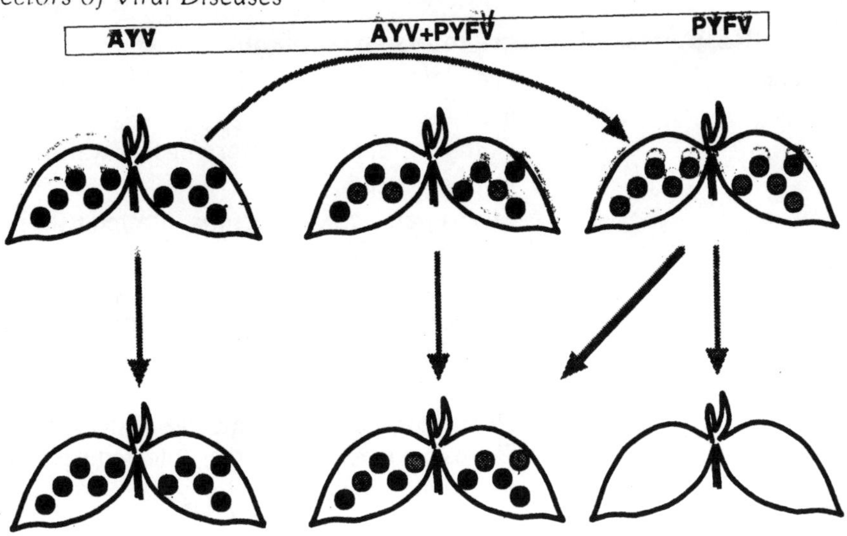

Fig. 3. Transmission of Parsnip yellow fleck virus (PYFV) requires the presence of helper Anthriscus yellows virus (AYV) for its transmission by vector aphid *Cavariella aegopodii*.

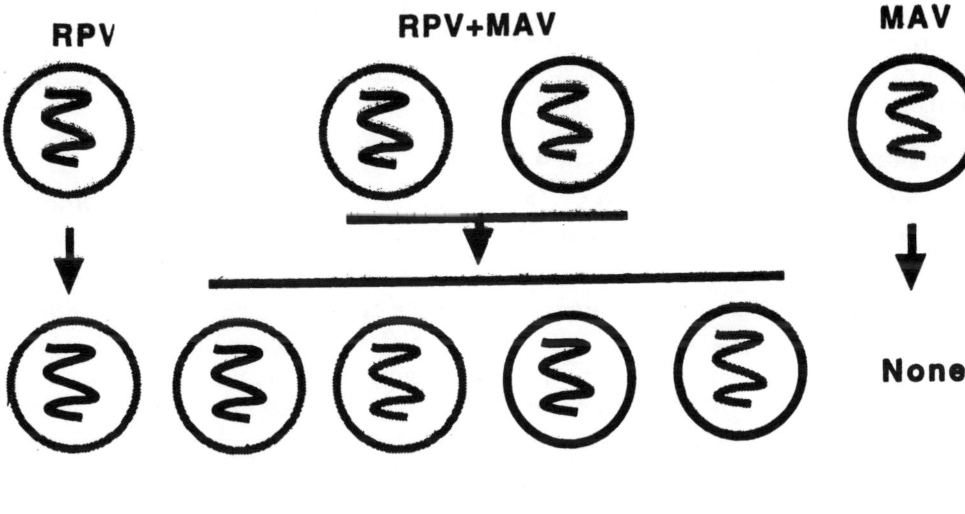

Fig. 4. Transmission of RPV and MAV strains of BYDV by *Rhopalosiphum padi*

helps in the transmission of PYFV (Fig. 3). RPV strain of barley yellow dwarf virus (BYDV) is efficiently transimitted by aphid *Rhopalosiphum padi* in a persistent manner, but not the MAV strain of BYDV. However, in mixed infections transencapsidation occurs which helps in the transmission of both the strains of BYDV with equal efficiency (Fig. 4). Similar helper transmission has been noticed for carrot mottle virus (CMoV) which is transmitted by *Cavariella aegopodii* when the plants are also infected with carrot red leaf virus, or by *Myzus persicae*

Table 4. Responses of insects to diseased plants.

Advantageous :

- increase is host range
  Dalbulus maidis on aster infected with MLO
- increased colonisation
  Myzus persicae on *Physalis* infected with potato leaf roll virus (PLRV)
- improved 'committed' phloem ingestion
  Schizaphis graminum on oats infected with barley yellow dwarf virus (BYDV)
- enhanced attraction
  Rhopalosiphum padi on cereals infected with BYDV, *Bemisia tabaci* on legumes infected with mungbean yellow mosaic virus.

Disadvantageous :

- reduced colonisation
  pea aphid (acyrthosiphon pisum) on clover infected with bean yellow mosaic virus
- Shorter life-span
  R. padi on cereals infected with BYDY

when the source plants are also infected with beet western yellows virus or potato leaf roll virus (Table 3). Such dependent transmissions must be common in India, but have remained undetected so far.

Responses of insects to diseased plants

Infection of plants by viruses or MLOs also influences the responses of insects. Some of these responses are advantageous to the insects and some disadvantageous (Table 4).

Thus, it is obvious that transmission of viruses by their vectors is a complex process involving many interactions including those influencing the behaviour of the insect vectors. Here we are briefly discussing the ecology of important groups of insect vectors of plant viruses and their management to minimise the economic losses caused by the viral diseases of plants.

Aphids (Order Homoptera : Family Aphididae)

Aphids constitute the largest group among the plant virus vectors. About 4000 species of aphids have been described (Eastop, 1977), but, only 200 species have been recorded as vectors of plant viruses (Kennedy *et al.*, 1962). Many aphids are known to transmit more than 30 viruses each.

The biology, feeding behaviour, host selection and parthenogenetic reproduction, world wide distribution make them ideal vectors of plant viruses. They are the most important group of vectors of plant viruses as they transmit over 65% of the known viruses (Table 1). In India 33 species of aphids have been identified as vectors of plant viruses as against one to three species of the other groups of insect vectors, and transmit 70% of over 230 viruses reported from the country (Muniyappa and Veeresh, 1982). Most of these viruses are transmitted non-persistently.

Aphids are more harmful as vectors of plant viruses than as direct pests. A single viruliferous aphid can cause more damage in 15 seconds by infecting a plant by virus(es) than a large number of nonviruliferous aphids feeding for prolonged periods on the same plant. Even aphids that do not colonize on certain plant species can transmit viruses during brief probings.

Under Delhi conditions maximum build up of aphid population and migration occurs in February (Fig 5) which also coincides with the rapid spread of aphid transmitted viruses in crop plants. Fortunately at that time most of the crops are near maturity, therefore, relatively less damage is caused by the viruses in *Rabi* crops. *Aphis craccivora* and *A. gossypii* are the two most dominant vector aphids in India. These aphids are found all the year round on the Indian plains. On high hills of the Himalayas, characterised by cooler temperatures, high wind velocity and high humidity, aphid population is not able to build up. This situation, coupled with the relative absence of vector aphids in North West Indian plains in November to mid January periods helped in the production of virus free potato seeds in the country through a system commonly referred as 'Seed Plot Technique' (Varma, 1993).

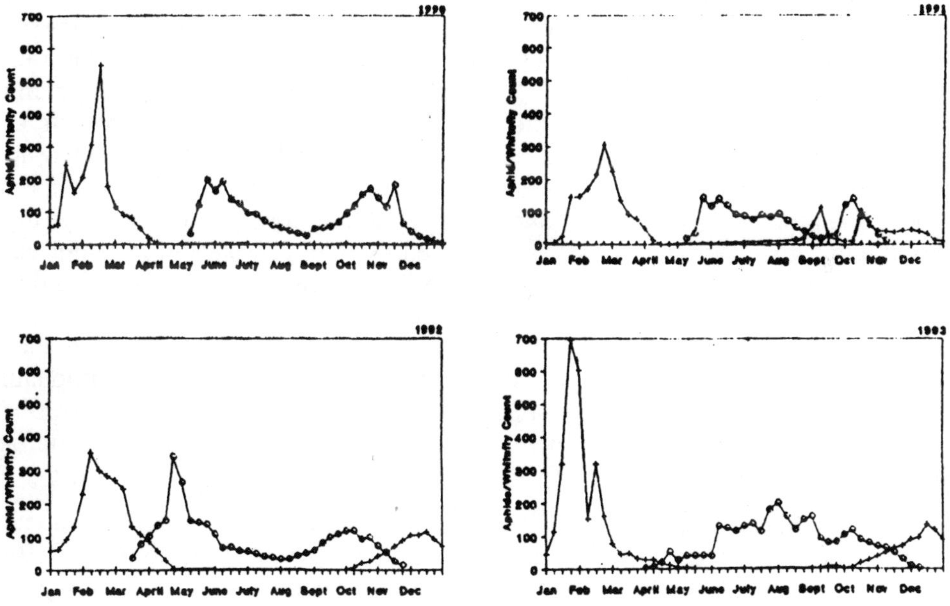

+ aphids; o whiteflies

Fig. 5. Population of aphids and whiteflies under Delhi conditions from 1990 to 1993

Whiteflies (Order Homoptera : Family Aleyrodidae)

Whiteflies are the most important group of insect vectors, particularly in the tropics and semitropics. Out of a total of 1156 species of whiteflies known (Mound and Halsey, 1978), *Bemisia tabaci, Trialeurodes abutilonea* and *T. vaporariorum* (Muniyappa, 1980) are known to transmit plant viruses. Of these *B. tabaci* is economically most important as it not only transmits a large number (over 70) of viruses, mostly belonging to geminivirus group, but also causes serious direct damage to crop plants. Some of the important viruses which are transmitted by *B. tabaci* are African and Indian cassava mosaic, bean golden mosaic, cotton leaf curl, mungbean yellow mosaic, papaya leaf curl, tobacco leaf curl, tomato leaf-curl, etc. It also causes diret damage as a pest as in cotton or by phytotoxicity resulting in disorders like squash silver leaf (Costa *et al.*, 1993).

B. tabaci has a wide host range. Over 320 plant spcies belonging to 63 families have been recorded to be its hosts. In India, *B. tabaci* is known to colonise 99 plant species belonging to 20 families (Pimpale and Summanwar, 1984). Host preference of whiteflies varies from region to region. In West Africa *B. tacaci* colonises cowpeas commonly but in North India they have not been found to colonise this host as other more preferable hosts are available (Varma, 1986). In a year whiteflies complete upto 15 life cycles (Azab *et al.*, 1971; Pimpale and Summanwar, 1984). The duration of life cycle is greatly influenced by weather conditions; the hot and humid conditions are most preferred.

The cropping systems also influence whitefly population. Until the early eighties the whiteflies were first trapped in the beginning of June and whitefly transmitted diseases (WTD) first appeared by the middle of June. This helped in raising mungbean yellow mosaic virus (MYMV) free mungbean crop during the summer months. But, by the begining of the 90s MYMV became a serious problem even in the summer crop (Varma *et al.*, 1992). During the last four years build up of whitefly population has been found throughout between March and October months under North Indian conditions. Maximum population, however, was in July-August period (Fig. 5). Although similar consistent data are not available for earlier years, recorded observations suggest that the whitefly population remained small during the winter and summer months under North Indian conditions. However, under South Indian conditions whitefly population remains high between December to May period (Saikia and Muniyappa, 1986). In Bengal also the whitefly population is more between August to February period (Ghatak *et al.*, 1986). These observations clearly indicate the influence of temperature and other weather conditions on the build up of whitefly population in different regions of the country.

Recently in the USA two distinct biotypes of *B. tabaci* have been identified on the basis of differences in esterases, and mating behaviour (Perring *et al.*, 1993). In India, also recent studies on esterases of whiteflies have indicated variations in different population of *B. tabaci* (Varma, Chandrashekhar and Srivastava, unpublished results). Further studies will show, whether these differences are due to distinct biotypes or a result of host plants.

Leafhoppers and planthoppers (Order Homoptera : Family Cicadellidae)

Leafhoppers and planthoppers transmit several economically important viruses. Nearly 15,000 species of leafhoppers belonging to 2000 genera are known but only 49 species representing 21 genera have been reported as vectors of plant viruses (Conti, 1985; Nault and Ammar, 1989; Nault and Madden, 1988; Nielson, 1985). Some of the important viruses transmitted by leaf-and-plant. hoppers are given in Table 5.

Table 5. Some plant viruses transmitted by leaf-and-plant hoppers.

Type of transmission	Virus	Vector
Semi persistent	Rice tungro	*Nephotettix nigropictus*
		N. virescens
	Maize chlorotic dwarf	*Graminella nigrifrons*
Circulative-nonpropagative	Maize streak gemini	*Cicadulina spp.*
Circulative-propagative	Maize mosaic rhabdo	*Peregrinus maidis*
	Potato yellow dwarf	*Agallia constricta*
	Rice stripe tenui	*Laodelphax striatellus*
	Wound tumor reo	*Agallia constricta*
	Rice dwarf reo	*Nephotettix impicticeps*

Beetles (Order Coleoptera : Families Chrysomelidae, Curculionidae, Meloidae and Coccinellidae)

About 74 species of beetles have been reported as vectors of plant viruses (Harris, 1981b). Earlier transmission of viruses by beetles was considered to be a mechanical process involving contamination of beetles mouth parts. But, now specific relationship between the beetles and the viruses they transmit has been demonstrated, particularly as the virus particles could be detected in the haemolymph of chrysomelid beetles (Slack and Scott, 1971). Regurgitant fluid of beetles contain potent nucleases, which selectively inactivate plant viruses. Thus the enzymatic function of ribonuclease determines as to which viruses could be transmitted by the beetles (Gergerich and Scott, 1988).

Nagaich *et al.* (1972) reported transmission of potato virus S and potato virus X by *Coccinella septempunctata* and *Epilachna ocellata*. These must be non-specific mechanical transmissions. Urdbean leaf crinkle virus is transmitted by *E. dodecastigma* (Beniwal and Bharathan, 1980), southern bean mosaic (SBMV) and black gram mottle (BGMoV) viruses are transmitted by *Madurasia obscurella* (Reddy and Varma, 1986; Krishnareddy, 1989). Both SBMV and BGMoV are retained by the vector for over ten days and are transmitted efficiently.

Thrips (Order Thysanoptera : Family Thripidae)

In recent years thrips have become important vectors of plant viruses (Ananthakrishan, 1980; Reddy and Wightman, 1988) as these transmit tomato spotted wilt virus (TSWV) which has a very wide host range and causes serious economic losses. Moreover, management of thrips is very difficult. In India *Frankliniella schultzei, Scirtothrips dorsalis* and *Thrips tabaci* are reported as vectors of TSWV (Muniyappa and Veeresh, 1982), but *F. palmi* is the most important vector of the virus causing bud necrosis of groundnut (DVR Reddy, personal communication).

Other important vectors (Class Arachnida; Order Acarina : Family Eriophyidae)

Mites possess mouth parts that are adopted for piercing and rasping which help in the transmission of plant viruses. Although mites are not insects these are briefly discussed here as a very serious disease - pigeonpea sterility mosaic (PSM) is transmitted by *Aceria cajani*. The organism causing PSM has not been identified so far, but its transmission characteristics are typical of viruses and virus-like organisms.

A. cajani oviposit the eggs on the vegetative tips of pigeonpea plants. The eggs 30x40 μ and are oval, milky white, transluscent and loosely fastened to trichomes. The female mites have a short preoviposition period varying between 24 and 27 hr and then lay 1 to 3 eggs per day. The viruliferous mite have been shown to spread the disease upto 1000m in the windward direction (Anonymous, 1982).

Management of insect vectors of plant viruses

Several approaches for the integrated management of insect vectos of plant viruses have been tried (Varma, 1993). These can be grouped as avoidance of vectors, chemical control of vectors, biological control of vectors, prevention of viral transmissibility by vectors, modification of cultural practices and use of crop varieties resistant to vectos. All these approaches are briefly discussed here.

Avoidance of vectors

This approach has been found very effective for the production of healthy seed potato on the plains of Northern India. As the aphid population usually starts to build up from the second half of January (Fig. 5), the potatoes sown in late September and harvested by the first week of January escape infection by aphid transmitted viruses. This 'seed plot technique' has been a grand success in increasing potato production in the country.

In North India, mungbean yellow mosaic virus (MYMV) is a major constraint in the production of mungbean during *Kharif* (rainy season), which is the main growing season. Short duration varieties of mungbean were developed which could be fruitfully cultivated during the *Zaid* (summer season) under irrigation so that losses due to MYMV were avoided due to the absence of whitefly population. In the recent years, however, build up of whitefly population is observed even in the summer season (Fig. 5), resulting in the spread of MYMV even in this season.

Chemical control of vectors

Chemical control of vectors is an important approach for the management of the insect vectors of viral diseases. For chemical control, timely application is very important. In the U. K. sugar beet yellows disease is managed very effectively by applying insecticides whenever a warning is issued on the basis of vector population. The success of such control measures depends on efficient monitoring of the vectors (Varma, 1976). Several insecticides, particularly systemic granular insecticides have been reported effective in reducing vector population (Varma *et al.*, 1992). Recently introduced synthetic pyrethroids have been reported to be effective in reducing the transmission of non-persistent viruses (Gibson and Cayley, 1984. Atiri *et al.*, 1987). But the use of insecticides to prevent transmission of non-persistently transmitted viruses by their vectors has only been partially successful. because the insects are able to transmit the viruses before they are affected by insecticides (Loebenstein and Raccah, 1980; Jayasena and Randles, 1985). Usually insecticidal control is better for persistent and semi-persistent viruses which require longer inoculation feeds by their vectors (Varma, 1983).

Soil application of carbofuran granules @ 1.5Kg a.i./ha, or disulfoton 1.5 Kg a.i./ha reduced the whiteflies population and leaf curl incidence in tomato to a great extent (Singh *et al.*, 1979). Dubey *et al.*(1986) reported that soil application of carbofuran 3G@ 1.25 kg a.i./ha with 3 sprays of dimethoate (0.1%) significantly reduced the leaf curl of tomato. Phosphamidon (0.02%) and methyl demeton (0.025%) were reported effective in reducing the incidence of yellow-vein mosaic of bhindi which is spread by whiteflies (Singh and Singh, 1986). Contact insectcides quinalphos (0.15 *a.l.*), chlorpyrifos and phenthoate (0.5% spray) were also effective in reducing the incidence of yellow vein mosaic (YVM) disease of okra (Ramachandran and Summanwar, 1986).

Aldicarb was the most effective and economical insetecticide in controlling whiteflies and yellow mosaic disease of blackgram specially when two or three applications @ 1.5 Kg a.i./ha) given (Subrahmaniam and Varma, 1986). Carbofuran (1 Kg a.i./ha) was also reported effective in reducing the yellow mosaic disease in greengram. Mulching with yellow polythene or plastic painted with aluminium paints followed by sprays with monocrotophos and formothion (0.1%) reduced the incidence of chilli mosaic spread by aphids (Reddy and Wightman, 1988).

Treatment of *B. tabaci* females with an antibiotic oxytetracycline hydrochloride adversely effected oviposition and growth and development of offspring on sweet potato (Costa *et al.*, 1993). Insecticides are extensively used to prevent the spread of viruses and their vectors all over the world. Proper timing for treatment, however, is very important.

Biological control of insect vectors

Biological control of insect vectors has not received proper attention. It was very successful in arresting the spread of carrot mottle dwarf virus in Australia and New Zealand through the introduction of small wasp *Aphidius* sp., which reduced the population of the vector *Cavariella aegopodii*. The spread of citrus greening disease, caused by a bacteria like organism, was successfully restricted through the control of its psyllid vectors *Diaphorina citri* and *Trioza erytreae* in the Reunion Island. Whithin 24 months of release of 30-50 eulophid ectoparasites *Tetrastichus dryi* and *T. radiatus* per hm, the population of *D. citri* vector was reduced and spread of greening disease checked. (Aubert and Quilici, 1984).

Paecilomyces farinosus parasitises *B. tabaci* and reduces the whitefly population on treated plants under field conditions (Nene, 1972; Dhar and Varma, unpublished results). Fungal pathogen *Verticillium lecanii* is also reported effective in killing the adult and larval stage of whitefly (Bedford *et al.*, 1993).

McAuslane *et al.*, (1993) reported that *Encarsia nigricephala, E. pergandiella, Eretmocerus californicus* and *E. transvena* were the most common parasitoids of whitefly at Gainesville, U.S.A.

Trioxys complanatus Wash. contributes significant role in reducing the population of vector aphid. *Praon exsoletum* and *Aphelinus asychis,* and *A ervi* are also effective parasites of aphids in Australia. Fungus *Zoophthora radicans* also reduced aphid population in Southern Queensland and Northern New South Wales, Australia (Lehane, 1982).

Prevention of viral transmission by vectors

Mineral or vegetable oil sprays have been very extensively used to reduce the transmission of non-persistent and semi-persistent viruses by aphids and whiteflies. Non residual 'Krishi' oil was effective in reducing the whitefly population and spread of tomato leaf curl virus (TLCV) under field condition by 90% (Sastry *et al.*, 1978; Singh *et al.*, 1973). Four sprays of power oil (1.0%) at 10 days intervals reduced the whitefly population and leaf curl incidence of chillies to a great extent (Singh *et al.*, 1979). Oil sprays were found very effective in reducing the transmission of watermelon mosaic virus - which is transmitted by aphids in a non-persistent manner (Raychaudhuri and Varma, 1983).

Modification of cultural practices

Suitable modification of cultural practices like, time of sowing, intercropping, agronomic practices can help greatly in minimising the incidence of viruses, particularly those with a narrow host range. Sterility mosaic 'virus' which is transmitted by eriophid mite, *Aceria cajani* causes serious economic losses in India. The host range of both the virus and its vector is restricted to pigeonpea but they perpetuate on volunteer plants and ratoon/overlapping crops. Removal of volunteers and prohibiting ratoon/overlapping crop even for one season, would break the disease cycle and prevent the losses caused by sterility mosaic (Varma, 1976). In Nigeria when tomatoes were grown in small plots more than 90% incidence of tomato leaf curl was observed, but when the crop was grown in large areas (c 200 ha) with clean cultivation the incidence of the viral disease was less than 0.01% (Varma, 1986).

Barrier crops

Barrier crops have been found useful in controlling the whitefly and aphid transmitted viruses. Subrahmaniam and Varma, (1986) observed that pearl millet was a better barrier crop than maize crop in reducing the yellow mosaic disease incidence of blackgram. Planting of six rows of either sunnhemp or maize as a border was found to reduce the incidence of tomato leaf curl virus (TLCV). The spread of TLCV can be minimized by barrier crops of maize, sorghum, pearlmillet.

Further, Singh (1989) recommended five or six rows of either of these crops all around the main tomato plot should be sown atleast 50-60 days before transplanting of tomato. These barrier crops checked incoming viruliferous whiteflies from entering the tomato crop. The incidence of chilli-mosaic transmitted by aphids, was also reduced by sowing barrier crops like sunflower, seasame, sorghum or pearl millet around the chilli plots (Singh, 1989).

Use of repellents and attractants

Reflective Mulching

Relfective surfaces laid on the soil around plants have also been found highly effective in controlling aphids and whiteflies.

For the management of watermelon mosaic virus-1 (WMV-1) and (WMV-2) various workers have tried polythene mulches with varying success in different cucurbitaceous crops (McLean *et al.*, 1982; Wyman *et al.*, 1979). Yellow polythene (YPM), transparent polythene (TPM) and straw mulch (SM) were tested by Vani *et al.* (1989) for the management of mosaic disease on muskmelon (*Cucumis melo* L.) and observed all the three types of mulches reduced the incidence of mosaic disease. The increase in yield by YPM was 36 per cent, TPM 74 per cent and SM 51 per cent.

Tomato leaf curl virus can also be reduced with yellow polyethylene mulching. Singh (1989) observed the appearance of yellow-vein mosaic disease of okra can be delayed with coloured polythene much. Singh (1989) observed soil mulching with sawdust or wood-chips or with aluminium foil repells the aphid vectors and checks the spread of cucumber mosaic virus. Aluminium foil and grey-plastic mulch have been found effective in reducing (upto 87.9%) infection caused by cucumber mosaic virus in cucumber.

Use of strips of reflecting aluminium foils as mulches protects plants from infection as vectors are repelled due to reflected ultraviolet light. This method is used to protect a variety of plant species like gladiolus, turnip, lettuce, sugarbeet, cucurbits etc (Varma, 1983).

Use of aphid alarm pheromone

Aphids have a built-in mechanism to resist the effects of predatory insects. By rotating the cornicle they secrete 'alarm pheromones' or *trans B* farnescene. A chemical evaporting from these drop-lets is detected by sensores on the antenne of aphids feeding nearby (Nault, 1973). Subsequently Phelan and Miller (1982) studied the effect of *(E) -B-* farnescene, undecanoic acid, dodecanoic acid and heptadecanoic acid on alate *M. persicae* landing behaviour and reported undecanoic and dodecanoic acids reduced all host-selection behaviours. A derivative prepared from aphid alarm pheromone (*E*)- *B*-farnescene and saturated strainght 14-carbon chain dialkyl ester of acetylene dicarboxylic acid was reported by Gibson *et al.* (1984) to be effective in inhibiting acquisition of the non-persistently transmitted potato virus Y by *M. persicae*.

Use of polymer webs to virus transmission by aphids

Harrewign *et al.* (1991) demonstrated that stylet penetration by aphids (*M. persicae*) into potato leaves could be prevented by covering the plants with polypropylene fleeces and polyethylene netting materials. Polypropylene fleeces (Lutrasil) provided complete protection of plants against potato Y and potato leafroll viruses. The aphids showed a peculiar behaviour on Agrifleece and Lutrasil. It was observed that the aphids tapped with their proboscis on the fibres of fleece and seldomly touched the leaf surface.

Use of resistant varieties

Use of resistant varieties of crop plants is the most important and practical approach for managing viral diseases in IPM system. Genetic resistance to plant viruses may operate through resistance to vectors, transmission by vectors/seeds, virus multiplication, symptom development, intercellular movement, hypersensitivity or immunity to virus infection.

Most (approximately 80%) varietal resistance to viruses is genetically, simple, with resistance and susceptibility segregating in a typical Mendelian system involving a single locus (Khetarpal *et al.*, 1990).

Resistance genes have been identified for nearly 130 host-virus combinations. Of these resistance genes, 47% are dominant, 44% recessive and the remainder incompletely dominant (gene dosage dependent). Similarly, of the three different hosts identified to have monogenic resistance to cucumber mosaic virus, two are dominant and one is recessive. Thus the mechanism of resistance varies not only from virus to virus, but even the same virus in different hosts (Varma, 1993).

Resistant sources have been identified in many crops for vectors and viruses mostly under field screening conditions. Sources of resistance to mungbean yellow mosaic viruses (MYMV), have been found in a large number of accessions of mungbean and blackgram (Varma, 1976), and cowpea (Reddy and Varma, 1986; Srivastava, 1989). Resistance to other viruses like groundnut bud necrosis (Amin, 1985) and sterility mosiac of pigeonpea (Anin *et al.*, 1988 have also been found. Resistance may break down through development of biotypes of vectors or new strains of viruses. Many high yielding cultivars of rice resistant to rice tungro virus (RTV) are widely grown in the Philippines. The resistance to RTV in some cultivars are overcome by the vector leaf hopper (*Nephotettix virescence)* (Dahal *et al.*, 1990). New strains arise frequently in viruses with divided genomes, like whitefly-transmitted geminivirus. For example, the resistance to MYMV in mungbean breaks down in less than six years, perhaps due to the occurrence of natural recombinants (Varma *et al.*, 1992).

Integrated management of viral diseases in sustainable agriculture

Integrated management of vector-borne viruses is of utmost importance, because the direct damage caused by vectors is also reduced. Of the various approaches that can be integrated for the management of different groups of viruses (Table 6), use of resistant varieties is most important.

Concluding remarks

Vector-virus-plant interactions are complex. In India very little attention has been given to understand these interactions, and to generate information on the biology of insect vectors. The weather, natural vegetation and the relay cropping systems provide ideal conditions for the perpetuation of the viruses and their vectors. In many instances an

Table 6. Management pratices which can be integrated for the control of four selected groups of viruses which affect grain legumes.

Management practice	WTG	*Virus* Poty	*Group* Luteo	Tospo
Avoidance of sources of infection				
Use of virus free true seed	x	+	x	+ Cont.....
Avoidance of weeds, volunteers, overlapping generations, other crops etc.	+	+	+	+
Avoidance and control of vectors.				
Avoidance	?	+	+	?
Chemical control	+	?	+	+
Biological control	?	+	+	?
Oil spray	?	+	?	?
Modification of cropping cultural practices				
Clean cultivation	+	+	+	+
Intercropping	?	+	+	?
Mulches	+	+	+	?
Resistance				
Resistant varieties	+	+	+	+
Induced resistance				
Cross protection	-	+	-	?
Systemic acquired resistance	-	?	-	?
Trasformed plants	?	+	?	+

+ effective approach; - ineffective approach; ? efficacy doubtful WTG Whitefly-transmitted germinivirus.

The groups of viruses were selected on the basis of their economic importance and tranmission characteristics. Potyviruses by aphids in a non-persistent manner, tospoviruses by thrips, luteo viruses by aphids and WTGs by whiteflies in a persistent manner.

(Source: Varma, 1993)

insect vector may not be a pest *per se*, but causes enormous losses by infecting plants with viruses. The vectors, therefore, can cause damage even when present in small numbers making their management even more difficult. To avoid the losses caused by insect vectors through the transmission of viruses use of integrated management is essential. This requires development of appropriate models for IPM, based on different components of management practices for insect vectors of plant viruses. Judicious and timely use of management practices like chemical control is possible when adequate data on the biology and ecology of the vectors is known. A beginning has been made in this direction. Biological control of vectors and use of biotechnological approaches for preventing transmission of viruses by insect vectors have to be added to the armoury of IPM for sustainable growth in agricultural production.

References

Amin, K. S., Pratibha Shukla, Singh, R. A., Nene, Y. L., Reddy., M. V., Zote, K. K., Mahmood, M. , Arjuman., G., Gangadharam, K., Bendre, N. I., Gupta, R. P., Chauhan, V. B., Rathi, Y. P. S. and Tripathi., H. S. 1988. Sources of resistance to pigeonpea sterility mosaic disease by multilocation infector-row technique. *In* National Symposium on 'Plant virus problems in India'. Division of Mycology and Plant Patholgy, I.A.R.I., New Delhi 10012 Abstract 9-10

Amin, P. W. 1985. Apparent resistance of groundnut cultivar Robut 33-1 to bud necrosis disease. *Plant Disease,* **69** : 718-719.

Ananthakrishnan, T. N. 1980. Thrips. *In* K. F. Harris and K. Maramorosch [eds], *Vector of plant pathogens.* Academic Press, New York. 149-164.

Anonumous 1982. ICRISAT Annual Report. 134 p.

Atiri, G. I., Thottappilly, G and Ligan, D. 1987. Effects of cypermethin and deltamethrim on the feeding behaviour of *Aphis craccivora* and transmission of cowpea aphid-borne mosaic virus. *Ann. appl. Biology.,* **110** (3) : 445-461.

Aubert, B. and Quilici, S. 1984. Biological control of the African and Asian citrus psyllids (Homoptera : Psylloidea) through eulophid and encyrtid parasites (Hymenoptera : Chalcidiodea) in Reunion Island. *In Proceedings of the 9th conference of the International Organisation of Citrus Virologist.* IOCV Riverside, CA, P 100-108.

Azab, A. K., Megahed, M. M. and El-Mirsawi, H. D. 1971. On the biology of *Bemisia tabaci* (Genn.) (Homoptera : Aleyrodidae) *Bull. Soc. Ent. Egypte,* **55** : 305-315.

Bedford, Ian, Markham Peter G., Helyer, Neil, Willians Coure M. de 1993. More than just another whitefly. *Horticulture Week.* (Aug. 6) : 24-28.

Beniwal, S.P.S. and Bharathan, N. 1980. Beetle transmission of Urdbean leaf crinkle virus. *Indian Phytopath.,* **33** : 600-601.

Conti, M. 1985. Transmission of plant viruses by leafhoppers and planthoppers. *In* L. R. Nault and J. G. Rodriguez [eds.], *The Leafhoppers and Planthoppers.* Wiley Interscience, New York. 289-307.

Costa, H. S., Ullman, D. E., Johnson, M. W. and Tabashnik. 1993. Antibiotic Oxytetracycline interferes with *Bemisia tabaci (Homoptera : Aleyrodidae)* oviposition, development and ability to induce squash silverleaf *Ann. ent. Soc Am.,* **86** (6) : 740-748.

Dahal, G., Hibino, H., gabinagan, R. C., Tiango, E. R., Flores, Z. M., Aquiero, V. M., 1990. changes in cultivar reaction to tungro due to changes in 'virulance' of the leafhopper vector. *Phytopathology,* **80** : 659-665

Dubey, G. S., Yadav, B. P. and Ramayan Singh, 1986. Chemical control of leaf curl of tomato. *In* National seminar on 'whitefly transmitted plant virus diseases' (25-27 June 1986). . Division of Mycology and Plant Pathology, I. A. R. I., New Delhi 110012. Abstract 32p.

Eastop, V. F. 1977. Worldwide importance of aphids as virus vectors. *In* K. F. Harris and K. Maramorosch [eds.], *Aphids as Virus Vectors,* Academic Press, New York 3-47.

Garrett, R. G. 1973. Non-persistent aphid-borne viruses. *In* A. J. Gibbs [ed.], *Viruses and Invertebrates,* Elsevier, New York. 476-492.

Gergerich R. C. and Scott, H. A. 1988. The enzymatic functions of ribonuclease determines plant virus transmission by leaf-feeding beetles. *Phytopathology,* **78** : 270-272.

Ghatak, S. S., Mukhopadhyay, S. and Varma, A. 1986. Virus Transmission characteristics of India whiteflies *In* National seminar on 'whitefly transmitted plant virus, diseases (25-27 June 1986). Division of Mycology and Plant Pathology, I. A. R. I., New Delhi 110012 Abstract 29 p.

Gibson, R. W. and Cayley, G. R. 1984. Improved control of potato virus by mineral oil plus pyrethroid cypermethrin aplied electrostatically *Crop Protection,* **3** : 469-478.

Gibson, R. W., Pickett, J. A., Dawson, G. W., Rice A. D. and Stribley, M. F. 1984. Effects of aphid alarm pheromone derivatives and related compounds on non-and semi persistent plant virus transmission by *Myzus persicae. Ann. Appl. Biol.,* **104**: 203-209.

Harrewijn, P., Ouden, H. den and Piron, P. G. M. 1991. Polymer webs to prevent virus transmission by aphids in seed potatoes. *Entomol. Exp. appl.,* **58** : 101-107.

Harris, K. F. 1979. Leafhoppers and aphids as biological vectors: vector-virus relationship. *In* K. Maramorosch and K. F. Harris [eds.], *Leafhopper Vectors and Plant Diseases Agents,* Academic Press, New York. 217-308.

Harris, K. F. 1981a. Horizontal transmission of plant viruses. *In* J. J. Jr. Mckelvey, B. Eldridge and K. Maramorsch [eds.], *Vectors of Disease Agents, Interactions with Plants, Animals and Man.* Praeger, New York. 92-108.

Harris, K. F. 1981b. Arthoropod and nematode vectors of plant Viruses. *A Rev. Phytopath.,* **19**: 391-426.

ICRISAT 1982. Annual Report. 134p

Jayasena, K. W. and Randle, J. W. 1985. The effect of insecticides and a plant barrier row on aphid populations and the spread of bean yellow mosaic potyvirus and subterranean clover red leaf virus in *Vicia fabae* in South Australia. *Ann. appl. Biology,* **107**: 355-364.

Kennedy, J. S., Day, M. F. and Eastop, V. F. 1962. A complex aphids as vectors of plant virus. Commonwealth Institute of Entomology, London. 1-114.

Khetarpal, R. K., Maury, Y. , Cousin, R., Burghofar, A. and Varma, A. 1990. Studies on resistance of pea to pea seed borne mosaic virus and new pathotypes. *Ann. appl. Biol.,* **116**: 297-304.

Krishnareddy,, M. 1989. Studies on yellow mosaic and leaf crinkle diseases of blackgram. Ph.D Thesis, Post Graduate School, I. A. R. I., New Delhi. 263 pp.

Lehane, Leigh. 1982. Biological control of lucerne aphids. *Rural Research,* **114**: 4-10.

Loebenstein, G. and Raccah, B. 1980. Control of non-persistently transmitted aphid borne viruses. *Phytoparasitca,* **8**: 221.

McAuslane, H. J., Johnson, F. A., Knauft, D. A. and Colvin, D. L. 1993. Seasonal. abundance and within-plants distribution of parasitoids of *Bemisia tabaci* (Homoptera:

Aleyrodidae) in peanuts. *Environ Entomol.*, **22** (5) : 1043-1050.

McLean, G. D., Burt, J. R., Thomas, D. W. and Sproul, A. N. 1992. The use of reflective mulch to reduce the incidence of watermelon mosaic virus in Western Australia. *Crop Protection,* **1**: 491-496.

Mound, L. A. and Halsey, S. H. 1978. *Whitefly of the world A systematic catalogue of the Aleyrodidae (Homoptera) with host plant and natural entry data.* British Museum (Natural Hostory) and John Wiley and Sons chichester. England. 340 pp.

Muniyappa, V. 1980. Whiteflies. *In* K. F. Harris and K. Maramorosch, K. [eds.], *Vectors of Plant Pathogens,* Academic Press, New York. 115-132.

Muniyappa, V., Chandrasekhariah, S. C. and Shivashankar G. 1979. Horsegram cultures tolerant to yellow mosaic. *Indian J. Genet. P1. Breed.,* **38** : 148.

Muniyappa, V., Reddy, H. R. and Shivashankar C. 1976. Yellow mosaic disease on horse gram (*Macrotyloma uniflorus* syn. *Dolichos biflorus)* 111. Evaluation of germplasm for the disease. *Curr. Res.,* **5** : 52-53

Muniyappa, V. and Veeresh, G. K. 1982. Arthropod vectors of virus and mycoplasma diseases o plants in India. *Colemania,* **2** : 19-38. (issued 14 July 1986).

Nagaich, B. B., Upreti, Ç. C. and Verma, K. D. 1972. Studies on transmission of potato viruses X and S by *Epilachna* and *Coccinella* beetle. *Sci. Cult.,* 38: 27-28.

Nault, L. R. 1973. Alarm pheromones help aphids escape predators. *Ohio Report,* **8** (1) : 16-17.

Nault, L. R. and Ammar, EL Desouky 1989 Leafhopper and planthopper transmission of plant viruses *A. Rev. Ent.,* **34** : 503-529.

Nault, L. R. and Madden, L. V. 1988. Phylogenetic relatedness of maize chlorotic dwarf virus leafhopper vectors. *Phytopathology,* **78** (12) 1683-1687.

Nene, Y. L. 1972. A survey of viral diseases of pulse crops in Uttar Pradesh, Pantnagar. U. P. G. B. Pant University of Agriculture and Technology. 191 p.

Nene, Y. L., Srivastava, S. K. and Naresh, J. S. 1972. Evaluation of blackgram (*Phaseolus mungo Roxb.)* and greengram *(Phaseolus aureus* Roxb.) varieties and germplasm for resistance to yellow mosaic virus of greengram. *Indian J. agri. Sci.,* **42**: 251-252.

Nielson, M. W. 1985. Leafhopper systematics. *In* L. R. Nault and J. G. Rodriguez [eds.], *The Leafhoppers and Planthoppers.* Wiley Interscience, New York. 11-39.

Perring, Thomas M., Cooper, Arthur, D., Rodriguez, R. J. Farrar, C. A. and bellows Jr. T. S. 1993. Identification of a whitefly species by genomic and behavioral studies. *Science,* **259**: 74-77.

Phelan, P. L. and Miller, J. R. 1982. Post-landing behaviour of alate *Myzus persicae* as altered by (E) -*B*- farnesene and three carboxylic acids. *Ent. exp. app1.,* **32:** 46-53.

Pimpale, T. D. and Summanwar, A. S. 1984. Studies on different stages in the life cycle and influence of season on the duration of different generations of the whitefly, *Bemisia tabaci* Genn. *Pestology,* **8** (12): 15-19.

Pirone, T. P. 1969. Mechanism of transmission of stylet-borne viruses. *In* K. Maramorosch [ed.] *Virus, Vectors and Vegetation.* Interscience Publishers, New York. 199-210.

Ramachandran, P. and Summanwar, A. S. 1986. Effect of systemic and contact pesticides on the spread of yellow vein mosaic (YVM) disease of bhendi. *In* National Seminar on 'whitefly transmitted plant virus diseases' (25-27, June 1986). Division of Mycology and Plant Pathology, I. A. R. I., New Delhi-110012. Abstract 11-12 p.

Raychaudhuri, M. and Varma, A. 1983. Effect of oils on transmission of marrow mosaic virus by *Myzus persicae* Sulz. *J. ent. Res.*, **7**: 107-111.

Reddy, D. R. R. and Varma, A. 1986. Management of golden mosaic disease of cowpea. *In* National Seminar on 'whitefly transmitted plant virus diseases' (25-27 June 1986). Division of Mycology and Plant Pathology, I. A. R. I., New Delhi-110012. Abstract 45 p.

Reddy, D. V. R. and Wightman, J. A. 1988. Tomato spotted wilt virus : thrips transmission and control. *Adv. Dis. Vect. Res.*, **5**: 202-220.

Saika, A. K. and Muniyappa, V. 1986. Epidemiology and control of tomato leaf curl virus. *In* National Seminar on 'whitefly transmitted pland virus diseases' (25-27 June 1986). Division of Mycology and Plant Pathology, I. A. R. I., New Delhi-110012. Abstract 30-31 p.

Sastry, K. S. M., Singh, S. J. and Sastry, K. S. 1978. Efficacy of Krishi oil in relation to the control of whitefly population and the spread of Tomato leaf curl virus (TLCV). *Pesticides,* **12** (1) ; 28-29.

Sharma, S. R. and Varma, A. 1975. Natural incidence of cowpea viruses and their effect on yeild of cowpea. *Indian Phytopath.,* **28** (3) : 330-334.

Singh, B. R. and Singh, Mahant. 1986. Control of yellow-vein mosaic of okra by checking its vector whitefly through adjusting dates of sowing, insecticidal application and crop barrier. *In* National Seminar on 'whitefly transmitted plant virus diseases' (25-27 June 1986) Division of Mycology and Plant Pathology, I. A. R. I., New Delhi 110012. Abstract 33 p.

Singh, S. J. 1989. Virus and mycoplasma diseases on vegetable crops. *Indian Horticulture,* **33** and **34** (4&1) (Jan.-March, April-June) 68-75.

Singh, S. J., Sastry, K. S. M. and Sastry, K. S. 1973. Effect of oil spray on the control of tomato leaf curl virus in field. *Indian J. agric. Sci.,* **43** (7) : 669-672.

Singh, S. J., Sastry, K. S. and Sastry, K. S. M. 1979. Efficacy of different insecticides and oil in control of leaf curl virus disease of chillies. *Zeit fur pflan and Pflanzenschutz,* **86** (5) : 253-256.

Slack, S. A. and Scott, H. A. 1971. Hemolymph as a reservoir for the cowpea strain of southern bean mosaic in the bean leaf beetle. *Phytopathology,* **61** : 538-540.

Srivastava, S. K. 1989. Further studies on virus of cowpea (*Vigna unguiculata* ssp. *unguiculata* (L) Walp.). Ph.D thesis. P. G. School, I. A. R. I., New Delhi-110012, 199 pp.

Subrahmaniam, K. and Varma, A. 1986. Management of yellow mosaic disease of blackgram through cultural practices and chemical treatment. *In* National Seminar on 'whitefly transmitted plant virus diseases' (25-27 June 1986), Division of Mycology and Plant Pathology, I. A. R. I. Abstract **43** p.

Sylvester, E. S. 1969. Virus transmission by aphids - a view point. *In* K. Maramorosch [ed.], ***Viruses, Vectors, and Vegetation.*** Interscience Publishers, New York. 159-173.

Thronbury, D. W., Hellmann, G. M., Rhoads, R. E. and Pirone, T. P. 1985. Purification and characterization of potyvirus helper component. *Virology,* **144**: 260-267.

Varma, A. 1976. Recent trends in control of plant virus disease. *Proc. Nat. Acad. Sci.,* India Section B (Biol. Sci.) **46**: 193-206.

Varma, A. 1983. Control of plant virus vectors an integral part of pest management. *In* R. A. Agarwal, G. P. Gupta, Prem Kishore and Dinesh Chandra [eds.], *Principles and concepts of Integrated Pest Mangement.* Division of Entomology, I. A. R. I., New Delhi-110012. 48-53.

Varma. A. 1986. Management of whitefly transmitted diseases. *In* National Seminar on 'whitefly transmitted plant virus diseases' (25-27 June 1986) Division of Mycology and Plant Pathology, I. A. R. I., Abstract 42 p.

Varma. Anupam 1993. Integrated management of plant viral diseases *In Crop Protection and sustainable agriculture.* Wiley, Chichester (Ciba Foundation Symposium 177): 140-157.

Varma. A., Dhar, A. K. and Mandal, B. 1992. MYMV transmission and control in India. *In* S. K. Green and D. Kim [eds.] *Mungbean Yellow Mosaic Disease Proceedings of an international workshop* held at Bangkak, Thailand (2-3 July 1991). Asian Vegetable Research and Development Center. P. O. Box 205, Taipei 10099. 8-27.

Varma. A and Rajamannar, M. 1989. Yellow type diseases - distribution identity and importance. *In* S. P. Raychaudhuri and Anupam Varma [eds.], *Plant diseases caused by Fastidious Prokanyotes.* Today and Tomorrows/Scholarly Publications, New Delhi-110005. 1-23 p.

Watson. M. A. 1972. Transmission of plant viruses by aphids. *In* C. I. Kado and H. O. Agrawal [eds.], *Principles and Techniques in Plant Virology.* Chapter 5. Van Nostrand Reinhold, New York, 131-167.

Watson. M. A. and Roberts, F. M. 1939. A comparative study of the transmission of *Hyoscynmus* virus 3, potato virus Y and cucumber virus by the vectors *Myzus persicae* (Sulz.), *M. circumflexus* (Buckton) and *Marcrosiphum gei* (Koch). *Proc. of Royal Soc. London.* B, **127** : 543-576.

Wyman. J. A., Toscano, N. C, Kido, K., Johnson, H and Mayberry, K. S. 1979. Effects of mulching on the spread of aphid-transmitted watermelon mosaic virus to summer squash. *J. econ. Ent.,* **72** : 139-193.

Recent Advances in Indian Entomology
Editor: O.P. Lal
APC Publications Pvt. Ltd., New Delhi, 1996, pp. 63-66

Recent Advances in the Control of Insects of Public Health Importance

V. P. SHARMA

Insects of public health importance are mainly the mosquitoes, sandflies, flies and the cockroaches. These insects transmit important and deadly diseases such as malaria, filaria, japanese encephalitis (JE), dengue fever (DF), dengue shock syndrome (DSS), dengue haemorrhagic fever (DHF), visceral and cutaneous leishmanisais, typhoid, enteric diseases, and a host of other viral infections of known and unknown etiology.

In India disease burden due to insects of public health importance is enourmous and this disease scenario is likely to stay for long as the main cause of high morbidity and associated mortality. In India the National Malaria Eradication Programme (NMEP) is responsible for the control of all vector borne diseases. Malaria control in the field is carried out by the state health departments. Based on the surveillance data collected by the states, NMEP reports about 2 million malaria cases annually, and 400-500 deaths due to malaria. There is incontrovertible evidence that the incidence of malaria is grossly underestimated. Malaria control in the endemic belts is becoming difficult to control because of technical obstacles such as inter alia the resistance in vectors to the insecticides, resistance in *P. falciparum* to the anti-malarials, inadequate knowledge about the biology of the vectors and the mismatched control strategies. Malaria has taken deep roots in the country and new ecotypes have been established. The problem of filaria is not even addressed adequately. Filaria surveys are still not complete, but based on the available figures there are about 26 million microfilaria carriers and 20 million cases of various types of disease manifestations. Control efforts by the National Filaria Control Programme (NFCP) are confined to urban areas only. Like malaria, filaria incidence is also increasing Viral diseases DF, DHF, and DSS are becoming a major cause of concern as the morbidity is very high, and particularly DHF and DSS cause deaths. There are no accurate statistics available in the country but the disease is spreading to cover almost all urban areas, bringing epidemics and now it is entering into the rural areas with the process of urbanization and piped water supply. Japanese encephalitis transmission is associated with rice paddy culture. There are several thousand deaths annually and the disease appears suddenly killing and maiming thousands in the country side. Vector control is therefore not of much value. Kala azar incidence is >70,000 cases annually causing protracted morbidity and eventually death in untreated cases. Kala azar transmission has also spread to eastern UP, West Bengal and has covered the entire state of Bihar. Diarrhea, Typoid, and enteric group of diseases account for about 30 % morbidity in India in the world. Therefore, the vector borne disease scenario in the country is quite depressing and calls for a holistic approach to bring all control technologies together for a selective mix of appropriate technologies for the strategic development of situation specific efforts in respect of the control of each disease.

Recent advances in vector control

In this brief session it is not possible to describe all the developments that are taking place in vector control. Today I would like to discuss some results of research carried out by the scientists of the Malaria Research Centre in addressing the problem of malaria control.

Vector biology

Our knowledge base on the vector biology has always been weak, and this was the main reason of control failures and widespread malaria resurgence inspite of heavy investments and nearly perfect field operations. For the first time *An. culicifacies*, the vector responsible for about 70% malaria transmission in the country, was colonized at the MRC. Soon thereafter it was possible to initiate many laboratory and field studies. One area that has received our attention is the cytotaxonomy of mosquitoes. Anophelines are characterized by the presence of sibling species atleast in some important taxons investigated. These sibling species can be identified by using many markers, but we have extensively used inversions in the polytene chromosomes. The vector *An culicifacies* comprises of 4 sibling species. Our laboratory studies substantiated in the field have revealed : (i) there are clear differences in the biology of the immatures with respect to their breeding habitats and the life budget features, (ii) response to DDT, HCH, and Malathion varies among the species complexes so that spraying may result in ealy or late appearance of physiological resistance, (iii) biting behaviour is quite specific both in respect to their preference for host and also the biting rhythm so that transmission dynamics of malaria is species dependent, and obviously this behaviour will profoundly effect control measures, particularly the introduction of the insecticide impregnated bed nets, and (iv) the vectorial efficiency of each species is also a characteristic feature governing the transmission of malaria. Natural distribution of sibling species in the country may be sympatric or allopatric. Species B. is found all over the country, species A is found from north to south, species C is found from west to east ad species D is found is some pockets mainly in the central India. Studies in the field on the transmission dynamics of *An. culicifacies* transmitted malaria revealed that malaria is prevanlent in areas with species A, C, & D but not in areas with B. It was therefore possible to reduce field operations accordingly and save insecticides in malaria control. This information is also being used in the introduction of impregnated bet nets, and in the malariogenic stratification for prospective planning of control measures.

Bioenvironmental Malaria control

Use of insecticides in malaria control has finally started producing diminishing returns. The situation could be retrieved for some time by using replacement insecticides but the practical life of most of the known insecticides is 5 to 7 years. It is therefore important that malaria control is based on technologies that are eco-friendly, sustainable, cost effective, and provide long term control. With these objectives a new strategy named the bioenvironmental control was conceived and launched in the country. This strategy comprises of : (i) vector control by environmental management and biological control methods, and (ii) parasite control by early case detection and treatment. In this approach to vector control health education, community participation, inter and intra sectorial coordination, invoking legal measures are the important component. Insecticides are not applied but held in reserve for emergency situations.

This experiment was launched at various locations in the country to demonstrate malaria control. In about three years successful malaria control was demonstrated in (i) rural areas in Haldwani, Shahjahanpur, Rourkela, Nadiad, and in the Car Nicobar island, (ii) industrial areas in Hardwar (BHEL), Rishikesh (IDPL), Mathura refinery (IOC), and (iii) urban areas in Madras, and Panjim (Goa). It was also demonstrated that the methods employed were simple, indigenous, low cost and can be integrated with the primary health care system. In comparing the cost of field operations it was estimated that the method of malaria control by the alternate strategy was cheaper than DDT spraying which is today the cheapest method of malaria control. Malaria control could also be linked with income generating schemes such as the edible fish production and social forestry.

During the course of the implementation of bioenvironmental control of malaria several new technologies were introduced. Expanded polystryrene beads were applied in the control of mosquito breeding in wells, overhead water tanks, sluice valve pits, septic tanks etc., fly ash was diverted to fill low lying areas, pot holes, ditches etc., drainage was improved in many rural areas to prevent mosquito breeding; in marshy areas euclyptus plantations were carried out to eliminate standing water and lower the water table; larvivorous fishes were produced in millions and billions to control mosquito breeding, and at the same time it was demonstrated that the use of fish in malaria control is not hazardous to the environment; bye-laws were amended, health education campaigns were intensified, community participation was encouraged, there was increased emphasis of interaction with the communities, government functionaries; and inter and intra sectorial coordination was elicited.

Insecticide impregnated mosquito bed nets

A recent development is the use of treated bed nets in malaria control. This strategy has evoked world wide interest. In India bed nets were first introduced in Sonapur PHC in the control of malaria. The results in terms of protection from mosquito bites and reduction in the epidemiological indices wee spectacular. The strategy was passed on to the state health and tribal welfare departments. At present bed nets are widely used in the endemic areas of Assam, particularly in the organized sector such as the tea estates, industries etc. Based on the experience gained in Assam bed nets were introduced in the mining belts of Orissa near Rourkela. Since the introduction of bed net programme malaria incidence has started dwindling, now touching low levels. Bed nets when introduced in areas with *An. culicifacies* and particularly in tribal populations, the results was not encouraging. We have stratified the country for the application of bed net programme. It is now possible to provide good protection from malaria to populations living in malaria refractory areas. Bed net programme is cost effective, appreciated by the communities, sustainable and can be incorporated with the PHC system.

Remote sensing and the geographical information system

We are exploring the possiblities of the use of remote sensed data in the mapping of mosquito breeding. Preliminary results have shown promise in mapping major mosquito production sites related to the quality of water. Although it may not be possible to apply this technology in case of vectors which prefer to breed in small places, but in many cases breeding of *An. culicifacies* could be predicted, say in the rice field areas, flood prone areas, large marshy areas and in some areas during the monsoon. A study has also been initiated to apply GIS in enhancing our understanding in establishing cause and effect relationship in malaria. Preliminary results based on data collected from Kheda and Car Nicobar have brought out primary and secondary causes of mosquito production leading to disease transmission. This line of investigation is still at the infancy stage, and in the coming few years it may be possible to plan malaria control on more scientific lines, particularly what may be required is to correct the main cause of problem, instead of spraying in time and space.

Neem based pesticides

We have demonstrated that neem oil is one of the best repellent for the mosquitoes, and the sandflies. Neem oil can be applied in low concentrations of 2 to 5% in either coconut oil or mustard oil on the exposed body parts. Repellent action lasts for upto 12 hrs. Neem oil can also be used in the mats and heated in any device marketed for producing vapours. In this case about 5-10% neem oil diluted in any solvent like acetone, alcohol or even the kerosene can be used. Either a used mat or a card board dipped in the diluted solulation can replace the commercial mats that contain insecticide as a repellent. Neem oil can also be burnt in kerosene lamps at 2-4 % concentrations to repel mosquitoes and the sand flies etc. This is the real low cost solution appropriate for rural India in terms of providing personal protection against the bite of haematophagoous insects.

Recent Advances in Indian Entomology
Editor: O.P. Lal
APC Publications Pvt. Ltd., New Delhi, 1996, pp. 67-82

Neem for Management of Insect Pests : Advantages and Disadvantages

R. P. SINGH and SHIVENDRA SINGH

Every year more than 20,000species of field and storage pests destroy approximately one third of the world's food production, valued at billions of dollars. Synthetic pesticides which are used for managing these pests suffer from serious drawbacks. A recent estimate by World Health Organisation (WHO) puts the number of acute poisoning due to use of synthetic pesticides at threee million with 220,000 deaths (Hoddy, 1993). It is also calculated that as many as 25 million agricultural workers in developing countries have episode of pesticide poisoning but few seek medical attention (Hoddy, 1993).

Human suffering is not the only problem associated with the use of synthetic pesticides. About 500 species of insect have been reported to have developed resistance to one or more insecticide and more than 12 species to practically all insecticides (Georghiou, 1990). Failure of synthetic insecticide to contain American bollowrm, *Heliothis armigera* is Guntur and Prakasham district of Andhra Pradesh resulted in enormous losses to cotton crop. Several poor farmers who were much dependent on the crop committed suicide. Cost is yet another problem associated with the use of synthetic insecticides. It is bound to increase due to depleting natural reserves. Certain crop pests which were managed with 2 or 3 sprays now require, 10-24 sprays. The problem gets further compounded because this leads to annihilation of natural biotic factors which otherwise keep the pest under control.

It is evident from the foregoing that synthetic insecticides are no solution for the management of insect pests. We should have a pesticide which is not only effective but also safe, available, renewable and at the same time economical

The fight between insect and plants started some 500 million year ago. The war between attacker and attacked is still on. Their presence is testimoney of the fact that insects have not been able to eliminate even single plant species. Plants are thus well-prepared to resist the attacker. For this plants have developed several ways and means (physical and chemical) to resist the attacker especially insects. Of these, chemicals elaborated by plants play a vital role in its defence mechanism.

Research carried out on neem *(Azadirachta indica)* during the last 30 years have unravelled the presence of array of chemicals which have adverse effect on insect and at the same time safe to human beings.

Historical aspects

Neem tree (*Azadirachta indica* A. Juss) is a native to India. In Sanskrit neem is known as "Arishta" which means relieving sickness. Its medicinal uses are mentioned in our earliest sanskrit writings, the earliest authentic record of its medicinal use is in "Artha Shastra" (Treatise on Polity) of Kautilya (4th century BC). Various parts of the tree (leaves, bark, flowers and fruits etc.) have been employed in ayurved by the famous Hindu physician Susruta. Besides being used as folklore medicine the tree has been utilised as source of insecticide since time immemorial. Burning of neem leaves to generate smoke to ward off mosquitoes, mixing of leaves with grains during storage and keeping dried leaves between folds of clothes in trunks to ward off insects is well known. Neem may be the first insecticide used by man agaist insect pest.

Management of insect pests

Search for alterntive to synthetic insecticides started following the publication of 'Silent Spring' by Rachel Carson in 1962. It was during the same year Pradhan et al., (1962) systematically investigated neem tree for its insecticidal property and reported that the active material in neem is concentrated in the seed kernel. An aqueous suspension prepared from seed kernel (NSKS) when sprayed at 0.1 percent concentration on standing crop at the Indian Agricultural Research Institute offered complete protection against locust swarm which invaded Delhi during the same year. This finding generated interest in neem the world over because no method was available whih could give complete protection against any field pest in continental area.

Ever since this findig that neem can deter feeding of such a highly polyphagus pest as *Schistocerca gregaria,* there has been intensive research on the chemical and insecticidal aspects of neem seed kernel. Evaluations done during the last 30 years have revealed that neem seed extracts and pure compounds isolated from seed have diverse biological effect on insects : antifeedant, oviposition deterrent, repellent sterilant, growth disrupter, ovicidal and other behavioural and physiological effects.

No plant or chemical is known which has such a diverse biological effect on insects. Also there is no report of any extract or pure compound which effect insect at all the steps of host selection (Fig. 1). The extra ordinary property of neem makes it an unique source of insecticide. Of the various biological effects growth and development disrupting activity of neem is the most important. In fact rarely have insects not succumed to the effect of neem. About 350 species of insects have been evaluated with neem and almost all have been found sensitive.

Some key pests of agriculture which have been evaluated with neem and found suspceptible are as under :

Desert Locust, *Schistocerca gregaria* and **Migratory Locust,** *Locusta migratoria* Laboratory tests with, aqueous suspension of seed kernel exhibited high antifeedant activity against *S. gregaria and L. migratoria* (Table 1).

The standing crops at Indian Agricultural Research Institute (IARI) sprayed with

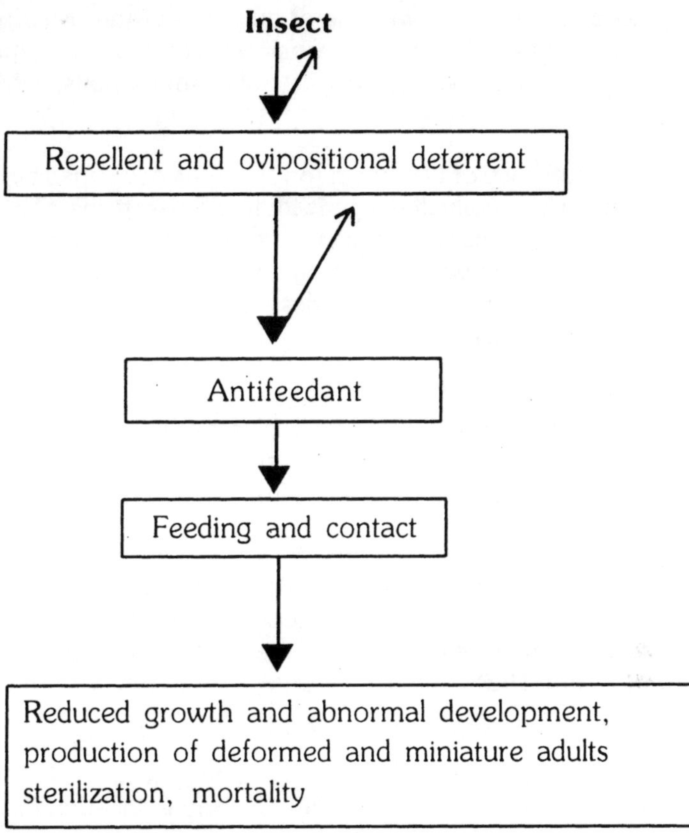

Fig 1. Biological effects of neem at various steps of host selection by insects

Table 1. Comparative antifeedant activity of neem seed kernel suspension against migratory and desert locusts.

Test insect	*Percentage leaf consumed at % conc*								
	0.5	0.25	0.1	0.05	0.01	0.005	0.001	0.000t	untreated
Migratory Locust	0	0	0	0	6	3.6	71.1	91.7	100
Desert locust	-	0	0	1	0	0	0	44.4	100

Source : Pradhan et. a1; (1962)

0.1 % concentration of NSKS at the rate of 300-600 litres/hectare were completely saved (Pradhan *et al.*, 1962). Ever since then NSKS is recommended for protection of crops against the desert locust. As soon as information regarding locust invasion is received, farmers are advised to be ready with NSKS.

Maize stem borer *(Chilo partellus)*

In a field trial conducted at IARI using 5 and 10% Kernel-based and 5% oil-based granules significantly reduced damage and increased yield of grain and fodder. The treatment was at par with 5% granule of quinalphos and *Bacillus thuringiensis*. The difference betweeen two neem kernel granules was not significant indicating that 5% granule was effective as 10% granule (Anonymous, 1983).

Cotton boll worms

In a field trial efficacy of two neem formulations developed by National chemcial Laboratory, Pune were evaluated alongwith three insecticide. The data gathered on percentage of fruiting bodies damage by boll worms revealed that the neem formulations were effective in reducing the damage caused by boll worms and compared well with synthetic insecticides. The yield of fibre was better or at par with synthetic insecticides.

Mango hopper *(Indioscopus clypealis)*

Two neem formulation - 'Nimba' and 'Neem emulsion' developed at IARI, were evaluated against the mango hooper, a serious pest of mango in northern India. Both the formulations gave effective control of the pest. (Srivastava et al., 1993).

Green leaf hopper *(Nephotettix spp)*, **leaf folder** *(Caphalocrocis medianalis)*

NSKS at 5% concentration significantly reduced the damage by leaf folder. The efficacy of NSKS was comparable to monocrotophos. NSKS also effectively reduced the population of green leaf hopper. The spray remained effective for 14 days (Rajasekaran et al., 1988).

Tobacco caterpillar *(Spodoptera litura)*

S. litura is currently being managed effectively with 2% NSKS. Extensive field trails conducted by Joshi et a1., (1990) at various locations in Andhra Pradesh revealled that NSKS is not only cheaper than conventional insecticide but also as effective as synthetic insecticide. Normally 4 sprays are needed to contain this pest. The first spray is given 20 days after seed germination. Second, third and fourth treatments are given at intervals of 10 days and if necessary a fifth spray is also given. The protection of the crop is due to the oviposition deterrent effect and IGR activity of the extract.

Diamond back moth *(Plutella xylostella)*

P. xylostella is a serious pest of vegetable crops and has developed resistance to almost all conventional insecticides. Vegetable crops like, chinese cabbage, cabbage and tomato were protected against the damage caused by *P. xylostella* when treated with neem seed kernel suspension, the extract was found as effective as fenvalerate in protecting the cabbage crop from *P. xylostella* damage, 60-70 days after planting. At 90 days after treatment neem oil registered marginally higher population than fenvalerate. However, neem oil was better than *B. thuringiensis*, Diflubenzuron and allitin treatments. (Chandra Mohan and Nanjan, 1990)

Gram pod borer *(Heliothis armigera)*, **spotted pod borer** *(Maruca testulalis)* **and pod fly** *(Melanagromyza obtusa)*

Singh et al., (1985) conducted field trial using neem oil and ethanolic extract of

NSK for control of the above three borers of pigeon pea and compared it with fenvalerate an effective insecticide recommended for these borers, only one spray @ 625 lit/hactare was given at 50% flowering stage. Neem oil at 2% was effective only against pod fly while ethanolic extaract at the same concentration considerably lowered the incidence of all the 3 borers.

Stored grain pests

Neem seed kernel powder (NKP), neem oil and deoiled kernel powder (DNKP) are the main products evaluated for protection of stored grain pests and almost all the pest have been found susceptible to these products. The efficacy of various neem products/derivatives against important stored grain pests is summarised here under.

Pulse beetles *(Callosobruchus maculatus* and *C. chinensis)*

Neem seed kernel powder at 1 to 2% concentration gave good control of pulse beetle in cowpea, pigeon pea, mung bean, chickpea and pea (Jotwani and Sircar, 1967) and lentil, grasspea, redgram, blackgram and green gram (Yadav, 1973).

Neem oil at 0.25 to 5% concentration has been reported (Ketkar, 1987; Yadav, 1985; Zeherer, 1984) to give excellent control of *C. maculatus and C. chinensis.*

Lesser grain borer *(Rhyzopertha dominica)*

Rhyzopertha has also been found very sensitive to neem treatments. Rice under storage can be effectively protected by mixing of dry neem leaves with it.

Neem oil at 0.005% significantly reduced damaged in paddy (Prakash et al., 1980). Neem seed kernel powder at 0.25-1% showed marked reduction in the population of the pest in wheat (Jilani and Haq, 1984). Treatments at a still higher concentration (1-2%) offered a very satisfactory control of the beetle in wheat (Jotwani and Sircar, 1965)

Rice weevil *(Sitophilus oryzae)*

Acetone extract of neem leaf is reported to cause 98.33% mortality of the rice weevil upto a period of three months (Mukherjee et al., 1990). Defatted neem kernel powder at 0.1% concentration admixed with shorgum grain completely protected the grains from the damage of the weevil, there was no feeding in the grains and the weevils died due to starvation (Mohan et al., 1990).

Neem kernel powder (1-2%) gave better protection of maize than ethylene dibromide-carbon tetrachloride (ED CT) (Chachoria et al., 1971).

Khapra beetle *(Trogoderma granarium)*

T. granarium is perhaps the most susceptible insect to neems treatments.

Neem seed kernel powder admixed @ 1 to 2 parts/100 parts (W/W) of wheat seed afforded very satisfactory protection of wheat seeds against the beetle (Jotwani and Sircar, 1965). DNKP admixed with wheat @ 0.03-0.05 parts/100 parts of wheat (w/w) caused 100% mortality of the first instar larvae (Singh and Kataria, 1986).

Angoumous moth *(Sitotroga cerealella)*

Neem's activity against Lepidopterous pests of storage is well exhibited against the Angoumous moth and rice moth. Neem oil at 0.1% reduced the damage by *S. cerealella* in wheat to a level comparable to 5% Malathion dust (Verma et. al., 1985). Multiplication of the moth was completely checked in paddy treated with 0.025% neem oil (Prakash et al., 1980).

Exposure of the rice moth to vapours of neem oil considerably reduced egg output and egg viability, thereby reducing the next generation (Pathak and Krishna, 1985). DNKP at 1-2% is reported to arrest completely the development of *C. cephalonica* in wheat flour (Jhansi and Singh (1993). Maize grains teated with 2% neem kernel powder suffered less damage than ED-CT (30ml/40 kg.) (Chachoria et al., 1971).

Biologically active constituents

Almost every part of the tree has long been used traditionally as medicine for a variety of ailments especially against skin diseases. Neem oil is perhaps the oldest oil used in medicine. It is because of this reason that chemists in the begining of this century were concerned with the pharmacological aspects of neem. Chatterji and Sen (1919) were the first to isolate an acidic principle which they named as "Margosic-acid". This was followed by isolation of a Sulphur free crystalline bitter acid called margosopicrin from oil (Watson, 1923).

Siddiqui (1942) isolated three compounds viz., nimbin nimbinin and nimbidin from oil. Nimbidin is antiarrhritic and anti-inflamatory and possess significant anti-ulcer potential. Sodium nimbidate was later found to be spermicidal.

Sixties saw remarkable discovery in neem chemistry and biological activity against insect pests. Pradhan et al., (1962) reported extra ordinary antifeedant activity of neem seed kernel extract against desert locust, *Schistocerca gregaria*. Meliantrol a triterpenoid alcohol was later isolated by Lavie et al., (1967) from neem oil and also from fruit of dharek, *Melia azedarach*. This compound inhibited the feeding of desert locust, *Schistocerca gregaria* at 8 ng/cm^2 filter paper. Almost simultenaeously Butterworth and Morgan (1968) isolated a tetranortriterpenoid, azadirachtin from neem seed kernel. This was later found to be present in neem leaves also. Azadirachtin was much stronger feeding inhibitor than meliantriol. The former deterred feeding of lucust at 1 ng/cm^2. Since then more than 150 constituents have been identified from different parts of the tree and their structures elucidated. Of these about 40 are reported to effect insects in a variety of ways.

Azadirachtin one of these coumpounds is considered as most important. Studies revealed that azadirachtin is a mixture of several isomers (A, B, C, D, E, F, G, H and I). Azadirachtin A and B are about 85% and 15% respectively; other isomers are only in traces.

Compounds reported to cause adverse effects on insects are as under :

Antifeedant compounds

Compound	Insect	Reference
Azadirachtin	*Schistocerca gregaria*	Butterworth and Morgan, 1968
Salannin	*Musca domestica*	Henderson et. al., 1967
Meliantriol	*Schistocerca gregaria*	Lavie et al., 1967
1, 3- deacetyl vilasinin	*Epilachna varivestis*	Kraus and Cramer, 1981
3-deacetyl salannin	*Musca domestica*	Schwinger et al., 1984
Salannol acetate Salannol	*Musca domestica*	Keller, 1985
Salannolactum (21) Salannolactum (23)	*Epilachna varivestis*	Klenk, 1985
Azadirone Azadiradione 14-epoxy azadiradione	*Epilachna varivestis*	Kraus et al., 1987
1-tigloy1-3-acetyl vilasinin	*Epilachna varivestis*	Kraus, 1986
Vepaol Iso vepaol	*Mythimna separata*	Sankaram et al., 1987

Insect Growth Regulatory (IGR) compounds

Compounds	*Insect*	Reference
Azadirachtin	*Plutella xylostella*	Ruscoe, 1972
Margosinolide Isomargosinolide		Kubo, 1986
Nimocinolide Isonimocinolide	*Musca domestica*	Siddiqui et al., 1986
Nimbolicinoic acid	*Callosobruchus analis*	Naqvi, 1987
Deacetyl azadirachtinol	*Heliothis virescens*	Kubo, 1986
22-23-dihydro-23B methoxy azadirachtin	*Spodoptera littoralis*	Kraus et al., 1987
Vepaol Isovepaol	*Spodoptera litura* *Mythimna separata* *Chilo partellus* *Myllocerus sp.*	Sankaram et al., 1987
Nimocinol	*Musca domestica*	Naqvi, 1987

Toxic compounds

Compounds	*Insect*	*Reference*
Azadirachtin Deacetyl salannin	*Dactynotus carthami*	Devakumar et al., 1986
Salannin A derivative of Salannin	*Lipaphis erysimi*	Singh et al., 1988

Sterilizing compounds

Compound	*Insect*	*Reference*
Azadirachtin	*Oncopeltus fasciatus*	Dorn et al., 1987
Azadirachtin	*Spodoptera litura*	Gujar and Mehrotra, 1984
Nimbocinolide	*Musca domestica*	Siddiqui et al., 1986

Advantages of using neem

There is growing conciousness among public and health authorities about the side effects of synthetic pesticides. Thus any method of management of insect pest should not only be acceptable to entomologist but also to public, health authorities and environmentalists. Neem besides being effective against insect pests has other attributes which make neem as an ideal source of insecticide.

1. *Toxicology*

We have been living with the tree for centuries without any record of any ill effect on our health. In fact the tree has always been considered as purifier of the surroundings.

Neem twigs are still being used as toothbrush and considered very good for oral hygine. Neem flowers are used in drinks and neem leaves are taken as pickle in many parts of the country without appraently any side effects. Mahatma Gandhi used to make 'Chutni' of neem leaves for eating along with food.

Laboratory studies on rabbits and guinea pig showed that an ethanol extract of seed installed into the rabbit eye caused no irritation and no significant skin-sensitivity resulted from injection of the extract into shaved skin of guinea pig

Albino rats and mice showed no toxic effects when fed nimbidin upto 2000 mg/kg or when administered intraperitoneally at 1000 mg/kg. Dogs fed nimbidin daily at 10 and 20 mg/kg showed no systemic effects. (Pillai and Santhakumari, 1984).

Neem rich 100, a formulation containing 30% neem oil, applied dermally to albino rats at daily doses at 200, 400 and 600 mg/kg for 3 weeks caused no overt signs of toxicity or abnormal behaviour. Treated rats exhibited higher food consumption, gained weight and showed no abnormal blood levels (Qadri et. al., 1984).

Application of 0.25 ml Margosan-0 (a commercial neem product) on the bare skin of guinea pig did not produce sensitization (Larson, 1986). Studies on five strains of *Salmonella typhimurium* showed that Margosan-0 is non-mutagenic. Neem extracts are also non-corcinogenic. As compared to synthetic insecticides, neem based formulations are safe as evident from the toxicological data. It is also not carcinogenic.

2. Safety to predators and parasitoides

All the literature on effect of neem formulations against predators and parasitoides show that the formulations are safe and maintain these biological control agents in the field. Some relevant references are given below :

Insect	*Extract*	*Effect*	*Reference*
A. EGG PARASITOID			
Telenomus remus	Seed kernel suspension	No adverse effect on adult emergence	Joshi et al., 1982
Trichogramma japonicum	Seed kernel suspension & oil	Safe	Jayaraj et al., 1983
Trichogramma chilonis	Seed kernel extract	No effect on adult emergence	Raguraman, 1993
B. LARVAL PARASITIODES			
Bracon hebator	oil	Safe	Jhansi and Sundra Babu, 1987
Bracon hebator	Replin & Neemguard	Safe	Sirinivasa Babu et al., 1993
Apanteles plutellae	Seed Kernei extract	No harmful effect	Bandra and Kudagamage, 1993
C. PUPAL PARASITIODES			
Lycosa pseudoannulata	oil	No adverse effect	Saxena et al., 1984
Tetrastichus	Seed kernel	Safe	Jayaraj et al., 1993
Orius sp.	Neem fraction	No effect	Sharma et al.,1984
Syrphid larvae and Coccinelid adults	Neem emulsified concentrate	No adverse effect	Srivastava and Parmar, 1985
Cyrtorhinus lividipennis	Seed kernel suspension, oil & aqueous extract of neem cake.	Safe	Mohan, 1989

3 Development of resistance

Development of resistance to insecticides is a predictable response by insects. Insects hve developed resistance to insecticides like organochlorines, O. P. compounds, carbamates etc. Infact more than 500 insect pests have developed resistance to one or more insecticides and 12 key pests have developed resistance to almost all the insecticides. In view of the above observations, it was feared that insects will soon develop resistance to neem products in near future. Till date, no report has been published on the development of resistance by insects to neem products. This is because neem extracts have a complex array of chemcials that act in conjuction and effect various physiological and behavioural parameters in insects at the same time.

Vollinger (1987) demonstrated that two genetically different strains of *Plutella xylostella* treated with a neem seed extract showed no sign of resistance in feeding and fecundity tests upto 35 generations. In contrast, deltamethrin treated lines developed resistance factor of 20 in one line and 35 in other. There was, also, no cross resistance between deltamethrin and neem seed kernel extract in the deltamethrin resistant lines. The activity of esterase and multi-function oxidase engymes did not change during the 35 generations. But continious use of pure comounds like azadirachtin, salannin etc. may lead to resistance for which precautions should be taken.

4. Availability and renewability

Another important factor in pest management after the material has been found to be effective, is the availability of raw material. Fortunately India has the largest number of neem trees in the world. It is an evergreen fast growing tree which thrives well in tropical climates having an annual rainfall of 400 to 800 mm and an extended dry season, which is typical to India. There are about 19-20 million trees in India and each tree on the average gives 15-50 kg of seeds. Assuming that an average size tree yields 30 kg seeds annually, the total seeds yield will be 570,000 tonnes which is worth Rs. 5700 million approximately. The collection of seeds alone can provide emloyment to nearly 0.6 million people in the country for 60 days. The neem tree is present in all the states of the country and grows on all types of soil, acidic to alkaline, it can tolerate temperature extreems from 0 to 49^0 C (Hegde, 1993). The neem tree is thus very well suited to Indian agroclimate and its availability is not a problem.

Regarding its renewability, the most active part of the tree, the seeds, are available year after year throughout the life of the tree. The leaves are also availble throughout the year.

5. Ecological impact

Neem extracts have no adverse ecological impact and are much safer than the conventionally used synthetic pesticides.

Neem oil extractive, a by product of neem oil industry, is an effective mosquito larvicide (Attri and Ravi Prasad, 1980). The extractive when tested against *Gambusia* fish, a voracious predator of mosquito larvae, showed no toxicity.

Neem leaves and kernel incorporated in the soil showed positive effect on weight and survival of earthworm, *Eisenia foetida* (Rossner and Zebitz, 1987). Application of neem cake to the soil retards soil urease activity and organic matter decomposition.

Grant and Schmutterer (1987) reported that soil treated with neem extract can be used to manage the ostracod *Heterocypris luzonensis* which feeds on N_2 fixing algae.

6. *Cost factor*

Cost of synthetic insecticides is yet another important factor affecting their use by poor and marginal farmers in developing countries.

The problem has been compounded due to development of resistance where increasing dosages of insecticides are used. This further adds not only to hazard but also the cost. Neem tree is fortunately readily available in most developing countries. With little training farmers can not only solve their pest problem with neem but can also supplement their income by selling surplus seeds. It has been calculated that three trees yield about 100 kg seeds which is worth Rs. 1000/- Crushing 100 kg seeds in local 'Kholu' will yield 25 kg oil valued at Rs. 1250/-. The crushing cost is about Rs. 250/- so the profit is Rs. 1000/- (Singh and Kataria, 1992). The advantage of selling oil rather than unprocessed seeds is that the farmer gets neem cake as a by-product which can either be sold for about Rs. 10/- per kg or used for coating urea to increase its efficiency. Cake can also be used as a pesticide.

7. *Medicinal uses*

Neem is also known for folklore medicine. Recent researches have brought to light that neem can be effectively utilized in several cases as medicine, thus besides combating pest problems poor farmers who cannot afford costly allopathic medicines can use it as a substitute.

Disadvantages

1. The most important drawback is that unlike synthetic insecticides which kill the insects instantly neem has slow effect and hence it is difficult to convince uneducated farmers in developing countries.

2. The cost of neem based products is nearly as much as synthetic insecticides. Thus if choice is given farmers still prefer synthetic insecticides. Crude extracts however are much cheaper.

3. Low stability under field conditions as compared to several synthetic insecticides is yet another drawback.

References

Anonymous. 1983. Neem in Agriculture. Research Bulletin No. 40. Indian Agricultural Research Institute, New Delhi-12.

Attri, B. S. and Raviprasad. 1980. Neem oil extractive as an effective mosquito larvicide. *Indian J. Ent* **42** : 371-374.

Bandra, K. A. N. P. and Kudagamage, C. 1993. Potential of neem for controlling cabbage leaf eating caterpillar complex. *Abstracts, World Neem Conference,* February 24-28, Bangalore, India.

Butterworth, J. H. and Morgan, E. D. 1968. Isolation of a substance that suppresses feeding in Locust. *J. Chem. Soc. Chem. Comm.* 23-24.

Carson, R. 1962. *Silent Spring.* Fancett Publication Inc. Greenwich Corpn.

Chachoria, H. S., Chandratre, M.T. and Ketkar, C. M. 1971. Insecticidal trial against stored grain pests of maize seeds with neem kernel powder carried out at Agricultural School, Manjuri, III. *In : Utilization of Neem (Azadirachta indica A. Juss) and its by- products* Ed., C. M. Ketkar, Final Tech. Report. *Modified neem cake Manurial Project,* II 53. pp. 8-9.

Chandra Mohan, N. and Nanjan, K. 1990. Effect of neem oil against diamond back moth in cabbage. *Neem Newsl.* **7** (4) : 35 - 36.

Chatterji, K. K. and Sen, R. N. 1919. *Ind. Med. Gaz.* **54** : 174 (through Siddiqui, S., 1942).

Devakumar, C., Saxena, V. S. and Mukherjee, S. K. 1986. Evaluation of neem *(Azadirachta indica* A. Juss) limonoids and azadirachtin against safflower aphid *(Dactynotus carthami). Indian. J. Ent.* **48** : 467-470

Dorn, A., Rademacher, J. M. and Shen, E. 1987. Effect of azadirachtin on reproductive organs and fertility in the large milkweed bug *Oncopeltus fasciatus. In : Natural Pesticides from neem tree and other tropical plants. Proc. 3rd. Intl. Neem Conference,* p. 273-288. (Eds., H. Schumtterer and K. R. S. Ascher), Nairobi, Kenya.

Fagoonee, I. 1987. Use of neem in vegetable crop protection in Mauritius. *In : Natural Pesticides from Neem tree and other tropical plants. Proc. 3rd. Intl. Neem Conference,* p. 419-429 (Eds. H. Schmutterer and K. R. S. Ascher) Nairobi, Kenya.

Georghiou, G. P. 1990. Overview of insecticide resistance. *In : Managing Resistance to Agrochemicals : Fundamental Research to Practical Strategies.* (Eds., M. B. Green, H. M. LeBaron and W. K. Moberg.) ACS Symopsium Series 421 pp. 18-41. American Chemical Society, Washington D. C.

Grant, I. F. and Schmutterer, H. 1987. Effect of aqueous neem seed kernel extract on ostracod (Class : Crustacea) development and population density in low land rice fields. *In : Natural pesticides from neem tree and other tropical plants.* (Eds, H. Schmutterer and K. R. S Ascher) G. T. Z., Eschborn, Germany p. 591-607.

Gujar, G. T. and Mehrotra, K. N. 1984. Inhibition of growth and development of tobacco catterpillar, *Spodoptera litura* due to azadirachtin and other neem products. *Indian. J. Ent.* **45** *(4)* : 431-435.

Hegde, N. G. 1993. Neem production and Development - Constrains at Grass roots level. *In : Souvenir, World Neem Conference, pp. 1-7, India.*

Henderson, R.,, McGrindle, R. and Overton, K. H. 1964. Salannin. *Tetrahedron letter* 3969-3974.

Hoddy, E. 1993. Neem : Natures bitter boon. *Plant Trees* **2** (5) : 17-18.

Jayaraj, S., Bharati, M. and Sundra Babu, P. C. 1993. Integrated Pest Management. *In*

: *Neem Research and Development* (Eds., N. S. Randhawa and B. S. Parmar) pp. 154-167, Society of Pesticide Science, New Delhi.

Jhansi Rani, B. 1984. *Studies on the biological efficacy of deoiled neem (Azadirachta indica* A. Juss) *Kernel against a few insects.* M. Sc. Thesis, P. G. School, IARI, New Delhi.

Jhansi Rani, B and Singh, R. P. 1993. Indentification of effective and inexpensive neem *(Azadirachta indica)* seed kernel extract for management of *Helicoverpa armigera* (Hubner). *World Neem Conference,* Bangalore, India, 24-28 Feb., Abstract, p. 7.

Jhansi, K. and Sundra Babu, P. C. 1987. Evaluation of the effect of parasite, fungus and neem oil individually and in combination against the spotted pod borer, *Maruca testulalis* (Geyer). *Andhra Agri. J.* **34** : 340-342.

Jilani, G. and Haq, H. S. 1984. Studies on some indigenous plant materials as grains protectants against insect pest of stored grain. *Pakistan Ent.* **6** : 24.

Jilani, G. , Noorullah, and Khan, M. I. 1986. 2nd Annual Report IFS project. Pest Management Research Institute, Pakistan Agricultural Research Council, Karachi, Pakistan.

Joshi, B. G., Ramaprasad, G. and Sitaramiah, S. 1982. Effect of neem seed kernel suspension on *Telenomus remus*, an egg parasite of *Spodoptera litura. Phytoparasitica* **10** *(1)* : 61-63.

Joshi, B. G. , Sriprakash, K. and Murthy, P. N. S. 1990. Efficacy of neem seed kernel suspension, neem seed bitter, *Pongamia* cake water extract, *Bacillus thuringiensis* Berliner, Nuclear polyhedrosis virus alone and in combination against tobacco caterpillar, *Spodoptera litura F.* in nurseris. *Abstracts of National Symposium on problems and prospects of botanical pesticides in integrated pest management,* p. 9, CTRI, Rajahmundry.

Jotwani, M. G. and Sircar, P. 1965. Neem seed as protectant against stored grain pests infesting wheat seed. *Indian. J. Ent.* **27** *(2)* : 160-164.

Jotwani, M. G. and Sircar, P 1967 Neem as a protectant against bruchid, *Callosobruchus maculatus (*Fab.) infesting some leguminous seeds. *Indian J. Ent.* **29** *(1)* : 21-24.

Keller, U. 1985. *Uhterschungen zur fra B - Und Eutwicklung-schemmen den Aktivital von Mellaceaneninhaltsstoffen, Epilachna-varivestis Indual-choice Ano-choice test und in Lang-Zeitversuchen.* Wissenchaftliche Arbeit, University of Hohenhiem, FRG.

Ketkar, C. M. 1987. Use of tree born non-edible oils like neem, karanga, sal and kusum as surface protectants for stored cowpea seeds against *Callosobruchus maculatus* and for greengram against *C. chinensis. Proceedings 3rd International Neem Conference*, p. 57, Nairobi, Kenya.

Klenk, A. 1985. *Isoleirung and Strucktraufklarung von Biologisch Aktiven Tetranortriterpinoiden aus Azadirachta indica* A. Juss. *Meliaceaea Ein nevur. Struckturvor Schlang fur Azarachtin.* Ph. D., Thesis, University of Hohenhiem, FRG.

Koul, O. 1984. Azadirachtin II. Interaction with the reproductive behaviour of red cotton bug. *Z. angew. Ent.* **98** : 221-223.

Kraus, W. 1986. New trends in natural product chemistry. *Proceedings 2nd. International Symposium and Pakistan US Botanical Workshop on Natural Product Chemistry.* p. 237-256, Karachi, Pakistan.

Kraus, W., Baumann, S., Bokel, M., Keller, U., Keller, A. , Klingela, M. , Pohne, H. and Schwinger, M. 1987. Control of insect feeding and development, by constitutents of *Melia azedarach and Azadirachta* indica In : *Natural Pesticides from Neem tree and Other Tropical plants.* Eds., (H. Schumtterer and K. R. S. Ascher) p. 111-125, GTZ, Eschborn, Germany.

Kraus, W. , Bokel, M. , Bruhn, A. ; Cramer, R. , Kliaber, I. , Klenk, A. , Nagi, G., Pohnl, H. , Siado, H. , and Volger, B. 1987. Sturcture determination by NMR of azadirachtin and related compounds from *Azadirachta indica* A, Juss. (Meliaceae). *Tetrahedron* **43** : 2817-2830.

Kraus, W. and Cramer, K. 1981. *Neue Tetraortriterpinoiden nik insect extra* B - *hemmender aus neem oil. Liebigs Ann. chem.* p. 181-189.

Kubo, I. , Matsumoto, A., Matsumoto, T. and Klocke, J. A. 1986. New insect ecdysis inhibitory limonoid, deacetyl azadirachtinol isolated from *Azadirachta indica*(Meliaceae) oil. *Tetrahedron* **42** : 489-496.

Larson, R. O. 1986. Development of Margosan -O, a pesticide from neem seed. *Proceedings 3rd. International Neem Conference,* p. 243-250, Nairobi, Kenya.

Lavie, D. , Jain, M. K. and Shapan - Gabrith, S. R. 1967. A loucust phagodetterent from two *Melia* spp. *J. Chem. Soc. Chem. Comm.* 910-911.

Mohan, K. 1989. *Studies on the effect of neemproducts and vegetable oils against major pests of rice and safety to natural enemies* M. Sc. (Ag.), Thesis, Tamil Nadu Agricultural University, Coimbatore, India.

Mohan, S. , Palanisamy, P. T. , Parvathy, K. , Rajasekhran, B. and Balsubramanian. 1990. Defatted neem kernel powder for rice weevil, *Sitophilus oryzae* Linn. control. *Neem Newsl.* **7** (1) : 1.

Mukherjee, V. , Singh, A. K. and Hameed, S. F. 1990. Evaluation of neem *(Azadirachta indica* A. Juss) products as wheat grain protectant against *Sitophilus oryzae* Linn. *Neem Newsl.* **7** (1) : 2-4.

Naqvi, S. N. H. 1987. Biological evaluation of fresh neem extract and some neem components with reference to abnormalities and esterase activity in insects. In : Natural pesticides from neem tree and other tropical plants (Eds., H. Schmutterer and K. R. S. Ascher), P. 315 - 330, GTZ, Eschborn, Germany.

Pathak, P. H. and Krishna, S. S. 1985. Neem seed oil, a capable ingredient to check rice moth reproduction (lepidoptera : Gelechidae). *Zeitschrift fur Angewandte Entomologie 100* : 33-35.

Pillai, N. R. and Santhakumari, G. S. 1984. Toxicity studies on nimbidin, a potential antiulcer drug. *Planta Medicina* **50** :146-148.

Pradhan, S. , Jotwani, M. G. and Rai, B. K. 1962. The neem seed deterrent to locust. *Indian Farming* **12** : 7-11.

Prakash. A. , Pasalu, I. C. and Mathur, K. C. 1980. Evaluation of plant products as garin protectants against insects of stored paddy. *Bull. Grain Tech.* **18** : 25-28.

Qadri, S. S. H. , Usha, G. and Jabeen, K. 1984. Sub-acute dermal toxicity of Neemrich-100 (*Tech.) to rats. International Pest Control* 26 : 18-20.

Raguraman, S. 1993. *Behavioural and physiological effects of neem (Azadirachta india* A. Juss.) *seed kernel extracts on insect egg parasitiod, Trichogramma chilonis* Ishii *and larval parasitoid, Bracon hebator* Say. Ph. D. Thesis, PG School, IARI, New Delhi.

Rajsekaran, B. , Jayaraj, S. , Saroja, R. and Raju, N. 1988. Field evaluation of neem products on the management of certain insect pest of rice. *Final Workshop of IRRI - ADB - EWC Project on Botanical pest control in Rice Based cropping systems.* IRRI, Manila, Phillipines.

Rembold, H. and Siber, K. P. 1981. Inhibition of oogenesis and overian ecdysteroid synthesis by azadirachtin in *Locusta migratoria* (R and F). *Z. Naturforsch* **C 36** : 466-469.

Rossner, J. and Zebitz, C. P. W. 1987. Effect of soil treatment with neem products on earthworms (Lumbricidae). *In : Natural Pesticides from Neem Tree and Other Tropical Plants*.p. 627-632 (Eds., H. Schmutterer and K. R. S. Acher.), GTZ., Eschborn, Germany.

Ruscoe, C. N. E. 1972. Growth and disruption effects of an insect antifeedant. *Nature (New Biology)* **236** *(66)* : 159-160.

Sankaram, A. V. B. , Murthy, M. M., Bhaskariah, K. , Subramanyam, M. , Sultan, N. , Sharma, H. C. , Leuschner, K. , Ramprasad, G. , Sitaramiah, S. , Rukmani, C. and Rao, U. 1987. Chemistry, biological activity and utilization aspects of some promising neem extractives. *In : Natural Pesticides from Neem tree and other tropical plants. Proc. 3rd. Int. 1. Neem Conf.* (Eds., H. Schmutterer and K. R. S. Ascher), pp. 127-148, GTZ., Eschborn, Germany.

Saxena, R. C. , Jilani, G. and Kareen, A. A. 1989. Effect of neem on stored grain insects. In : *Focus on phytochemical pesticides*, Vol. I. *The Neem Tree* (Eds., M. Jacobson) , pp. 97-III, Boca Ratou : CRC Press, Florida, USA.

Schwinger, M. , Ehhammer, B. and Kraus, W. 1984. Methodology of *Epilachna varivestis* bioassay of antifeedants demonstraed with some compounds from *Azadirachta indica* and *Melia azedarach*. *In : Natural Pesticides from the Neem Tree and other tropical plants* (Eds., H. Schmutterer and K. R. S. Ascher), pp. 181-198, GTZ. , Eschborn, Germany.

Sharma, H. C. , Leuschner, K. , Shankaram, A. V. B. Gunasekar, D. , Marthandamorti, M. , Baskariah, K. , Subramanyam, M. and Sultana, N. 1984. Insect antifeedants and growth inhibitors from neem *Azadirachta indica* and *Plumbago zeylonica* In : *Natural Pesticides from the Neem Tree and other Tropical Plants* (Eds. , H. Schmutterer and K. R. S. Ascher), pp. 291-320, G. T. Z., Eschborn, Germany.

Siddiqui, S. 1942. A note on the isolation of three new bitter principles from nim oil. *Curr. Sci.* **11** : 278-279.

Siddiqui, S. , Faizal, S. , Mahmood, T. and Siddiqui, B. S. 1986. Two new insect growth regulator meliacins from *Azadirachta indica* A. *Juss (Meliaceae). J. Chem. Soc. Perken. Trans.* I :1021-1025.

Singh, R. P. 1986. Neem (*Azadirachta indica* A. Juss) an effective substitute to synthetic insecticides. *Prodceedings of National Conference on key pests of Agricultural Crops* (Ed. , Y. K. Mathur) pp. 322-333.

Singh, R. P. , Singh, Y. and Singh, S. P. 1985. Field evaluations of neem (*Azadirachta indica* A. Juss) seed kernel extracts against the pod borers of pigeon pea *Cajanus cajan* (L.) Mill sp. *Ind. J. Ent.* **47** : 111-112.

Singh, R. P. and Kataria, P. K. 1986. Deoiled neem kernel powder as protectant of wheat seed against *Trogoderma granarium* Everts. *Ind. J. Ent.* **48** *(I)* : 119-120.

Singh R. P. and Kataria, P. K. 1992. Benefits and uses of neem tree. *In : Proceedings of Promotion of non-wood forest produce through social forestry.* (Eds. , N. G. Hegde and J. N. Daniel pp. 153-165, Pune, India.

Singh. R. P. , Devakumar, C. and Dhingra, S. 1988. Activity of neem seed kernel extract against the mustard aphid, *Lipaphis erysimi*. *Phytoparasitica* **16** : 225-229.

Srinivasa Babu, K. , Murthy, M. N. S. and Ramesh Babu, T. 1993. Effect of botanical insecticdes on certain parasitoids. *Abstract, World Neem Conference*, Bangalore, India.

Srivastava, K P. and Parmar, B. S. 1985. Evaluation of neem oil emulsifiable concentrate against sorghum aphids. *Neem Newsl* **2** (I) : 7.

Srivastava, R. P. Singh, R. P. and Masarrat Haseeb. 1993. Evaluation of formulated neem (*Azadirachta indica* A. Juss) kernel extract against mangohoppers and mealy bugs. *Abstract, World Neem Conference*. Bangalore, India.

Verma, S. P. , Singh, B. and Singh, Y. P. 1985. Studies on the comparative efficacy of certain grain protectants against *Sitotrogra cerealella Oliver* *Bull. grain Tech*. **21** *(I)* : 37-42.

Vollinger, M. 1987. The possible development of resistance against neem seed kernel extract and deltamethrin in *Plutella xylostella*. *Proceedings 3rd Intl. Neem Conference*, Nairobi, Kenya.

Yadav, T. D. 1973. Studies on the insecticidal treatment against Callosobruchus maculatus (Fab.) and *C. chinensis* (Linn.) damaging stored leguminous seeds. Ph. D. Thesis, Agra University, Agra.

Yadav, T. D. 1985. Antiovipositional and ovicidal toxicity of neem (*Azadirachta indica* A. Juss) oil against the spp. *Callosobruchus*. *Neem Newsl*. **2** *(II)* : 5-6.

Zeherer, W. 1984. The effect of traditional preservatives used in northern Togo and of neem oil for control of storage pests. *Proceedings 2nd. International neem Conference*, Rauichholzhausen, pp. 453-460.

Recent Advances in Indian Entomology
Editor: O.P. Lal
APC Publications Pvt. Ltd., New Delhi 1996, pp. 83-85

Recent Advances in Toxicology

SHASHI VERMA and G. RAVI

Toxicology is a science related to the effects of chemicals or physical agents on the functioning of the living organisms. Toxicology of pesticides in agriculture means proper utilisation of minimum amount of pesticides to get maximum benefit and leaving the least hazardous effects.

It is well known that pesticides play an indispensable role both in increasing agricultural productivity and in ensuring a stable food supply of high quality. The injudicious use of pesticides has led to the development of resistance and resurgence in insects and mites, disturbed the balance of the ecosystem and some other problems directly or indirectly related to environmental pollution.

In order to determine the most effective control programme for a complex of pests, it is useful and necessary to view a crop field as an ecosystem of interacting components. The major components of the biotic part of an agricultural ecosystem are the crop, its pest, beneficial organisms which suppress the pest population or promote growth. The dynamics and interactions among these populations will be strongly infulenced by the abiotic environment which include solar radiation, temperature, humidity and availability of water and nutrients in the soil. Moreover the cultural practices employed in growing the crop will have various interactions with biotic and abiotic factors (Fig-1).

Recent work in toxicology has been reported on the methods to get maximum benefit from pesticides with least harmful effects in the environment. These are as follows :-

1. Formulation of new pesticides with improvement in safety to the environment alongwith selective and specific toxicity to pests.
2. Stability of formulations in storage.
3. New formulations based on the controlled release or controlled persistence of pesticides.
4. New formulations based on the admixture with synergistic chemicals to promote toxicity against pests.
5. Formulation of low price pesticides.
6. It is not the pesticide which matters, but it is important to get improved application technology, improved treatment timings etc.
7. After the application of insecticides, it is necessary to know the degradation to non-toxic components by the improved estimation technology in order

to know the maximum period of insecticide degradation for establishing the safe interval between last treatment and harvest to avoid hazards.

8. In toxicity testing, the use of structural activity relationship is becoming an increasingly important predictive tool. The reliability of this tool is, of course dependent on the accuracy of the data bases.
9. Studies on the biochemical reaction of the insecticide at the site of action for developing specific and selective insecticides.
10. Development of new compounds other than that of chemicals for the control of pests e.g. plant and animal origin pesticides, insect growth regulators, juvenile hormones, juvenile mimics, antijuvenile hormones etc.
11. The great emphasis is also needed on the compatibility of biological control with chemical control in order to develop insect pest management programme for all the economically important crops.
12. Importance is needed much more on the proper statistical analyses and interpretation of results.

The main aim is to get an ideal pesticide for killing the target organisms without affecting other beneficial ones. It has become increasinly evident that biological response to substances may be caused atleast in part by metabolites of the substances. The fundamental goal of toxicological research is to provide a rational basis for recommending acceptible safe level of human exposure to harmful agents.

The information in this paper is restricated to methodology of residue analysis and its interpretation, because other aspects of toxicology are going to be or already have been dealt by other scientists.

It is often more difficult to judge the validity of an analytical study than to perform the analysis. So the present outline describes the types of prerequiste information considered essential for actual determination procedures and also all of those steps preliminary to determination i.e. sample collection, extraction, clean-up and final test preparation. The emphasis is upon validation of procedures and reporting in sufficient detail of the procedure. These are as follows ·

i) *Sample Selection* : Method of collecting samples must be given in detail to enable one to distinguish between random and haphazard samples. Representative sample of larger mass should be indicated.

ii) *Sample collection and storage* : This step is prior to an analytical study and offers many opportunities for loss of integrity of smaples which must be described in full detail. The type of sample tools and containers should be written. The temperature for storing the samples should be mentioned.

iii) *Sample extraction* : It sounds very simple but it is very essential to mention the name of the solvent, the time taken for homogenizing the sample with solvent, the number of cycles of solvent in soxhlet extraction, solvent to sample ratio, the temperature at the time of extraction and concentration of the extract.

iv) *Clean-up procedures* : Mention should be made for the methods of using adosorbents and elutes for clean up of the insecticidal extracts.

v) *Validation studies* : Prior to analysis for unknown samples, all the procedures must be validated in terms of recovery, reporducibility, sensitivity and accuracy.

a) *Recovery studies* : The first requirement is for the spiking or fortification procedure itself to be given in the detail. Samples should be fortified at a minimum *ppm* level.

b) *Limits of detection or sensitivity* : The limit of detection must be taken as not the smallest amount causing response, but rather the smallest amount to which the analytical criteria employed for quantitative identification.

c) *Qualitative determination* : At least one criterion unique to the compound and the criteria for the identification of interfering substances must be mentioned.

d) *Quantitative determination* : Not all the methods show linera calibration curve. If interpolation is utilised, then its procedure must be fully described, whether estimation has been done by bioassay, chemical methods, GLC, HPLC, NMR, IR etc.

e) *Stripping extracts* (Blank) : Samples of the same material that will subsequently be analysed in the study, but known to lack detectable levels of the compounds of interest, must be put through the entire procedure including storage for the same length of time in the same environment as the test samples. Extraction and clean-up procedures applied to these blanks must be same as will be used for real samples.

f) The extracts of the different treated samples should be concentrated under the same conditions to the same level to avoid the error in the loss of insecticide as much as possible.

In conclusion it can be said that knowledge of analytical chemistry alongwith biological background is very essential in all aspects of toxicological work. Becuase the qualitative determination in the past have given way to quantitative analysis to a level of counting molecular cells in the present era by both biologists and chemists. The comparisan can be done as follows :-

i) At a time, when toxicologists were limited in sophistication to mearly counting the number of dead and alive organisms after exposure to chemicals, the chemists were making qualitative analysis at the per cent level.

ii) As toxicologists progressed beyond the above level and began to look for visible changes in organs, the analytical chemists were detecting at ppm level.

iii) The use of light microscopy and biochemical meausres was the next step by toxicologists, the analytical chemists analysed at ppb level.

iv) Now toxicologists are using electron microscopy to assess structural changes and are detecting chemical changes at the molecular level of all organisation, chemists are detecting at *ppt* level and even below.

Future aspects in toxicology can be taken in the areas of metaboilism and molecular interactions between toxins and target molecules for understanding the mechanism of toxicity and immunity.

Recent Advances in Indian Entomology
Editor: O.P. Lal
APC Publications Pvt. Ltd., New Delhi, 1996, pp. 86-101

Neem-potential and Use in Indian Agriculture*

N. S. RANDHAWA and B. S. PARMAR

Neem, *Azadirachta indica* A. Juss, an evergreen tree belonging to Meliaceae family, had its orgin in the Indian sub-continent. It is currently grown in Africa and other continents also where it spread through the Indian immigrants as well as other means.

The tree can be propagated by direct seeding or raising the plants in nurseries by direct seeding and transplanting or by planting cuttings. A spacing of 1.8 m x 1.8 m or wider is recommended (De Jussian, 1963; Gorse, 1986). Tissue culture has been deemed unrealistic. It takes around 10-12 years to become a full tree when it attains a height of 10-18 m. The tree starts bearing fruits when 3-5 years old and is fully productive around 10 years. The fruit yield per tree varies from 30-50 kg. After depulping and washing, the seeds remain vaiable for only 1-2 months in most cases and are, therefore, collected when fully ripe and are sown as soon as possible.

Extensive adaptability

Neem grows well from sea level to over 670 m (2000 ft), under semi arid to sub humid conditions receiving 450 to 1150 mm annual rainfall and with temperatures upto 49° C. Excessive cold and frost during sapling and seedling stages are harmful for its growth. Therefore, neem trees can be seen in a large belt extending from the southern most tip of India at Cape Comorin to Punjab and from the lush green high rainfall areas of Kerala to the semi arid parts of eastern, western, central and northern Indian states. It grows well on all soils but being a very competitive tree for water and soil nutrients, its growth is hindered on marginal soils. Deep sand and well drained soils are most suitable for its growth and it can even grow on slightly saline and alkali or 'usar' soils, though may, at times, show susceptibility to moderate salinity. Waterlogged and impeded drainage conditions are not suitable and so are zinc and potassium deficiencies; causing leaf chlorosis. There are contradictory reports on its compatability with other plant species in its environment and due to this, its potential in agro-forestry is a subject of debate that needs to be researched.

A neglected native tree

The tree has acquired considerable significance internationally and its various aspects have been discussed at four exclusive international conferences besides a

*Paper-Presented at the Symposium on Resources for Sustainable Agriculture: The Use of Neem and Other Plant Material for Pest Control and Rural Development, Honolulu, Hawai (1991)

number of other seminars and workshops. The national and international attention has been focussed primarily on its value as a botanical pesticide. No organised effort has been made to cash on its multifarious properties and undertake its holistic development. Several exotic introductions have received far more attention of forest departments as well as the research institutions than this native tree. In spite of recorded evidence of its multiple uses spread over several centuries, there has been no effort to promote its propagation. For over three decades the oft repeated statements are : "14 million neem trees with a potential to produce over 0.4 m.t. of neem fruit and 20-25% collection of fruits" (Ketkar, 1976 cites a 1959 survey; KVIC, 1988). Although there is a mention of its propagation at 1.8 m x 1.8 m row and tree spacings, however, in view of the larger spread of its crown it ought to be 10 m x 10 m or even more. Even at the wider spacing bringing just 1 million hectare of India's vast wastelands, can accommodate 60-100 million trees.

The present paper focusses on the status of neem research and its utilisation in India. An integrated research and development programme on its propagation and diversification of its utilization are suggested.

Biomass production and utilization

The yield of neem varies between 10 and 100 tons biomass (dried material) per hectare per year depending upon rainfall, site characteristics and spacing and 40 tons of biomass (equiv. 12.5 m^3 solid wood) can be achieved under average management conditions (Michel-Kim and Brandt, 1981). About 50% of the biomass is reportedly contained in the leaves, about 25% in the fruit, and 25% in the wood (1 m^3 =800 kg).

Fuelwood

Although there are a few references about the potential of neem as a source of fuelwood, but there is very little quantitative information on the thermal efficiency of neem wood as compared with the well recognised fuel wood species. In the Deccan plateau region of India, full grown neem tree (12-15 year old) is expected to provide 400-500 kg fuelwood (Pers. Comm. from Dr. S. Chinnamani, Ex-ADG (Agro-forestry), ICAR, New Delhi).

Bose and Bandoyopadhyay (1986) reported that growing neem for fuelwood in an alkaline soil of village Dhanwas, Distt. Gurgaon, Haryana (pH 9.7-10.4, EC 0.2-1.4 m mhos cm^{-1}, sandyloam) needed soil amelioration (1.5 kg gypsum, 5 kg cowdung in each pit of 0.45 x 0.45 m, row to row 3 m and between plants 2 m). For every rupee invested in the above afforestation programme, a benefit of rupees 1.24 was reported. The net benefit for a 10 year horizon was rupees 22,509 per hectare (around US $ 1800 on 1988 basis).

Timber

The wood is used by villagers for furniture, cart and farm implements. It is tough and resembles cuban mahogany and is highly resistant to insect attack. The heartwood is red and sap wood greyish-white with sp. gr. 0.69 (Rajput *et. al.* 1985). Chauhan and Bist (1987) have reported that neem is suitable for making plywood for general purposes, for concrete shuttering works, preservative treated plywood and fire retardant treated plywood. It is equally suitable for making blackboards.

Pest control and allied chemicals

The earliest observation that *Melia* was avoided by locust (Herculais, 1888-94) was subsequently substantiated by several workers on *Melia* (Carlothays, 1910) as well as on *Azadirachta indica* (Mann and Burns, 1927). Neem derivatives have now been reported as pesticide (including antifeedant, metamorphosis disruptor, growth disruptor/juvenilizing, ovicidal/oviposition deterent, antifecundity, antireproduction etc.) and pesticide adjuvants against over 250 insect species all over the world.

The pioneering work on the bioactivity of neem derivatives was conducted at the Indian Agricultural Research Institute, New Delhi. Sinha (1960) reported the gustatory repellent property of alcohol extract of neem cake. It was followed by the publication of Pradhan *et a1* (1962) reporting the antifeedant property of neem seed kernel against the desert locust, *Schistocerca gregaria*. Since then, the bioactivity of neem derivatives has been reported in India against 106 species of insects belonging to 10 orders of Class Insecta (Table 1), 12 spcies of nematodes and 9 species of fungi (Table 2). Several products are stated to be commercially marketted in the country and a few more are in the pipeline for commercialization in the near future (Table 3). Leading business houses namely M/s Godrej Soaps, Tata Oil Mills Company and National Organic Chemical Industries Ltd. have recently announced plans to produce neem based pesticides for the domestic market or for export (The Times of India, New Delhi, February 18, 1991).

Source of nutrients and modifier of nutrient release

Due to its antimicrobial action, neem serves as a slowly decomposing material,

Table 1. Orders of class insecta evaluated with neem products in India.

S. No.	Order	No. of crop + Storage insects
1.	Orthoptera	5
2.	Dictyoptera	1
3.	Lepidoptera	29 + 3
4.	Hemiptera	23
5.	Diptera	12
6.	Coleoptera	16 + 12
7.	Hymenoptera	1
8.	Isoptera	1
9.	Thysanoptera	2
10	Siphonaptera	1

Table 2. Nematodes and fungi against which neem products have been tested.

S1. No.	Common Name	Scientific name
NEMATODES		
1.	Spiral nematode	*Helicotylenchus indicus* Siddiqui
2.	Rice root nematode	*Hirschmanniella oryzae* Luc. and Goodey 1962
3.	Ring nematode	*Hemicriconemoide* sp.
4.	Lance nematode	*Haplolaimus indicus* Sher.
5.	Root knot nematode	*Meloidogyne incognita* (Kofoid and White, 1919; Chitwood, 1949)
6.	-do-	*M. arenaria* (Neel, 1889; Chitwood, 1949)
7.	-do-	*M. javanica* (Treub. 1885; Chitwood, 1949)
8.	Reniform nematode	*Rotylenchulus reniformis* Linford and Oliveira, 1940)
9.	Stunt nematode	*Tylenchus filiformis* Butschli
10.	-do-	*Tylenchorhynchus brassicae* Siddiqui, 1961.
11.	-do-	*T. elegans Siddiqui*, 1961
12.	Burrowing nematode	*Radopholous similis* (Cobb, 1893, Thorne, 1949)
FUNGI		
1.	Tomato rot	*Aspergillus niger* Van Tiegham
2.	Groundnut rust	*Puccinia arachidis* Speg.
3.	Rice sheath rot	*Sarocladium oryzae* (Sawada) W. Gams and U. Hawksworth
4.	Onion bulb rot and wilt	*Fusarium oxysporum* f. sp. Cepae Sch1.
5.	Chickpea root rot	*F. solani* (Mart) App. and Wellenw.
6.	Cotton root rot	*Rhizoctonia solani* Kuhn
7.	Soybean root rot	*Macrophomia phaseolina* (Tassi) Goid.
8.	Coconut wilt	*Ganoderma lucidum* (Leys) Karst.
9.	Betelvine wilt	*Phytophthora caspici* Leonian

Singh and Kataria (1991)

Table 3. Commercially produced and pipeline neem based products in India (1991).

Product	Activity	Manufacturer
Wellgro	Repellent Fungus inhibitory Antiviral Plant nutrition N-Loss prevention	M/s ITC Ltd.-ILTD Division, Rajahmundry
Repelin	Against leaf eating caterpillars Safe to pollinators	-do-
Neemark	Antifeedant/repellent Nematicidal Synergistic	M/s West Coast Herbochem Pvt. Ltd., Bombay-400 025.
Neemguard	Repellent, Metamorphosis disruption, Synergistic Safe to natural enemies of insects.	M/s Akshay Chemical, Bombay-400 052. (Marketed by M/s Gharda Chemicals Ltd., Bombay-400 050.)
Moskit	Mosquito repellent	M/s Investment & Commercial Enterprises, Bombay-400022. (MCDA Agro Pvt. Ltd., Bombay.)
Jeevan Crop Protector	Antifeedant/repellent Disturbs growth & reproduction	MCDA Agro Pvt. Ltd. Bombay-400 001.
Neemrich	Neemrich I-Warehouse pests Neemrich II- Antifeedant	Developed by National Chemical Laboratory, Pune.
Vapacide	Antifeedant	Developed by Regional Research Laboratory, Hyderabad.
Neem based Emulsifiable Concentrate, Dust, Water Dispersible Powder, Granule.	Pesticide	Developed by Indian Agricultural Research Institute, New Delhi-110 012.
Nimba	Pesticide	-do-

leading to a slower release of the nutrients contained in it. This attribute of the neem based organic manures has been usefully exploited in the country. The slow release of nutrients has been reported to be due to the presence of extractable principles of neem seed and cake (Sinha, 1960; Sinha and Gulati, 1968). The effect was vindicated by successful incorporation of acetone extract of the cake into urea in field trials (Bains *et al.*, 1971). The utility of neem products namely seed kernel and its extractives, seed crush, deoiled cake, oil etc., as rewarding adjuvants for nitrogenous fertilizers has been subsequently demonstrated in several field trials.

The mechanism of N regulation by these products is attributed to nitrification retardation by selective suppression of the activity of NH_4^+ to NO_2^-; the subsequent conversion of NO_2 to NO_3 by *Nitrobacter* sp. being not affected. Depending upon the precision of coating, upto 46% nitrification inhibition till 3 weeks has been reported with the cake coated urea. The cake or its extractives have also been incorporated at reduced concentration, in the matrix of the polymers, often used for coating urea (Mishra and Chhonkar, 1978), to yield products with N release extended upto 160 days or so. Ammonia volatilization loss is also reduced in case of neem coated urea super granules. The retarded N release has been reflected in increased crop yields (Table 4).

The 'ghani' extracted neem oil which contains bitters has been reported to cause a remarkable inhibition of nitrification of ammonium sulphate at 30% of N dosage as compared to the hexane or petroleum ether extracted oil that is devoid of such

Table 4. Crop response to neem cake application as fertilizer adjuvant (Devakumar and Mukerjee, 1985).

	TREATMENT			RESPONSE		
Crop	No. of trials	Fertilizer (kg N/ha)	Cake (kg/ha)	Yield (t/ha)	Increase (%)	Reference
Paddy	12	50	16.5	4.12	-	Ketkar (1976)
	325	100	30	3.33	5.4	Ketkar (1976)
	6	100	44	4.55	10	Anonymous (1983)
	2	100	22	3.47	25	Nayak *et al*,(1981)
Paddy-Wheat (P-W)	2	100	44	3.88(P) 3.26(W)	12 24.8	Subbaiah *et al* (1979)
Wheat-Paddy-Wheat	2	50	22	3.35(W) 2.67(P)	12.9	Ketkar (1976)
Sugarcane	9	400	140	128.9	3.6	Ketkar (1976)
	2	75-150	33-44	61.8	12.9	Prasad (1976), Sharma *et al*, (1980)
Wheat	2	100	10	4.59	8	Murugappan and Mishra (1979).
Maize (Fodder)	2	100	22	57.74	6.7	Ketkar (1976)
Finger-millet	2	45	44	3.55	5.0	Subbaiah *et al*. (1982)
Potato	2	125	27.5	29.67	19	Ketkar (1976)

bitters (Patil, 1972). That the oil owed its activity to the presence of the bitters (lipid associates) has been established by separately evaluating these at 30% of N level with urea and ammonium sulphate. The crude extract of neem was also comparable with karanjin (a furano-flavonoid component of *Pongamia glabra)* at 5% N dosage (Sahrawat and Parmar, 1975; Sahrawat, 1982). In addition to nitrification retardation, neem cake has been shown to retard soil urease activity (Reddy and Mishra, 1983) and organic matter decomposition (Mishra *et al.*, 1972).

Neem based slow mineralising fertilizers are commercially produced in the country. Recently, two such products, namely 'Jeevan soil conditioner' claimed for improving soil health, releasing nitrogen slowly to match the plant needs and 'Neemin' a fertilizer coating with neem extract for slow nitrogen release are stated to be marketted by M/s MCDA Agro Pvt. Ltd. Bombay-400001 and M/s Godrej respectively.

Cattle and poultry feed

The cake obtained by seed pressing or crushing contains varying amount of the bitters. There are contradictory reports regarding the utility of this cake as a cattle or poultry feed.

To overcome the toxicity of the cake whenever observed, its admixture with various edible cakes or other feed concentrates has been suggested. Since it is a rich source of crude protein, amino acids (Table 5) mineral and other components, the cake devoid of the bitter lipid associates can serve as an excellent feed for livestock and poultry.

Neem leaves being rich in digestible nutrients, also serve as cattle feed either

Table 5. Amino acid composition (%) of processed *Melia indica* seed meal protein.

Alanine	5.3	Histidine	3.4	Proline	5.9
Arginine	6.3	[Leucine		Serine	7.6
Aspartic acid	6.8	Nor-Leucine	8.0	Threonine	5.3
Cystine	3.1	Isoleucine]		Tyrosine	3.4
Glutamic acid	11.8	Lysine	3.9	[Valine	
Glutamine	1.6	[Phenyl alanine		Methionine]	4.5
Glycine	7.4	Tryptophane]	5.3	Unidentified	3.9

Mitra and Mishra (1967)

as such or in admixture with other feeds. Goat and camel browse neem leaves well.

Medicinal applications

Medicinal uses of the neem tree are manifold and have been known since times immemorial. The use of various parts of the tree as a remedy for constipation, diabetes, indigestion, itches, pyorrhoea, sleeplessness, stomach ache, ulcers and skin diseases is well documented.

Recently, there has been a considerable interest in India about the contraceptive property of neem oil and other extractives. Neem oil has been reported to possess strong spermicidal action against rhesus monkey and human spermatozoa both *in vitro* and *in vivo*. It has also anti-implantation effect when applied intravaginally pre or post-mating and pre or post implantation. The advantage of contraception using neem oil is that the fertility can be restored in around 30 days on cessation of the treatment. The oil is also reported to exhibit anitovulatory activity in both the reflex and spontaneous ovulators. The oil administered orally possessed significant male infertility activity with no interference in the spermatogenesis which is reportedly retarded with oilcake feeding. In females, both oral and intravaginal oil applications caused antifertility effects, considered to be due to the foetal resorption.

Recently, Riar *et al.* (1990) have identified a steam volatile, odorous fraction of neem oil coded as NIM-76 inhibiting spermatozoal motility *in vitro* at 0.25 mg/ml for rat and 25 mg/ml for human spermatozoa. No undesirable symptoms on its application in vagina were observed.

Based on above observations a commercial product 'sensal' described as a safe, reliable, intravaginal, pre-coital, spermicidal contraceptive, having also antiseptic and medicinal properties is being manufactured by M/s Excelsior Enterprises, Kanpur-208004. It contains neem oil 98% W/V, camphor 1.5% W/V, 'ketki', cintronella and lemongrass oils 0.5% W/V.

Neem oil has also been reported to be very useful in the management of secondary hyperglycemia (diabetes). Its 5-10 drops taken daily act as antihyperglycemic and blood purifier. It is commercially available in gelatin capsules manufactured by M/s Kee-Pharma, New Delhi under the name 'Nimbola'.

An ointment prepared from neem oil and bark alongwith traces of 'haldi' in vaseline acts as antiseptic, fungicidal, astringent, insect repellent, softner of eczyma crust and adsorbent, thereby promoting wound healing in animals by first intention, thus minimizing formation of scar.

Miscellaneous

Neem products are made into toiletries such as soap, toothpaste, oil etc., glue, cotton paint, silk paint and tannin. The seed pulp is also a promising substrate for generating methane gas. The oil can be used for light and heating and is even edible.

Toxicology

Neem is reportedly non-toxic to man or animals and non-polluting to the environment. However, a syndrome similar to Reye's syndrome appeared in children who were given large doses of neem oil (Sinniah *et al.*, 1982, 1985). Since Reye's syndrome is poorly understood, a thorough testing of neem extracts before using for medicinal purposes was suggested. The extracts have been reported to be toxic to guinea pigs and rabbits (Sadre *et al.*, 1984). The insectivorous fish Gumbusia spp., and tadpoles died at 0.4% concentration (Jotwani and Srivastava, 1981). The seeds may sometimes carry *Aspergillus flavus*, producing aflatoxins (Pereira and Wohlgemuth, 1982). However, Bhatnagar and MeCormick (1988) reported that the neem leaf extracts did not affect the growth of *A. flavus* and *A. parasiticus* but blocked aflatoxin biosynthesis by interfering in polyketide formation at concentrations above 10% V/V. the effect diminished in heated leaf extracts. The volatiles from extracts were, however, ineffective. Azadirachtin and neem oil have no mutagenecity in the Ames test (Warthen and Uebel 1980; Elizabeth Hoddy, 1985).

Economic potential

In India, neem is mostly raised as an isolated ornamental shade tree or as avenue plantation along the roads. In spite of the unorganized character of these plantations, the stated 14 million trees with only 20-25% collection are supporting 141 seed collection and 70 oil production centres in the country. At present, except for a pure stand of 50,000 neem trees in Saudi Arabia (Ahmad *et al.*, 1989), one does not come across in literature any bigger organised plantation of neem as monoculture or as various kinds of neem based intercropping systems. A concerted effort to undertake plantation of a pure stand of neem in just 1 million hectare of the country's vast wastelands can provide 60-100 million neem trees.

An average yield under normal management is reported to be around 40 tons of biomass per hectare per year in the ratio of 50: 25: 25 as leaves, fruits and wood respectively (Michel Kim and Brandt, 1980). However, going by the yield of fruit and wood reported under Indian conditions, the ratio works out to be around 39: 6: 55. This discrepency needs to be reconciled.

At the 1988 price of neem products (Table 6), fruits alone can fetch about Rs. 2500 to 3500 per hectare per year. The economic value of the oil from these seeds can be of the order of about Rs. 7,000 to 12,500 (assuming a yield of 30 kg fruits per tree, 0.25 kg oil per kg of fruits and 60-100 trees per hectare) and the cake around Rs. 1400-2300. This plantation is expected to yield about 18-31 tons of wood which at Rs. 500 or so per ton can yield about Rs. 9000 to Rs. 15000 per hectare per year. The 21-35 tons of leaves per hectare per year can be a bonus as feed stock for small ruminants (goats) or to be used for other purposes.

Table 6. Price fluctuation of neem pruducts between 1977-1988.

Product	Price range (Rs./Kg.)
Fruit	0.41 - 1.25
seed	0.83 - 2.00
oil	5.10 - 15.16
cake (expeller)	0.33 - 0.93
cake (solvent extracted)	0.44 - 0.93

Ketkar (1989)

Economic competitiveness of neem, therefore, compares very favourably with the prevailing land use economics in the semi arid and sub humid regions of the country.

Suggestions for future work

1. *Survey, identification and indexing of elite trees*

In view of the multifarious uses of neem derivatives, it may be desirable to identify the specific traits for which the trees need to be raised. Such traits may be in terms of quality of timber, thermal efficiency as firewood, bioactive ingredients, oil content, biomass, feedant value of leaves or cake for animals or source of nutrients for plants etc. Indexing and catalogueing of this variation in neem germplasm can prove handy in studies on tree genetics and specie improvement.

2. *Characterisation of agro-ecological conditions most-suited for specific traits*

Wide variation in the quality and quantity of products obtained from neem grown in different agro-climatic conditions of the country, has attracted the attention of the neem researchers as well as the neem users. There is enough evidence to state that the environment for neem growth and even that the neem processing (drying, storage, handling etc.) affect the content and characteristics of its various products. Therefore, it will be proper to identify the conditions most suitable for a trait and concentrate on obtaining or processing neem based products of that trait under the optimum environmental conditions.

3. *Raising the threshold of useful traits*

The threshold of useful traits such as azadirachtin content from 1-3% to 7-9%, biomass etc. may be raised through natural selection, exotic introductions or genetic or other biotechnology manipulations. Such approaches offer interesting possibilites that need to be exploited.

4. *Evaluation of germplasm for tolerance to soil and water related stresses*

Introduction of neem in problem wastelands offers a vast potential in India. A pre-requisite to such introductions will be a thorough evaluation, through multi-disciplinary programmes, of the available neem germplasm for tolerance to various soil and water related contraints (water stress, humidity, evapo-transpiration, temperature, rainfall etc.), so endemic to our wastelands. Successful screens can be employed for a large scale cultivation in such areas.

5. *Development of tissue culture and nursery management techniques for large scale propagation of Neem*

Biotechnology can find a useful application in quick multiplication of the desired neem clones through meristem culture. Tissue culture can be exploited to particularly obtain the potential bioactive materials like azadirachtin in larger amounts for use in pest control. Nursery management technology to enable a large scale propagation of the material needs to be developed.

6 *Development of agro-technology for management of Neem plantation*

Low input management technologies for a large scale propagation of neem under the normal conditions as well as under the soil and water related stress conditions and other problem environments, needs to be developed. In this regard, a special attention needs to be paid to its cultivation in the available wastelands including the problem soils. Also, the use of marginal quality water for its irrigation can be explored. Raising of neem plantations as monoculture or in the company of various agricultural, horticultural or pastoral crops or as wind breaks and avenue plantation, or in any other need based cropping system needs to be investigated.

7. *Improvement of harvest index by induing dwarfing or quick growth through grafting and genetic manipulations*

The harvest index may be increased by increasing the population of plants per unit area by developing fast growing germplasm or by other genetic inputs for maximum clean bole height and for maximum branching at low height or by developing pruning techniques which promote growth and yield of loppings and increase the lopping cycles.

8. *Evaluation of Neem based agro-forestry systems for carrying capacity of small ruminants*

The daily dietary requirement of natural/cultivated forages for an adult sheep or goat is taken around 3% air dried forage of its body weight (about 30 kg). This works out to about 0.9 to 1.0 kg of neem leaves per day. Since one tree is stated to yield around 350 kg leaves, it can roughtly sustain one goat or sheep per year. Assuming that neem is planted at a plant to plant and row to row distance of 10 x 10 m or more, about 60-100 trees can be raised in one hectare of wasteland. This can thus sustain over 60 goats per hectare.

9. *Promotion of on-farm production of Neem products*

In order to reduce the reliance on off-farm inputs to agricultural production, farmers should be trained to prepare their own products for use. Though standardisation of products at that level may be difficult, at least standard sequence of steps for observance at farmer's level can be offered. In this context. specific farmer level pest control practices need to be identified and offered as a package of practices. Particular significance can be attached to its use in pest control in high value crops like cotton, tobacco, horticultrual and vegetable crops. The products can also be used for safe storage of farm produce at the farmer's or community or even national level

10. *As component of integrated pest management*

Prolonged and non-judicious use of a few insecticides has led to the development of insect resistance against them. In India, the resistance has been reported in case of 27 insects including 14 in public health, 7 in agriculture and 6 in storage. Adaptive research trials on neem and other botanicals for promoting their use either alone or as a part of the integrated pest management system need to be undertaken. Their particular use in the food security chain of 20 m.t. or even more of foodgrain buffer stocks in India needs a serious consideration specially in view of their multi-thronged mode of action against insect pests.

11. Standardisation of Neem products

Neem contains a large number of active chemical constituents. For a large scale use of such products, development of standard products is vital. Standardisation of neem products based on scientifically agreed activities will go a long way in the promotion of these materials. As per the current information, it may be worthwhile to standardise the bioactive extractives on the basis of the azadirachtin content and the oil on its salanin content. Similarly, oil for soap making or edible purposes etc. may be standardised on the basis of certain vital attributes for the purpose.

12. Development of resistance against photo and thermal degradation

The light and heat lability of neem products is another area that needs attention. This is often an undesirable aspect particularly when outdoor applications are involved and repeated applications are necessitated on account of this deficiency. In view of the fact that the supply of botanicals will always be in limited amounts, appropriate technologies for improving the shelf and application lives of these products need to be developed. Suitable stabilizers for the purpose, therefore, need to be searched.

13. Storage and conservation of germplasm

There is a need to improve the poor viability of the seeds as well as their limited storage life. Low temperature storage of seeds has been reported in U. K. to improve the storage life of the seeds without any detrimental effect on their viability. Such informations need to be investigated and practised in India.

14. Potential use in modern medicine

Neem has traditionally played a key role in the Indian system of medicine. The use of neem products towards this end needs to be sustained and further promoted. Some of its recent applications particularly the use as a cheap and safe contraceptive material needs to be exploited in the national family planning programmes.

15. Establishment of Neem-based rural industrial infrastructure

Since every component of the neem tree can be put to multiple uses, organised plantations of neem on government wastelands, village community lands and marginal agricultural lands as part of social forestry and farm forestry programmes can help to establish neem based rural industrial infrastructures thereby generating employment and income for the village communities.

16. Basic chemical studies

Neem chemistry is yet another area requiring more attention in India. The synthesis of azadirachtin or other actives is still a distant dream though a significant headway has been reported in this direction. Smaller moieties from azadirachtin have been reported to possess bioactivity particularly insect growth inhibitory and feeding deterrent effects. There is a possibility of striking at simpler active analogues through structure optimisation studies and research efforts in this direction need to be accelerated.

17. Institutionalization of multidisciplinary teams and development of team projects

In view of the multidisciplinary nature of neem research, it is desirable that team projects involving different disciplines are formulated and institutionalised. The project should amalgamate experts from the basic, production, protection and social sciences and also devise ways and means of implementing important findings to transform research into practice. Only when information on all aspects of neem is generated, will help this native tree to truely prosper and benefit mankind.

18. Development of linkages

Successful implementation of a neem based programme will be seriously handicapped if appropriate linkages amongst researchers, extension agencies, the developmental organizations including farmers and other programme implementing bodies and the processing, marketing and consumer agencies etc. are not established. It is vital that the needs and view points of all are properly analysed to develop a sound-neem based programme. The stigma that the use of neem based products is orthodox and not modern, needs to be removed.

19. Effect of Neem on environment

The useful effects of neem on the environment have been often stated by the elderly people but the information lacks evidence in the modern format of science. Generating these informations will boost the raising of neem plantations and the use of neem products. Informations also need to be generated on the effect of neem plantations on the natural resources base e.g. land, water, etc. Even the effect of neem burning in purifying the environment needs scientific verification.

20. Policy issues

In spite of the fact that on the basis of its multifarious uses, neem qualifies to be considered as a national tree, it has not received the desired patronage from the Government, policy makers, administrators and consequently the public and private organizations. This drawback needs to be corrected and requisite amendments made. Neem promotional programmes need to be developed and undertaken as a national priority. These activities can be channelled through the establishment of a "Neem Marketing Bureau" on the lines of "Pyrethrum Marketing Bureau" in Kenya.

Special incentives can be provided for growing neem in wastelands, for promoting neem use in situations where the synthetics pose hazard and other well identified areas.

It may be appropriate to constitute a national group to overview progress in the diverse areas of neem research and its utilisation and to coordinate the team effort aimed at promoting this tree.

Acknowledgement

Financial support provided by EWC in meeting part of the expenditure for participation in the International Symposium is acknowledged with thanks

References

Ahmed, S. , Bamofleh, S. and Munshi, M. (1989). Cultivation of neem (*Azadirachta indica*, Meliaceae) in Saudi Arabia. *Economic Botany* **43** (1) : 35-38.

Bains, S. S., Prasad, R. and Bhatia, P. C. (1971). Use of indigenous materials to enhance the efficiency of fertilizer nitrogen for rice. *Fertil. News* **16** (3) : 30-32.

Bhatnagar, D. And MeCormick, S. P. (1988). The inhibitory effect of neem (*Azadirachta indica*) leaf extract on aflatoxin synthesis in *Aspergillus parasiticus*. *JAOCS* **65** (7) : 1166-1168.

Bose, R. K. and Bandoyopadhyay, S. K. (1986). Economics of energy plantations in alkali soils of Indian semi-arid regions. *Biomass* : 51-60.

Carlothays (1910) Eljardin botanico do Buchos-Ayers Jacobo Peuser, edit 1910 Nota : 173..

Chauhan, B. R. S. and Bist, J. P. S. (1987). Plywood from Indian timbers: *Azadirachta indica* (Neem). *J. Timb. Dev. Assoc. XXXIII:* 47-55.

Devakumar, C. and Mukerjee, S. K. (1985). Nitirification retardation by neem products. IARI *Neem Newsletter* **2** (2) : 11-14.

De Jussieu, A. (1963). *Azadirachta indica* and *Melia azedarach* Linn. Silvicultural characteristics and planting methods. *Rev. Bois Forets Trop.* No. 88, 23.

Gorse, J. E. (1986). Personal communication to Michael D. Benge. Cited from 'Cultivation and propagation of the neem tree', *The Neem Tree* vol. 1:1-16, (Ed. Martin Jacobson) CRC Press, Inc. Florida.

Herculais, Kuckel D., (1888-94). "Lee, Acridiens, Vulgo Sautere Ues, et Leurs Invasions on Algeric 1888-94".

Hoddy, Elizabeth (1985). Neem, the miracle tree. *The Economic Times, Bombay,* dated the 15th September, 1985.

IARI (1983). *Neem in Agriculture.* Res. Bull. No. 40, IARI, New Delhi.

Jotwani, M. G. and Srivastava, K. P. (1981). *Neem pesticide of the future II Protection against field pests. Pesticides* **15** : 40.

Ketkar, C. M. (1976). *Utilization of Neem (Azadirachta indica* Juss) *and its by-products.* Final Technical Report, Directorate of Non-edible oils and Soap Industry, Khadi and Village Industries Commission, Bombay.

Ketkar, C. M. (1989). Neem, Vast potential untapped. *Indian Express,* Bombay, the 19th January, 1989.

KVIC (1988) Bumper neem fruit crop this year. IARI *Neem Newsletter.* **5** (3) : 39-42.

Mann, H. H. and Burns, W. (1927). The locust attack of 1926-27 in Sind, Kathiawar and Gujarat. *Agric. J. India* **22** (5) : 325-332.

Michel-Kim, H. and Brandt, A. (1980). The cultivation of neem and processing it in a small village plant. *Proc. 1st Int. Neem Conf.,* Rottachegern, : 279-290, GTZ, Eschborn.

Mishra, K. C. and Chhonkar, P. K. (1978). Possible utilization of neem cake for inhibiting nitrification in soil. *J. Indian Soc. Soil Sci.* **26** : 90-92.

Mishra, M. M., Neelakantan, S., Narayanan, K. G. and Vyas, S. R. (1972). Studies on the decomposition of organic matter in soil. 2 Effect of Sesbania and bajra stubble plus neem cake. *J. Res. HAU* **2** : 150-155.

Murugappan, V. and Mishra, M. M. (1979) N-Lignin and margosa (neem) seed cake blended urea as nitrogen source for wheat. *Plant Soil* **52**: 131-134.

Nayak, R. L., Chatterjee, B. N. and Das, M. (1981) Regulation of nitrogen supply to lowland rice (*Oryza sativa*). *J. Indian Soc. Soil Sci.* **29**: 557-558.

Patil, N. D. (1972) Neem (*Azadirachta indica* L.) oil as nitrification inhibitor. *Fertil. News* **17** (9): 37-38.

Pereira, J. and Wohlgemuth, R. (1982). Neem (*Azadirachta indica*) of West African origin. *J. Appl. Entomol,* 94: 208.

Pradhan, S., Jotwani, M. G. and Rai, B. K. (1962). The neem seed deterrent to locust. *Indian Fmg.* **12** (8) : 7-11.

Prasad, R. (1979). Increasing fertilizer nitrogen efficiency with slow release fertilizer and nitrification inhibitor. *Fertil. News* **24** (9) 25-32.

Rajput, S. S., Shukla, N. K. and Gupta, V. K. (1985). Specific gravity of Indian timbers. *Journal of T. D. A. XXXI* (3).

Reddy, V. R. M. and Mishra, B. (1983). Effect of altered soil urease activity and temperature on ammonia volatilization from surface applied urea. *J. Indian Soc. Soil Sci.* **31** : 143-145.

Riar, S. S., Devakumar, C., Ilavazhagan, G., Bardhan, J., Kain, A. K., Thomas, P., Singh, R. and Singh, B. (1990). Volatile fraction of neem oil as a spermicide. *Contraception* **42** (4) : 479-487.

Sadre, N. L., Deshpande, Vibhavari. Y, Mendulkar, K. N. and Nandal, D. H. (1984). Male antifertility activity of *Azadirachta indica* in different species. In Natural pesticides from the neem tree and other tropical plants (Eds) H. Schmutterer, and K. R. S. Ascher, *Proc. IInd Intern. Neem Conf.* R auischholzhausen, 1983. 473-482. GTZ Press, Eschborn, West Germany.

Sahrawat, K. L. (1982). Comparative evalution of karanjin and extracts of karnaja (*Pongamia glabra* Vent) and neem (*Azadirachta indica L*) seeds for retardation of nitrification of urea in soil. *J. Indian Soc. Soil Sci.* **30** : 156-159.

Sahrawat, K. L. and Parmar, B. S. (1975). Alcohol extract of neem (*Azadirachta indica* L) seed as a nitrification inhibitor. *J. Indian Soc. Soil Sci.* **23** (1) : 131-134.

Sharma, R. C., Grewal, J. S., Sharma, A. K. and Sharma, H. C. (1980). Relative efficacy of calcium ammonium nitrate, dimethyl urea, urea and urea mixed with neem cake for potato. *Indian J. Agric. Sci.* **50** : 152-157.

Singh, R.P. and Kataria, P.K. (1991). Insects, nematodes and fungi evaluated with neem (*Azadirachita indica* A. Juss.) in India. IARI *Neem Newsletter* **8** (1) : 3-10.

Sinha, N. P. (1960). Studies on the better utilization of neem (*Azadirachta indica*). *M.Sc. Thesis,* P. G. School, IARI, New Delhi, India.

Sinha, N. P. and Gulati, K. C. (1968). Nitrification studies of solvent extracted neem seed cake and groundnut cake in Delhi soil. *Indian J. Agri. Sci.* **38** : 244-250.

Sinniah, D. *et al.* (1982). Reye-like Syndrome due to margosa oil poisoning : Report of a case with postmortem findings. *Am.J. Gastroenterol* **77**: 158.

Sinniah, D. *et. al.* (1985). Investigation of an animal model of a Reye like syndrome caused by margosa oil. *Pediatric Res.* **19** : 1346.

Subbaiah, S., Ramanathan, K. M., Francis, H. J., Suresh Kumar, R. and Kothandaraman, G. V. (1982). Influence of nitrogenous fertilizers with and without neem cake blending on the yield of finger millet. *J. Indian Soc. Soil Sci.* **30** : 37-43.

Subbaiah, S., Suresh Kumar, R. Ramanathan K. M. and Krishnamurthy, K. K. (1979). Effect of neem seed crush blended urea on the yield of finger millet variety Co. 7 in two different soils. *Madras Agric. J.* **66** : 53-55.

Warthen, J. D. Jr. and Uebel, E. C. (1980) Effects of *azadirachtin* on house cricket *Acheta domesticus*. *In* "Natural pesticides from the neem tree (*Azadirachta indica* A. Juss)". (Eds) H. Schmutterer, K. R. S. Ascher and H. Rembold GTZ. Eschborn: 137-148.

Recent Advances in Indian Entomology
Editor: O.P. Lal
APC Publications Pvt. Ltd., New Delhi, 1996, pp. 102-108

Monitoring of Pesticide Residues

N. P. AGNIHOTRI

The population of India is growing steadily at a rate of about 2.14% and is likely to touch 1 billion mark by the turn of this century from the current level of 883 million. To feed this population, accordingly, food production will have to be raised to 210 million tons from current level of 178 million tons. The additional food requirement can only be met by adoption of mordern crop practices like increasing area under hybrids and other high yielding varities with concomitant increase in fertilizers and water resourses. Bu this maximisation of yield is only possible if adequate plant protection measures are adopted. Because intensive agriculture with increase in water and nitrogen use inadvertetly lead to build up of pest population on account of congenial microclimate for insect growth. Further, warm and moist conditions which prevail in India, particularly during rainy months favour the build up of a variety of pests. Under these circumstances the use of pesticide is indispensible if the current level of food production has to be maintained or increased further.

Although pesticides have been accepted as an integral part of modern agriculture and gains accrued have been generally acknowledged by all. However, the residues that remain on the crop and particularly on the edible commodities have been of concern because of their toxic nature.

Fate of pesticide in plant and soil

The crops treated with pesticide invariably contain small amount of its residues at harvest. The amount of residues that may be present depends on the nature of the pesticide, type of crop, cultural practices and various other edaphic and climatic factors.

There is no denia of the fact that most pesticides are inherent poisons and when sprayed on the crop, the entire plant becomes toxic and spray that falls on the ground contaminates the soil as well. However, this toxicity is not permanent and the pesticides start breaking down as soon as they are applied by the action of biotic and abiotic factors. In India, under tropical conditions the dissipation of pesticides is much faster than reported under temperate climate because of the hot and moist condition which favour microbial and chemical degradation. In addition, light plays an important role in degradation of pestcides that remian on the surface of plant and soil. The intensity of sun light is very high in India because its geographical situation. Even the most persistent pesticides are broken down in matter of days and weeks when exposed to direct sun sunlight.

Whether pesticide residues are significant or not depends on the degree of exposure to pesticide residues and their toxicological properties. Thus, the hazards associated with pesticide residues depend mainly on two factors :

i. Nature and level of residues

ii. Toxicity to man and other life forms

Among these two factors, the former is of greater importance. The mere presence of pesticide residues in food does not indicate that it is hazardous. To be toxic, residues have to be present in large enough quantities considered unsafe or toxic. the detection of pesticide residues below the safe level is of academic importance only. There are many dietry components like heavy metals and toxins which are invariably present in our food but are not considered toxic despite their known toxicity because of their low concentrations. On the other hand, there are many compounds which are essential to health at low concentrations, for example, Vit A. thus, the concentration of the toxic material is the most important single parameter while considering its hazards.

Analysis of pesticide residues

Pesticides consitute large number of chemicals belonging to various groups and classes of compounds. Their rate of applications are generally low usally ranging from 15 g to 5 kg ai/ha. Comsequently, on application the initial deposits on plant/soil are in ppm or ppb range. Some compounds like HCH, synthetic pyrethroids, endosulfan, DDT contain isomers with varying degree of toxicity. Therefore, it is preferred that isomers are analysed separately. Following application, concentration of pesticide decreases further with time due to losses through degradation, volatilization, leaching and absorption by plants and animals. Moreover, pesticides undergo alteration and degradation in the environment, the metabolities thu formed, if toxic, have to be analysed along with the parent compound. Thus, the analysis of pesticide residues is a specialised field requiring sophisticated equipments and trained personnels.

Analysis of pesticide residues can be divided into four steps : i) Sampling, ii) Extraction, iii) Clean-up and iv) Estimation.

1. *Sampling* : A sample must be the true representative of the total medium under investigation. Statistical methods are available which can be followed to minimise errors during sampling.

2. *Extraction* : Pesticide residues are separated from sample matrix by using suitable solvent or solvent mixture depending on the solubility of the pesticide, the nature of the sample and final method of estimation. Usually, a lot of co-extractives (impurities) are also extracted along with the pesticide residues which may interfere with final analysis. However, careful selection of solvent can minimise these co-extratives. Solvents commonly used are : for fruits and vegetables, acetone or acetonitrile; for water, 15% methylene chloride in hexane; for dry products (< 2% fat, mixture of water and acetone or water and acetonitrile (75 : 25); for soil,

acetone or methanol; for meat and fish, water and acetonitrile (35 : 65) and for oil seeds, carbon tetrachloride.

3. Concentration : During extraction, mostly pesticide residues are collected in large volume of solvent which has to be reduced before further processing. Concentration is done by evaporating the solvent at low temperature by using either Kuderna Danish Evaporator or Rotary Vacuum Evaporator.

4. Clean-up : During extraction, many impurities like lipids, pigments, and other polar and non-polar compounds are also co-extracted along with pesticide residues which need to be removed to avoid interference in final estimation. Plant pigments, lipids and proteinaceous materials and other polar impurities are best removed by adsorption chromatography. Adsorbents used are charcoal, celite, silica, alumina, magnesia and florisil as are charocoal, celite, silica, alumina, magnesia and florisil as single or in combination. Sometimes mixture of adsorbents are preferred. Fatty materials are removed by petroleum ether-acetonitrile partitioning. Besides, chemical treatment involving strong alkali or acid are used for chemicals which are acid or alakali reisstant. For example, estimation of DDT in fatty materials like butter and oils.

5. Estimation : Since pesticides include a wide range of chemicals with different structures, there are a number of ways by which these can be estimated. These can be broadly classified into three categories : (a) Bioassay and enzyme inhbition methods, (b) Spectrophotmetric methods and (c) Chromatographic methods.

(a) Bioassay methods are fairly sensitive and do not require much sophisticated equipments. However, these methods are not specific and hence cannot be sued for identification of compounds or where mixture of compounds or their metabolites are present in the sample. Although the procedure is simple to follow the facilities required for broad spectrum bioassay programme are expensive and time consuming. An elaborate set up is required for constant rearing and maintenance of biocultures. For these reasons, bioassay methods are mostly used for screening of pesticides in development stages only.

Enzyme inhibition methods are highly sensitive but can be used only for those componds which are antienzymes. Carbamatic and organophosphorus insecticides are specific inhibitors of acetylcholine esterase and can be estimated colorimetrically by measuring relative inhibition of the enzyme. However, these methods are non-specific and suffer the same shortcomings as bioassay method.

(b) Sepctrophotometric methods include visible, ultravoilet and infrared absorption. Among these methods colorimetry has been, at least, in the past very popular because of its simplicity, rapidity, reproducibility and cost-effectiveness. UV methods, although much more sensititive than colorimetric methods require elaborate clean-up and are time consuming. IR methods are rapid and specific but are not sensitive enough to be recommended for pesticide residue analysis.

(c) Chromatographic methods include : (i) Thin layer chromatography (TLC), (ii) Gas liquid chromatography (GLC) and (iii) High pressure liquid chromatography (HPLC)

TLC is used for rapid screening and identification of pesticide residues. It is a simple and relatively inexpensive technique used for separation and identification of compounds at microgram level.

GLC is the most widely accepted technique in pesticide residue analysis. It provides required sensitivity, speed, accuracy and reproducibility. A wide range of detectors are available with can be utilised depending upon the type of the compound to be analysed and the sensitivity required. Mixture of pesticides and their metabolites or compounds with similar structures can be separated and analysed quantitatively in single estimation. GLC has all the advantages except that the compounds are heated at high temperature and during this process some compounds (For example carbamates) are either degraded or converted into their metabolites. For these type of thermolabile compounds and other non-volatile and polar compounds GLC is not suitable.

HPLC also has been used widely for analysis of pesticide residues particularly of those compounds which are non-volatile, polar and thermolabile. This method is complimentary to GLC. It has all the advantages of GLC except that sensitivity is slightly lower and the cost of operation is much higher.

Determination of safe level of pesticide

To determine the safe level of pesticide below which the risk to health is too small to be of concern, Codex Alimatarius Commission was established in 1964 under the aegis of FAO/WHO to fix the maximum residues limits (tolerance limits) of pesticide of food commodities. The recommendations are based on :

i Reports of the supervised field trials, to determine the maximum residues likely to occur on food through the application of minimum effective dosage of pesticide to control pest.

2. Detailed toxicological studies with parent compound and its toxic metabolites to determine toxic effect on man and animals.

On the basis of detailed toxicological studies '*No effect level*' is ascertained. This is the level of residues which causes no descernible effect on most sensitive of species. To ensure additional safety, NEL is again divided by a factor of 10-100. This level of intake is referred to as acceptabale daily intake (ADI) which is defined as the amount of a chemical if consumed every day for the entire life time with practical certainty that no harm will result even after life time exposure. It is expressed as mg/kg/day. For example, NEL for diazinon on different test animals was determined to be 0.1 mg/kg/ on day rat, 0.02 mg/kg/day on dog, 0.05 mg/kg/day on monkey and 0.02 mg/kg/day on man. The ADI for diazinon was fixed at 0.002 mg/kg/day.

The *Maximum permissible level* (MPL) is calculated taking into consideration ADI, Average body weight (50 kg for developing countries), weight of the treated food or class of food (kg).

For example, ADI for pyrethrin is 0.04 mg/kg/day and if treated class of food is 0.4 kg and average body weight is 50 kg, the MPL is calculated as follows :

Acceptable daily intake = 0.04 mg/kg/day

= 0.04 mg/kg/day X 50 kg (bdoy wt)

= 2 mg/day/man

Treated food = 0.4 kg

Maximum permissible limit = 2 mg/0.4 kg

= 5 mg/kg or 5 ppm

Maximum residue limit (MRL) : The permitted concentration of residues in/on food derived by taking into account both the range of residues actually remaining on food when offered for consumption following good agricultural practices and is never greater than the MPL for the food in question, expressed as ppm (mg/kg).

For example, if under normal agricultural practice the residue of pyrethrin is 1.5 ppm then the MRL of pyrethrin will be fixed at 1.5 ppm eventhough the MPL is 5 ppm.

Actual intake of residues through food

to ascertain the actual intake of residues through food.in light of their toxicity *Theoretical daily intake* (TDI) is calculated. The TDI of residues is calculated be treated into consideration daily human consumption of the treated crop and the MRL value for that crop. The process is repeated for each crop which could possibly be treated with the pesticide in question. The intake of pesticide from different crops consumed are calculated and the values are added up to determine TDI.

In practice the level of intake is invaribly much lower than expected in theory for the following reasons :

i. Not all crops are treated with each pesticide.

ii. Pesticides are not always applied at the highest recommended appliction rates.

iii. A large portion of residues in raw agricultural commodity is discarded as inedible portion (peels, leaves, stem etc.)

iv. Residues dissipate during storage and cooking

v. It is highly unlikely that a person will consume in the same day all the commodities in which pesticides have been used.

Safety to consumers

For the safety of consumers, a Registrtion Committee has been formed under "Insecticide act 1968" Manufacturer or formulator of pesticide who wishes to register a pesticide for use has to satisfy the Registration committee that pesticide allowed will not have residues on food above its prescribed MRL.

Besides, *Safety intervals* (The time required between the last spray and the harvest) are prescribed for each crop on which use of pesticide is prescribed. This is of importance particularly on vegetable and fruit crops which are harvested several times at short intervals.

Lastly, there is a provision in legislation under *prevention of Food Aultration Act 1954* that if food commodities are found to contain pesticide residues above their prescribed MRLs, farmer/grocer can be prosecuted.

Status of pesticide residues in food commodities in India

The limited surveys that have carried out at Indian Agricultural Research Institute, New Delhi, Pubjab Agriculture University, Ludhiana, Central Plant Protection Training Institute, Hyderabad and lately by different Centres of All India Coordinated Research Project on Pesticide Residues, New Delhi, clearly show the presence of pesticide residues in various food commodities and at times above their legally permitted levels. But, most of the time these residues consist of either DDT or HCH or both. the residues of organophosphorous, carbamate and synthetic pyrethroid insecticides were either missing or much below their prescribed MRL values.

Status of pesticide residues in crops from supervised field trials

Over 250 plant protection schedules consisting of organochlorine (lindane, HCH, endosulfan etc.), organophosphorous (dimethoate, disulfton, acephate, malathion, methyl parathion, fenitrothion, monocrotophos, mephospholan, phorate, phosphomidon etc), carbamates (carbofuran, carbaryl, aldicarb etc.) and synthetic pyrethroids (fenvalerate, cypermethrin, deltamethrin, flucythrinate, fluvalinate, cyhalothrin etc.) have been critically evaluated on various field crops including cotton, pulses (green gram, balck gram, chick pea, pea, pigeon pea), oilseeds

Toxicological data of pesticides.

Pesticide	LD 50 (mg/kg)	A. D. I. (mg/kg)	M. R. L.* (ppm)
ALDRIN	67	0.0001	0.2
CARBARYL	500-850	0.01	5.0-10.0
CARBOFURAN	8-14	0.003	0.2-0.5
CYPERMETHRIN	4123	0.006	0.1-0.2
2,4-D	500	0.3	0.2
DDT	113-118	0.005	7.0
DITHICARBAMATE	1500-3000	0.02	0.5-5.0
ENDOSULFAN	80-100	0.0075	0.2-2.0
FENVALERATE	451	0.06	0.2-2.0
LINDANE	88-270	0.01	0.5-3.0
MALATHION	1000-1375	0.02	0.5-8.0
MONOCROTHOPHOS	7-11	0.0006	0.2-1.0
METHYL PARATHION	14-24	0.001	0.2-1.0

* MRL for Vegetables and fruits

(mustard, sunflower, ground nut), cereals (wheat, rice, maize, sorghum) and vegetables (potato, okra, tomato, brinjal, cabbage, cauliflower, cowpea, cucurbits) from residue point of view in the Division of Agricultural Chemicals during last 25 years. It has been found that if pesticides are used at recommended rate of application following good agricultural practices, the residues are within safe limits.

Conclusion

The use of pesticides is inevitable to sustain cucrent level of food production or increase it further for future needs. It is posssible to use pesticides effectively and safely while deriving all benefits and yet, keep residues in food and environment well within the safe limits. Persistent organochlorine pesticides should be replaced by the biodegradable ones. Training for safe and effective use of pesticides can hardly be overemphasized.

References

Agnihotri. N.P. 1978. In : Residues in Indian environment (Eds. Edwards, C.A., Veeresh, G.I. and Kruiger, H.R.T.) University of Agricultural Sciences, Bangalore. pp 343-350.

Agnihotri, N.P. and Chattopadhyay, S. 1992. Pollution hazards through agro-chemicals. *Soil Fertility and Fertilizer Use* **5** : 157-174.

Jadhav, G.D. 1986. In : Proc. of Symp. on "Pesticide residue and environmental pollution" S.D. College, Muzaffarnagar (U.P.) pp 72-75.

Kalra, R.L. and Chawla, R.P. 1983. Final report of PL-480 project "Pesticide residues and monitoring of pesticidal pollution" Punjab Agricultural University, Ludhiana.

Lakshminarayana, V. 1980. Paper presented at the meeting of the "Pesticide residue problem in the country and to formulate a need based programme to be implemented" Punjab Agricultural University, Ludhiana.

Ray, P.K. 1989. Progress report of "Measurement of Ganga water quality Heavy metals and pesticides" Sponsored by Ganga Project Directorate, Ministry of Environment and Forestry, Govt. of India, I.T.R.C., Lucknow.

Recent Advances in Indian Entomology
Editor: O.P. Lal
APC Publications Pvt. Ltd., New Delhi, 1996, pp. 109-114

Pest Surveillance and Monitoring in IPM in Tamil Nadu

M. GOPALAN and S. SURESH

The prophylactic and calender based pesticide application has been slowly replaced by need based pest control because of limitations like pesticide resistance, resurgence of pests and environmental pollution by the former two methods. In this context, fixing ETL for key pests of various crops has assumed much significance. The intensity of a pest or disease is largely dependent on size of inoculum, susceptible nature of crop and favourable environment for further build up. For getting a good control of the pest and diseases, timely monitoring and application of different control agents are essential, which is obtained by a good pest surveillance and monitoring system. Lowe (1976) stated that pest surveillance forms an integral part of any and all control measures applied to a crop and that no pesticide should be applied without suffiecient justification which can only be obtained through surveillance. Green revolution has not produced the expected spectacular results in rice due to serious pest and disease problem (Chatterjee *et al.*, 1976). Hence judicious and need based applications have to be give based on surveillance information.

In strict sense, surveillance would mean vigilance on the build up of pests and pathogens for forewarning/forecasting their flare up based on weather parameters and other biotic and abiotic factors. It is done under two types: single action programme, in which a survey is made only once to estimate damage caused by a pest or pathogen while in continuous action programme, it is done over a span of time at regular intervals.

The organisation for IPM implementation in Tamil Nadu has been given below:

Organisation for IPM implementation in Tamil Nadu.

TNAU, Coimbatore		*State Extension Directorates*
State Level	Director, CPPS	Directors of the concerned departments
District Level	Plant Protection Scientists of Colleges, Research Stations. KVKs and Plant Clinic Centres	Joint Director of Agriculture and Deputy Director, Plant Protection
Division Level	-do-	Asst. Director of Agriculture
Village Level	-do-	Agricultural officers Village Level WorkersLink

Between the university and extension functionaries

The Tamil Nadu Agricultural University collaborates with the State Extension Departments for discussion and finalisation of the plant protection strategies including IPM in many ways.

State Plant Protection Committee

This is a body to decide once in a quarter the policy matter on plant protection in the state in which the Director, Centre for Plant Protection Studies, TNAU is also a member who contribute for policy decisions.

Scientific Workers Conference

This is an annual conference conducted by Tamil Nadu Agricultural University to decide research findings for adoption, on-farm trials, for confirmation and formulation of strategies for agricultural production for the ensuing year which includes plant protection. Feed back information about earlier recommendations is also analysed.

State Level Plant Protection Workers Meet

This meet is conducted annually when the subject matter specialists at the level of Joint Director and Deputy Director in the State Development Departments, Plant Protection scientists of TNAU including Krishi Vigyan Kendras (KVKs) and Plant Clinic Centres (PCCs) assemble and discuss exclusively the plant protection research findings of integrated pest and disease management for adoption and the on-farm trials for confirmation. The information and data collected through pest surveillance programme, the experiences and emerging problems in plant protection are analysed and strategies for IPM are evolved.

State Research Council, Regional Research Council and Extension Education Council

The feed back information from all these meets are utilized for formulating/ modifying the research programme in plant protection

Pest and Disease Clinics

The plant protection scientists in all Colleges and Research Stations advice the farmers on IPM strategies whenever needed.

Pest Monitoring in IPM

The concept of IPM is mainly based on epidemiology of the insect pests and pathogens and their effective monitoring. Timely monitoring helps in developing strategies for successful management of pests besides avoiding unnecessary application of pesticides. The TNAU and the State Department of Agriculture have jointly launched a pest and disease monitoring programme from 1984-85 onwards in all the districts of Tamil Nadu covering nearly 7250 points where fixed plot and roving surveys are being conducted. The farmers are advised to take up pest management measures only when the pest population level exceeds economic threshold level (ETL). The need based application of pesticides based on ETL has reduced the cost of the cultivation and avoided environmental harzards.

The weekly data on insect pests and diseases incidence along wih the weather parameters collected from fixed plot and roving surveys are reported to the Joint

Directors of Agriculture in each district. These data collected from the farmers fields and research centres in the district are pooled, interpreted jointly by the plant protection scientists of the University and Deputy Director of Agriculture (Plant Protection) every weeks and the action to be taken is decided. Messages are then finalised for dissemination to the farmers immediately and also flashed through radio and press if the situation warrants. If any outbreak occurs or if any pest exceeds ETL, the concerned staff are alerted to take up immediate control measures (Red Card) (Fig1).

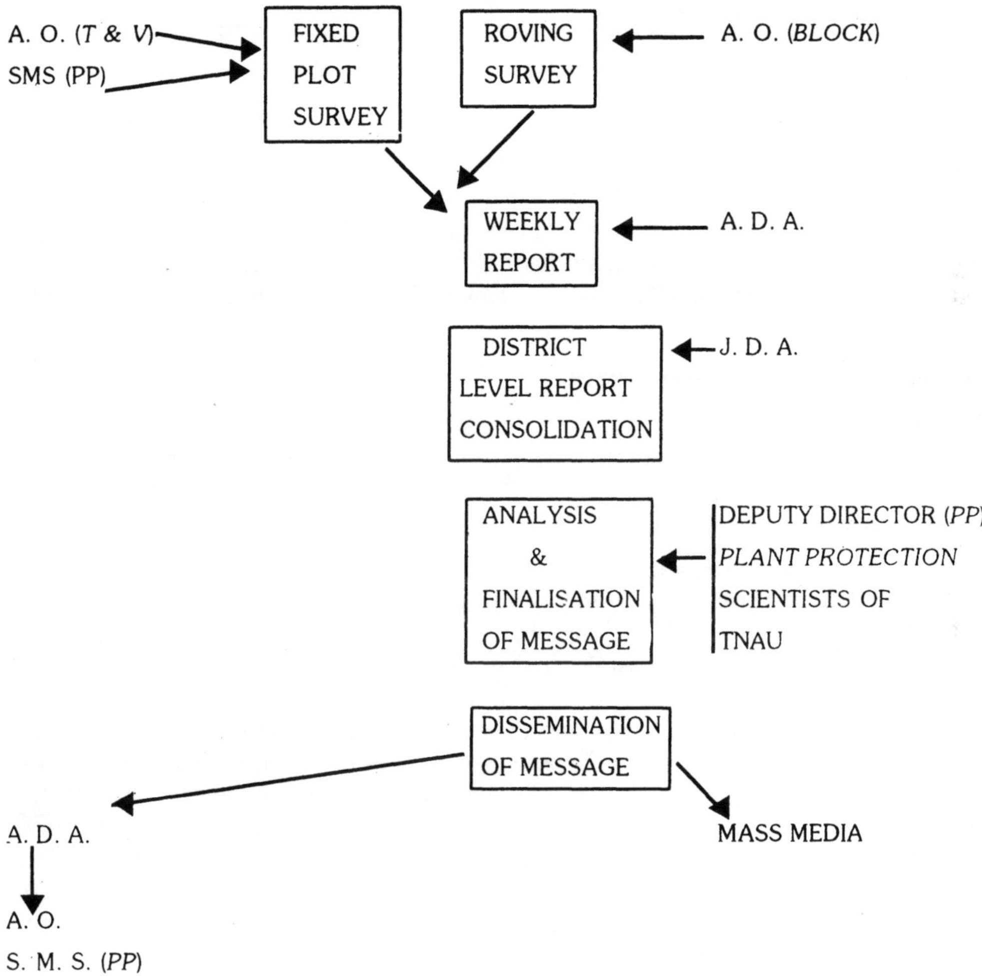

Figure 1. Pest Surveillance Programme

Special surveys

Problems related to pests and diseasses appear all of a sudden in different agroclimatic zones in the state. Under such situations with a view to managing the outbreak of insects and diseases, special surveys are undertaken as a collaborative venture between the TNAU and State development departments connected with agriculture. Surveys undertaken on rice tungro virus disease, cotton whitefly, coconut root wilt, cucumber mosaic virus on banana, nematodes of grapevine, citrus and betelvine are some of the examples in this direction.

Case studies

Rice Tungro Virus Disease

Rice tungro virus disease was first reported in India during 1967 (Raychaudhari *et al.*, 1967). It was reported from Tamil Nadu state during 1972 (Narayanasamy, 1972). The disease is a major problem in Orissa, West Bengal, Uttar Pradesh, Bihar, Kerala and Andhra Pradesh. This disease was of very minor importance in Tamil Nadu till 1980s. During the first week of November 1984, sudden outbreak of the disease in most of the rice growing tracts was reported. The causes for the malady were investigated by the Scientists of TNAU. The disease incidence ranged from 10 to 96 per cent in the districts of North Arcot, Chingleput, South Arcot, Thanjavur, Madurai, and Tirunelveli districts in all the cultivars raised viz. IR 50, ADT 36, Co 34, Co 36, Co 37, Co 43, TKM 9, IR 20, Ponmani, NLR 9672, ADT 31 and Ponni.

Factors responsible for epiphytotic

Scenario in 1984

The optimum weather condition for oviposition and nymphal hatching reported as 32°C maximum temperature and 26°C minimum temperature were prevalent during the peak period of catch of GLH in light trap.

Very high vector population was observed during the period in light trap catches during July and September, 1984. Two species of green leaf hopper vector *Nephotettix virescens* and *N. nigropictus* were prevalent in the state among which the former was the most predominant. Indexing of GLH vectors collected from Tamil Nadu Rice Research Institute, Aduthurai situated in the major rice growing areas of the state, showed high percentage of viruliferous vector population from July to November (20 to 35%). The analysis of weather data, RTV incidence and GLH vector population revealed that RTV incidence occurred in a cycle, viz., 1979, 1984 and 1992. In tungro years, sudden increase in vector population 60 to 70 times more than non-tungro years, combined with high incidence was noticed. The viruliferous vector population was also higher during the peak period than in non-tungro years. There was a positive correlation between the increase in vector population and outbreak of RTV combined with high percentage of viruliferous vector (Vidhyasekaran, 1988)

Continuous cropping of susceptible varieties especially in the norhern districts of Tamil Nadu like North Arcot, Chingleput districts and presence of inoculum in the harvested stubbles and weed hosts during the period were also the main reason for spread.

Scenario in 1992

During 1992 again there was severe epiphytotic of rice tungro in certain northern districts of the state during May-June, 1992. The reasons for epiphytotic were identified as : Growing susceptible varieities like TKM 9, IR 50, BR 5, ADT 31; Presence of RTV inoculum in the neighbouring Andhra Pradesh State; Continuous cultivation of the crop in some northern districts of the state; and Sudden increase in GLH population in these localities with high viruliferous vector population.

Immediately a comprehensive IPM strategy was intensified in almost all rice growing areas of the state.

- ★ Ploughing the severely affected crop and raising green manure
- ★ Avoid nursery near the lamp post.
- ★ Neem cake application to nursery as basal @ 5 kg/8 cent nursery
- ★ Carbofuran 3G at 10 DAS @ 2 kg/8 cent nursery
- ★ Removal of weed hosts in the vicinity of the field in bunds and border of the field
- ★ Synchronized sowing and planting in the unaffected districts
- ★ Removal of stubbles after harvest of the first crop and destruction
- ★ Continuous monitoring of GLH vector and vector indexing at weekly intervals in all the research stations of the state and reporting to the Director, CPPS
- ★ In the planted crop, spray monocrotophos at 500 ml/acre at 10 DAT
- ★ If the GLH population persists at 1/hill, spraying repeated after 15 days

This strategy has yielded successful results by coordinating with the Extension Functionaries of the state.

The scenario of RTV epiphytotic during 1984 and 1992 in the state and the vector population and viruliferous nature (Table 1) showed interesting results. During 1984, the GLH outbreak occurred in the northern districts during July-Aug. with more than 1.8 lakh combined with high incidence of RTV during Aug. Sep. The incidence spread to other parts of the state subsequently with GLH population going beyond 1.27 lakh during Aug.-Sep. with corresponding outbreak of RTV in the southern parts of the state

During 1992, the increase in GLH population occurred again in Northern districts during Aug.-Sep. (with more than 1.051) combined with viruliferous vector (35%) leading to RTV incidence up to 90 per cent. But due to immediate plant protection strategies and IPM technologies in the remaining districts all over the state, the RTV incidence was arrested and did not spread to the Southern parts, even though the GLH population was more than 2 lakhs during Sep.-Oct. 1992, while the viruliferous vector was below 4.0 per cent only.

Table 1. GLH population *(Nephotettix sp.)* their viruliferous nature and RTV Incidence in Tamil Nadu - 1984 & 1992.

	Aduthurai					Tirur			
	1984		1992			1984	1992		
Month	GLH (No.)	V V (%)	GLH (No.)	V V (%)	RTV (%)	GLH (No.)	GLH (No.)	VV (%)	RTV (%)
May	-	-	109	0	0	-	1948	1.0	0
June	-	-	180	0	0	926	895	5.0	17.5
July	-	-	92	0	0	181196	5192	31.0	96.0
August	-	-	137	0	0	30343	105012	35.0	94.8
Sept.	121483	0	80818	2.0	0	2407	1637	14.0	28.0
October	39645	40	260585	4.6	0.	906	500	12.0	25.0
Nov.	234	46	408	0	0	1278	7597	2.0	5.0
Dec.	561	26	166	0	0	10221	1930	1.0	0

GLH, green leafhopper; VV, viruliferous vector, RTV, rice tungro virus disease

The surveillance programme has been lauched with the aim to exploit and realise the unrealised yields of crops which were otherwise being taken by ravages of pests and diseases. Besides this, rational use of pesticides, preservation and augmentation of natural enemies, residueless produces will go a long way in helping the farming community and improvement of his live conditions. In course of time, it is also possible to develop pest and disease production model for each crop in each district.

References

Chatterjee, S. M. J. P. Kulshreshtha, S. Rajamani and P. S. Prakasa Rao. 1976. Insect pests of rice and their control. *Pesticide Informn.* **2** (3) : 33-36.

Lowe, J. A. 1976. Pest surveillance with special reference to rice and jowar. *Pesticide Informn.* **2** (3) : 28-30.

Narayanasamy, P. 1972. Influence of age of rice plants at the time of inoculation on the recovery of tungro virus by *Nephotettix impicticeps* (Ishihara).*Phytopath. Z.*, **74** : 104-114.

Raychaudhuri, S. P., M. D. Mishra and A. Ghosh. 1967. Preliminary note on the transmission of a virus disease resembling tungro of rice in India and other virus-like symptoms. *Plant Dis. Reptr.* **51** : 300-301.

Recent Advances in Indian Entomology
Editor: O.P. Lal
APC Publications Pvt. Ltd., New Delhi, 1996, pp. 115-126

IPM Research and Development in India: Progress and Priorities

A. K. RAHEJA

Significant changes in agriculture have taken place in the country in the last 2-3 decades. Food production has been increased by introduction of modern technology, the use of high yielding varieties and optimisation of agricultural inputs such as chemical fertilisers and pesticides. Associated with these changes are however, the concerns about the problems of sustainability and environment quality. The problems of pest control and concern about the continued use of pesticides in out environment has been the subject of intense debate and discussion. Agenda 21 of the United Nations Conference on Environment and Development (UNCED) at Rio de Janeiro in June, 1992 identified Integrated Pest Management & Control in Agriculture as one of the requirements for promoting sustainable agriculture & rural development.

There is no doubt that IPM has become a popular idea. But the point is whether there has been a real change or is it only a new label for the familiar practice of depending solely on pesticides? How far has IPM percolated to the farms or even beyond the community of researchers? How far have we been successful in curbing the use of pesticides and in creating an environment favourable to implementation of IPM? This paper examines some of these issues. The priorities in research and development of IPM are also discussed.

Present scenario

In India, crop losses due to all pests range from 10-30% per year depending on crop and the environment. Complete crop failure may occur in case of serious attack. Annual crop loss due to all pest in India was estimated at Rupees 6,000 Crores in 1983 which at today's prices could easily exceed Rs. 20,000 crores. Of this, losses from weeds are considered to be about 33 per cent, diseases-26 per cent, insects and rodents-26 per cent and birds, nematodes etc., account for the rest. Pest control in India is still largely dependent on the use of synthetic chemicals. More than 70 per cent of pesticides used in India are insecticides, 12-15 per cent fungicides and about 4-5 per cent are weedicides. In most instances, spray schedules are based on calenders with little consideration of real necessity and little regard to various detrimental side effects. This has resulted in degradation of environment, while our pest problems seem greater than ever. At first, we
environment, while our pest problems seem greater than ever. At first, we
experienced minor hints of trouble, but these have rapidly became alarming. There

are more and more reports of resistance of pests of pesticides, the number of pest outbreaks have increased and many innocuous insect species have attained the status of serious pests in recent years.

There has been a steep growth in pesticide usage in the country and the estimated demand for pesticide has risen to 82,000 metric tonnes (technical grade) in 1991 (Fig. 1). cotton and paddy account for almost 70% of the total pesticide consumption (table-1). From the state-wise consumption of pesticides shown in table-2, it is interesting to note that 33.6% of the total consumption of pesticides in India is concentrated only in Andhra Pradesh. The information contained in tables 2 and 3 clearly indicates that insect outbreaks have been maximum in the areas of crops where pesticide consumption is maximum. The indiscriminate use of pesticides at least in cotton and paddy is of such a magnitude that a re-appraisal is necessary. The two white-fly epidemics consecutive during 1983-84 and 1984-85 and the havoc brought by the American bolloworm, *Helicoverpa armigera* on cotton crop in recent years in Guntur and Pradasham districts of Andhra Pradesh are, therefore, not mere coincidences but indications of self-defeating features of injudicious use of broad spectrum pesticides.

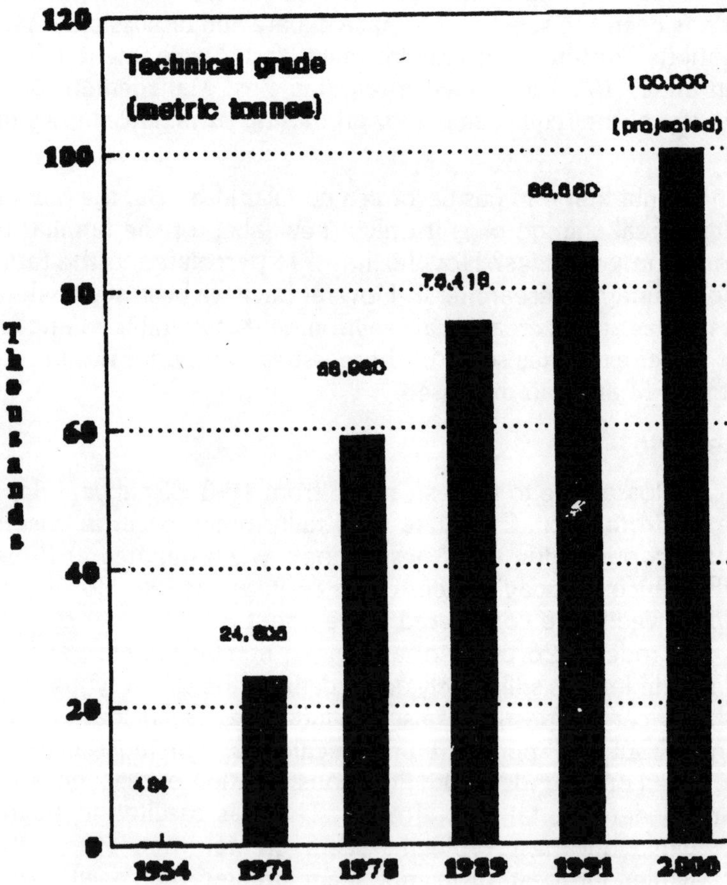

Figure :1 Pesticide Consumption in Indian Agriculture

Table 1. Pesticides market share (%) in relation to total cropped area and pest outbreaks recorded.

Crop	Pesticide share (%)	Cropped area share (%)	Major pests outbreaks
Cotton	51.6	5	Whitefly, *Heliothis*, spider mite.
Paddy	17.2	24	Brown plant hopper, gall midge, army worm.
Fruits & Vegetables	13.2	3	Spider mite, mango hopper.
Plantation	7.8	2	
Cereals, millets, oilseeds & pulses	6.7	58	Shoot fly (sorghum), *Heliothis* (pulses)
Sugarcane	2.3	2	
Others	1.2	6	

Note : Modified from Farmer's Journal (1988)

Table 2. statewise pesticides consumption and pest outbreaks.

State	% consumption	Pests outbreaks
Andhra Pradesh	33.6	Whitefly, *Heliothis*, BPH gall midge, spider mite.
Karnataka	16.2	Spider mite, groundnut leaf miner
Gujarat	15.2	Spider mite,
Punjab	11.4	Spider mite
Maharashtra	5.1	Whitefly
Others	18.5	

Note : Pesticide consumption figures are based on information compiled by Srivastava and Patel (1988).

Contributions by Indian scientists on the components of IPM are well documented. Biological control of pests, the development of resistant crop varieties, the manipulation of cultural practices to reduce pest incidence, and the use of botanicals, such as neem are practices well known in Indian agriculture. Additionally, crop diversity, and the fact that most agricultural holdings are relatively small and intensively managed are all factors that favour the implementation of IPM at the farm level. Indian scientists and extensionists are all aware of the problems that can result from over use of pesticides, and the concept of economic threshold for pesticides use is well organised in research community.

An integrated strategy for the management of a number of pests have been possible by, (i) breeding new varieties with built-in resistance, (ii) modifying agronomical practices to evade or reduce pest incidence, (iii) recommending safe and efficient methods of pest control through pest surveys and monitoring and (iv) biological control of pests with the help of their natural enemies like parasites, predators and pathogens. Easily adaptable and economical viable integrated pest management strategies have been developed for the control of major pests in rice, cotton, pulses, sugarcane etc. Large success in biological control of crop pests have been in the conservation of biologically useful organisms through either selective use of pesticides or their avoidance. Control of Pyrilla and top borer of sugarcane, mealy bug of coffee, lepidopterous pests affecting cotton, tobacco, coconut, sugarcane, etc. are a few examples where success has been achieved through the release of biocontrol agents. A major achievement has been the development of mass rearing technology for biotic agents such as *Trichogramma* spp., Chrysoperla spp. and nuclear polyhedrosis viruses (NPV) of *Helicoverpa* and *Spodoptera.* Spectacular success has been achieved in biological control of two aquatic weeds, viz., the water hyacinth *Eichhornia crassipes* and the water fern, *Sylvania molesta*.

Some of the successful IPM Programmes in the country are listed in table-3.

Research policy and infrastructure

The Indian Council Agricultural Research (ICAR) is fully committed to the promotion of the IPM concept. The "Development of Integrated Pest Mangement practice to optimise plant protection" has been included as a priority for the Eight Plan.

India has a large and execellent infrastructure including well traned scientific manpower. Research support for Agriculture is provided by the Indian Council of Agricultural Research (ICAR) through its various research institutions and All India Coordinated Research Projects and the State Agricultural Universities (SAU).

Plant protection research is an in-buit component of crop improvement research and its various disciplines are incorporated in the Crop Research Institutes as well as in the All India Coordinated Crop Improvement Projects of the Indian Council of Agricultural Research. However, there are certain special areas concerning pests and diseases or beneficial insects where the ICAR has implemented separate All India Coordinated Research Projects. These Coordinated Research Projects are :

i) Biological Control of Crop pests and weeds

ii) Nematode pests of Crop pests

iii) Rodent pests

iv) White Grub

v) Seed borne diseases (now merged with National Seed Project)

Table 3. Successful IPM programmes.

Crop / Pest Problems	Management tactics
Rice : Brown plant hopper, stem borer, gall midge	Resistant varieties, pest monitoring, Conservation and mass release of biotic agents, cultural methods.
Sugarcane : Tissue borers, pyrilla, scale insects	Resistant varieties, augmentation and release of parasites (*Trichogramma, Epiricania, Strumiopsis inference, Isotima javensis*, GV etc.), need based pesticides.
Cotton : Bolloworms, jassids, white flies	Mass release of biotic agents (NPV, *Chrysoperla, Trichogramma* spp., *Bracon* spp.), need based pesticides.
Groundnut : White-grubs	Cultural and mechanical means, seed treatment with safe pesticides.
Cruciferous vegetables : Diamond-back moth, other lepidoptera	Biotic agents, trap cropping, need based use of safe chemicals, *neem,* B.t.
Mango : leaf hoppers, mealy bugs	Tree banding, fungal pathogens.
Citrus, Grapes, Guava : Mealy bug	Mass release of *Cryptolaemus montrouzieri* and *Laptomastrix dactylopii*
Coconut : Rhinocerous beetle	Use of baculovirus.
Aquatic weeds : Water fern and water hyacinth	Introduction and mass release of exotic weevils, *Cyrtobagous salvinae* (for water fern) and *Neochitina* spp. (for water hyacinth)

vi) Agricultural ornithology

vii) Honey been research and training

viii) Agricultural Acarology

ix) Pesticide residues.

In addition, a National Centre for Integrated Pest Management has been recently established with the aim of evolving environmentally sound pest management strategies for pest and disease problems in major crops.

Promotion and implementation of IPM programmes

While tremendous advances have been made in the last 3-4 decades in agricultural research, the breadth of change is disappointing and still a large number of farmers do not appear to be affected by the new technology. The great majority of farmers in the country have access to only a limited rural infrastrure. Therefore development of appropriate IPM packages must be our first priority.

Role of researchers

In order to achieve the wide-spread adoption of biologically based technologies, much critically important research, education and demonstration work is still needed and unless and until the farmers, extension workers and even researchers and policy makers acquire greater confidence in the use of these approaches, the long term goal of creating a truely sustainable agriculture will continue to elude us. The efficacy and dependability of biological approaches has yet to be adequately demonstrated to the satisfaction of farmers and others. Greater emphasis must be given to the improvement and wide scale adoption of these methods.

It is very unlikely that chemicals will be replaced completely in the management of pests of agricultural crops - at least not in the foreseeable future. IPM, therefore, is the corner stone of the present day environmentally conscious agricultural order. Short of achieving maximum productivity without chemicals, our objective must be to maximise the biological half of the IPM equation.

The initiative for development of what is now described as biointensive pest management will have to continue to come from researchers and policy makers. As far as the farmers are concerned idealogical reasons are unlikely to change their attitude/ideas on chemical control. Also chemical companies are unlikely to stop the production of broad spectrum pesticides by their own initiative.

Eventhough research on number of novel approaches and 'hitech' methods including the use of transgenic plant has become fashionable,, pest mangement programmes for most farmers of the country will have to depnd overwhelmingly on making the best use of conventional methods, including the use of conventional pesticides, as part of IPM.

Role of Government and private sector

Agriculture is a state subject in India and extension per se is the responsibility of the State governments. However, the appropriae IPM technologies developed by ICAR/SAUs are validated through "First line extension programmes", e.g. Operational Research Projects, Lab-to-land programmes etc. The technology thus validated is then taken up for further extension by the concerned departments of Agriculture/Horticulture of the State governments.

Pilot projects on IPM were initiated by the government in the 5th Plan (1975-80) in a modest way. Pest surveillance activities which were first started in 1980 soon after India became a signitory to FAO - Intercountry IPC Rice programmes have now been extended to all the major crops.

In addition, the Government of India, Ministry of agriculture has established 26 centrally funded IPM centres in the country under the Directorate of Plant Protection, Quarantine and Storage to provide critical inputs needed for implementation of IPM programmes and to act as catalysts and model stations. These centres have been mandated to educate and farmers through training and demonstrations. These IPM centres can cover only about 5% of the total area under crops.

At present the Central Government is giving about 580 million rupees as subsidy for distribution of pesticides under various crop schemes. In order to promote IPM, it has been proposed that this expenditure should be reduced substantially and the saving diverted to IPM schemes.

A few private commercial insectaries in the country are now successfully producing biotic agents. A number of sugar factories are mass rearing egg parasites and distributing these to farmers for control of sugarcane borers. Also, there are some socially concious voluntary organisations promoting in general, better agriculture and IPM. Overall, however the role of private sector in popularising the adoption of IPM is very small.

The Central and State governments must also take the lead in changing the pest control picture through measures that would make chemical control less attractive, by legislation, registration, taxation etc. and through stimulating other control methods i.e., funding research, providing training at all levels, improvement of extension services and incentives to private companies.

Future thrusts

VIII Plan priorities

The major thrust of plant protection research in the VIII Plan of the country will continue to be on promotion of IPM concept in all major crops. This will include emphasis on :

i) biological control of insect and plant pathogens, specially mass multiplication and development of appropriate systems for dissemination of biological control agents; and use of novel methods such as insect growth regulators, pheromones and kairomones, etc.

ii) survey, surveillance and monitoring in order to develop medium and long term pest forecasting and forewarning systems;

iii) a national support system for screening of germplasm against major pest and disease problems;

iv) research on pesticides with particular emphasis on compounds of botanical origin, and

v) monitoring the development of pest resistance to pesticides (IRM).

vi) Establishment of a centre for bio-systematics of agriculturally important organism.

Critical research needs

The knowledge and information required to implement IPM strategies will have to be generated by basic research by carefully examining the factors affecting population regulation in various agro-ecological zones. Investigations are needed in population ecology, systematics, behavious, physiology, and biochemistry. The maintenance of a broad basic research effort can only yield suggestions for novel approaches to meet the challenge of rational pest management.

The emphasis in future research needs to be shifted from crop-based research to pest-oriented research. This view has been further reinforced by the recent outbreaks of polyphagous pests like *H. armigera* and whitefly with attack a number of different hosts.

i) *Insect systematic*

This has been a negelcted area of research in the country. In fact, taxonomists as a group appear to be becoming extinct and very little support is being given to maintenance and curating of taxonomy collections/museums in ICAR Institutes and Agricultural Universities. Research activity in insect systematics should not only be encouraged and but also adequately supprted. The development of comprehensive bio-systematic data base must be given top priority.

ii) *Biological control*

The need for improving and expanding the use of biotic control agents is well recognised. Technology for their mass propogation, harvesting, packaging, storage, distribution and release or application is now viewed as high priority.

The management of natural enemies (predators, parasites, pathogens) and selected beneficial organisms (antogonists, competitors and allelopaths etc.) and their products to reduce pest population and their effects is one of the most important components of our pest management strategy. The emphasis therefore is to develop pest management systems that enhance the survival and effectiveness of biological control agents by selective and minimal use of chemical pesticides.

Two aspects of biological control research that require attention are systematics and genetic improvement of biotic agents (through biotype selection, conventional crosses and genetic engineering) to enhance key attributes such as pesticide resistance and climatic tolerance. Research must also be continued on the biology, ecology and behaviour of biotic agents for using them more effectively in pest control.

Proper identification and quarantine facilities are vital to the continued success of biological control programme. Slight mistakes in identification or escape of exotic insects from quarantine can lead to disastrous results. There is a tendency these days to emphasize technology at the expense of the classical biological control approach. A balance between innovative technology and proven methods of biological control is essential for advancement of this field.

Considerable success in the country has been achieved in the use of microbials for the control of insect pests. Among the microbial pathogens, the baculovirouses, especially nuclear polyhedrosis and granulosis viruses are the most promising. The spore forming bacteria, *Bacillus thuringiensis* (Bt), and *Bacillus popilliae* and also some species of entomogenous fungi have been utilized with varied degree of success. A number of Bt formulations are now commercially available in the country. Microbial insecticides are considered as possible safe and effective replacement for chemical insecticides. However, compared to chemcials, these are poorly understood, little used and insufficiently supported.

Technology for mass production of pathogens such as NPV of IHelicoverpa and *spodoptera* has been developed but needs much refinement. There is also the need for development and optimization of application techniques to place the microbials at their most favourable infection sites. The critical research needs would also include identification of new potential candidate micro-organisms and their development into microbial pesticides; use of classical or molecule genetics to produce novel pesticides or enhance the effectivenss of those already in use; and consistent availability of easily produced, effective, quality assured preparations and maintenance of field infectivity of natural and artificially dispersed microbials.

The development of computer-based decision-making technology bringing together climatogical, microclimate, economic and biological data that will allow decision-making for maximizing the usefulness and effectiveness of biotic agents is very essential.

iii) *Host selection and feeding behaviour*

There is an urgent need to identify and understand the role of visual, chemical, mechanicl and biological signals involved in different steps of an insect's selection behaviour. Nutritional ecology is another area requiring attention. The role of chemicals mediating the feeding behaviour is not fully understood and little is known about their action threshold in crops. We must broaden and refine our basic knowledge of plant-insect relations. Host selection has been studied in only a few economically important insects. there is perhaps no example in which all the behaviour cues involved in host selection have been identified. also, there is very little information available on the manner in which multiple sensor cues reinforce each other and are integrated by insect's central nervous system. This presents both a challenge and opportunity to basic scientists for solving entomological problems of agricultural importance.

iv) *Breeding for resistance*

The green revolution of the seventies was a classical example of break through in wheat production because of the breeding for disease resistant varieties. Breeding for resistant varieties has also been adopted with certain amount of success in other crops such as paddy, sorghum, pulses etc. Breeding of high yielding varieties with in-built pest and disease resistance is one of the top priority areas of research. Special emphasis should be given to biotechnological approach to exploit somaclonal variation and genetic engineering to evolve new multiple resistant varieties.

While there is no question about the importance of resistance breeding in preventing insect and disease damage, this approach must be tempered by heavy doses of caution. Breeding for high yields and other desirable traits has fostered a widespread trend towards large scale monocultures. Intensive cultivation of high yielding varieties over contiguous areas has facilitated the pests of hiterto minor importance to acquire a status of major pests, for example whitefly in cotton and gall midge on paddy and sorghum. while the traditional agro-ecosystem was genetically diverse, there is now a danger of genetic erosion due to depletion of the gene base of many crop plants. There is also the problem of the evolution of biotypes which break the resistance. It is also unlikely that important polyphagous pests of our crops, such as *H armigera* can be adequately controlled by plant resistance only.

v) *Insect growth regulators and behaviour modifying chemicals*

A number of natural and synthetic hormonal chemicals are known. These include insect hormones, hormone analogues, and a diversity of synthetic compounds that mimic the action of insect hormones and/or interfere with hormonal regulated process of insects. These compounds cause diverse disruptive effects on insect development, moulting, metamorphosis, and reproduction which they are either applied directly to the insect or added to insects diet or medium in extremely small quantities. Some of the synthetic compounds, particularly few more active juevnile hormones, and chtin inhibitors have been commercially formulated. The challege lies in translating the disruptive effects of these compounds into effective pest management tools.

Insect pheromones and kaironones offer great promise in pest management. While pheromones can be employed in survey, monitoring, trapping and mating diruption of insects, kairomones can be used to increase the efficacy of parasitoids and predators. Research effort in future would have to be directed to evaluation of physical, chemical and biological factors influencing pheromonal traps and formulations. An essential requirement is the development of statistical modelling of pheromonal catches in relation to field populations and its application to pest forecasting.

vi) *Biotechnology*

Biotechnology will play a major role in developing new approaches to pest control. Our research should aim at using the modern tools of biotechnology to increase the efficiency of the established microbial agents by enhancing their virulence, increasing/decreasing their host range and modifying their tolerance to environment. the development of genetically engineered microbial pesticides offers exciting possibilities for the future. Another exciting area of research is the development of transgenic crop plants with insect resistant genes from Bt.

vii) *Pesticides*

Management of pesticide use is an integral part of IPM. Our priority should be to refine our pesticide application recommendations so as to reduce their use to the absolute minimum. Research must also be intensified on other pesticide related problems such as residues, resistance etc. The challenge to the industry lies in developing novel molecules and products that are less injurious to the environment, and more active against pests and of course are also not more expensive to use. Basic research into developing biodegradable pestidides by better understanding of the biochemical processes in the intoxication and detoxication process could be initiated by some of the high-tech research laboratories. Similarly, such research could also examine the degradation of pesticides by micro-organisms. Very little attention has been given to pesticide application techniques in the past and this deficiency must be corrected. Most spray application methods coventionally used though quite effective are wasteful and inefficient. Much of the pesticide is not deposited on the target resulting in contamination of the non-target areas, thus increasing the incompatibility of the chemicals with other mothods of control.

(a) *Insecticide resistance management :*

In India, in the last few years, the outbreaks of *H. armigera* in Andhra Pradesh, Tamil Nadu and more recently in Punjab and Haryana have caused great alarm. A very high degree of resistance to pyrethroids has been found in *H. armigera* populations in all the major cotton growing on pigeonpea and chickpea.

In order to ensure long-term effectiveness and provide guidelines for judicious use of all classes of insecticides, sound resistance management strategies are needed. This need is of course greatest in crops like cotton, where numerous insecticide applications throughout the season exert significant selection pressure on several arthropod species.

(b) *Botanical pesticides :*

A lot has been recently written about using "neem" any other plant products as pesticides. Neem has also been the subject of many a national and international symposia and seminars. while there is no doubt about the potential of neem derivatives as an important tool in pest management programmes, we are still far away from the stage of their effective utilization. There is, therefore, an urgent need to critically review the technology and develop inter-institutional and multi-disciplinary approach to solve the problems associated with the use of botanical pesticides. Cheap and simple methods for isolation of pure active compounds such as azadirachtin must be developed. Research is also needed to further elucidate the biological activity of neem compounds and their effect on insect growth and behaviour. Research on development on formulations of neem is being carried out in a number of laboratories in the country and we also now have a few commercial formulations available in India. There is need for standardization of these formulations in terms of the active ingredient and their bio-efficacy. It would appear that complexity of neem compounds precludes their synthetic production in the near future. Industry, however, can play a major role in development of suitable and stable formulations. A number of commercial neem formulations have been registered and are now available in the country. Their field efficacy for various pest problems is yet to be fully evaluated.

viii) *Vigilance against introduction of new pests :*

This aspect has gained special significance with the introduction of the new liberalised National Seed Policy in India which along with inadequate quarantine facilities and procedures has greatly increased the chances of introduction of new weeds, pathogens and insects into the country. Of course, weeds and pathogens have a greater chance of entry than insects. The arrival in the country of the psyllid pest, *Heteropsylla cubana* now affecting large tracts of *Lucaena* plantations in many States is one example. There is obviously a need for greater vigilance to prevent such occurrences. Similarly, we 'ave been greatly alarmed at the spread of the tomato leaf miner, *Liriomyza trifoli* in the country. It appears to have been imported in the country through chrysanthemum cuttings. Some of the other insects on which is essential include the cotton boll weevil, *Anthornomus grandis* and the Mediterranean fruit fly, *Ceratitis capitata.*

ix) *Integrated pest management :*

Implementation of IPM is only possible if criteria necessary for appropriate decision making are developed. The utilization of the concept of "economic threshold" and "economic injury level" still encounter numerous difficulties. A system management approach is essential for an effective IPM strategy. The research must be intensified to understand the interactions between crop, pest, climate, and natural enemies to fine tune our control approach to a particular pest problem in an agro-eco-system.

The survey and surveillance programmes must be strengthened and research is needed to develop effective methodology for sampling collection and analysis of such data. Development of forecasting and forewarning systems with corresponding advisory services are pre-requisites for effective implementation of IPM programmes. Here, the role of plant protection advisory services must be emphasised. Trained and specialized extension personnel are needed to undertake monitoring and scouting to provide forecasting and forewarning services and also to ensure timely action.

Concluding remarks

Pest will continue to be a major constraint and a serious threat to yield stability. Increases in crop yields in India (and many other countries) over the past two decades have reached a stage of rapidly diminishing returns. Even varieties that possess high yield potential will be unable to express this trait without pest control. Good pest management can reduce production variability and will be more lasting and even less expensive than the reliance on continuous breeding programmes. This is both an opportunity and a challenge to the ingenuity of the research workers - basic and applied, in the field of Entomology and IPM.

Recent Advances in Indian Entomology
Editor: O.P. Lal
APC Publications Pvt. Ltd., New Delhi, 1996, pp. 127-139

Problems of Insect Pests and Pesticide Hazards

O. P. LAL

Agriculture is the back bone of our country to provide food to our growing population and the raw materials to numerous industries like those engaged in the area of sugar, cotton, tea, coffee, jute, tobacco, vegetable oil and food processing, etc. Hence any factor reducing the yield is of great concern of one and all.

The struggle between man and insect began long before the dawn of civilisation and is still continuing with greater challenges as man wants to protect his health and provide an adequate food supply. No. doubt, pesticides have provided good control of many insect pests, diseases and weeds in the field of agriculture increasing our production and at the same time protected our health by controlling vectors which transmit diseases like malaria, filarisasis, dengu, encephalities, etc. Although pesticides continue to play a significant role in agriculture and public health, but the indiscriminate and over use of pesticides has been ecologically unsound and led to a number of problems such as insect pest resistance, outbreak of secondary pests, adverse effects on non-target organisms, toxic pesticide residues and direct hazards to the user. Hazards due to pesticides are more prevalent in developing countries than the developed countries. Though the developing countries use less than one quarter of the world's pesticidies but they suffer three quarter of total pesticide fatilities. About 25 million agricultural workers are poisoned by pesticides and some 20,000 die of pesticide poisoning in developing countries every year. Hence pesticides should be used with extreme caution.

Table 1. Estimated losses due to pests, diseases and weeds in field and storage in India in 1973-74.

Pest		Percentage Loss	Rupees in Crores
Weeds		33	1980
Diseases		26	1560
Insects		20	1200
Miscellaneous (Pests, storage and rodents, etc.)		21	1260
Total	1973-74	100	6000
	1989-90	100	20,000

Source : Pesticide Information Volume 1, No.1, April - June 1985, pp 12.

Crop losses in India

Today when the human population is crossing its limits, the damages caused by the enemies of plants is posing serious problems before farmers, scientists and the administrators. Pesticides Association of India conducted a study about 20 years back and concluded that the annual loss of agricultural production in India was Rs. 6000/- crores (Table 1). At present, this loss would be more than Rs. 20,000 crores as reported in Pesticide Information, Volume 15, No. 2, July - September, 1989, pp. 12.

The estimated crop losses during 1973-74 and 1989-90 (Pooran chand, 1992) are given in Table 2.

Table 2. Estimated crop losses during 1973-74 and 1989-90 in India.

Crop	1973-74			1989-90		
	Production (Million Tonnes)[1]	Percentage loss	Loss value (Million Rs.)*	Production (Million Tonnes)[2]	Percentage loss**	Loss value (Million Rs.)
1. Paddy	43.7	18.6	15443	74.1	18.6	55120
2. Wheat	22.0	11.4	3750	49.7	11.4	14150
3. Bajra	7.0	10.0	672	7.8	10.0	1287
4. Jowar	8.9	10.0	630	10.5	10.0	1732
5. Pulses	9.7	7.0	2037	12.61	7.0	4840
6. Oil seeds	8.6	25.0	4350	16.75	25.0	41800
7. Sugarcane	14.0	15.0	2844	222.6	15.0	13360
8. Cotton	1.0	22.0	9770	1.1	22.0	20000
9. Tea and Coffee	0.54	24.0	106	0.87	24.0	840*
10 Fruits	18.0	20.0	2160	-	20.0	8640*
11 Potatoes	4.6	11.0	305	11.87	11.0	3444
12 Miscellaneous	-	12.0	6353	-	12.0	19359
Total	-	-	48420 or 500 crores	-	-	184272 or 18400 crores

* Estimated value.

** Percentage loss was assumed same in 1989-90 as it was in 1973-74.

Source: 1. Pesticide Information, Volume 1, No. 1, April-June, 1975 pp. 12.
2. Economic survey, Govt. of India, 1990-91.

The extent of pesticide use in India

The extensive use of pesticides in India began after World War II with the discovery of DDT and HCH followed by organophosphates and such organochlorines as chlordane, heptachlor, dieldrin and endrin. Endrin has been banned in India due to its high toxicity. However, upto 1966 most of the pesticides used were in public health sector while it was less in agriculture. But, from 1967 onwards, when high yielding varieties, highly responsive to fertilizers and irrigation, were evolved, the pesticide consumption increased by leaps and bounds which is given in Table 3.

Table 3. Agricultural production and use of pesticides in India.

Year	Total production of food Grain (Million Tonnes)	Pesticides used in Agriculture (Thousand Tonnes)	Pesticides in public health (Thousand Tonnes)	Total pesticides consumption (Thousand Tonnes)
1960	80	2.0	21.0	23.0
1968	74	7.4	9.6	17.1
1970	108	10.2	8.8	19.0
1975	121	43.4	15.2	58.6
1980	126	40.6	15.5	56.1
1981	129	48.0	16.8	64.8
1982	133	58.0	19.3	72.3
1983	129	64.0	17.8	81.8
1984	151	68.0	20.6	88.6
1985	153	72.0	19.2	91.2
1990	176	81.5	—	119.2
1995	191	83.2	—	—

Source: Lal, 1996

However, the total pesticide consumption has gone upto 119.2 thousand tonnes in 1990 (Mehrotra, 1991). The cropwise consumption of pesticides is given in Table-4 which shows the severity of pests on different crops requiring much expenditure in plant protection.

If we look at the consumption of pesticides in different states of India (Table 5), Andhra Pradesh is found using maximum pesticides followed by Tamil Nadu, U.P., Punjab and West Bengal. But, on the basis of per hectare cropped area, the highest amount of pesticide is used in Tamil Nadu (1.8 kg), followed by Punjab (0.87 kg), Andhra Pradesh (0.83 kg) and Haryana (0.8 kg).

The increase in pesticide consumption may be due to developing insecticide resistant strains and over-utilization of pesticides due to lack of proper knowledge about the hazards and requirements of pesticides.

At present DDT along with HCH constitute bulk of the pesticide used in the country. Since its introduction in the country more than 270 thopusand tonnes of DDT has been used till to date, more than 220 thousand tonnes in public health and 50 thousand tonnes in agriculture. Compared to this, 600 thousand tonnes HCH has been used, about 525 thousand tonnes in agriculture and 75 thousand tonnes in public health. The large scale use of DDT and HCH have led to their ubiquitous presence in the Indian environment including human and animal tissues (mehrotra, 1989b).

Pesticide production units

There are more than 350 small scale and 50 large scale formulators of pesticides in India. Under the Insecticide Act 1968, the location should be controlled. Most of pesticide formulation units are in small scale sectors and hence they take only the consent of the Pollution Control Board that it is not located near a food processing or pharmaceutical factory. However, big factories like Union Carbide (Bhopal), should be in remote area with a proper on site emergency plan.

According to the Industrial Toxicology Research Centre, Lucknow, the total production capacity of pesticides is high in cities of Hyderabad, Bombay, Calcutta, Allwaiye and New Delhi. It is absolutely necessary to excercise a control in these cities and if possible even disperse the manufacturing units to rural areas.

Table 4. Cropwise consumption of pesticides at farmers levels.

Crop	% of * cropeed area covered	% of pesticide used	Cost (1980)** Rs.in crores	Cost (1986)** Rs.in crores
1. Cotton	5	54	139	250
2. Rice	24	17	102	100
3. Vegetables and fruits	3	13	39	100
4. Plantation	2	8	15	40
5. Pulses and oilseeds	10	2	15	30
6. Sugarcane	2	3	19	40
7. Others	5	3	14	20

Source : * Kapadia and Mohla, 1980.

Table 5. Consumption of pesticides in different states of India.

State/Union Territory	Consumption of pesticides (Technical Grade) 1974-75		Consumption of pesticides (Technical Grade) 1986-87	
	Total (in tonnes)	Per hectare of cropped area (in kg)	Total (in tonnes)	Per hectare of cropped area (kg)
1	2	3	4	5
1. Andhra Pradesh	10,030	0.7	12,000	0.83
2. Assam	155	0.1	650	0.42
3. Bihar	3,000	0.3	3,000	0.3
4. Gujarat	800	0.1	5,000	0.63
5. Haryana	2,400	0.5	3,800	0.80
6. Himachal Pradesh	315	0.3	300	0.3
7. Jammu and Kashmir	78	0.1	500	0.64
8. Karnataka	2,472	0.2	4,200	0.34
9. Kerala	586	0.2	1,200	0.40
10. Madhya Pradesh	3,422	0.2	4,000	0.23
11. Maharashtra	3,500	0.2	4,100	0.23
12. Manipur	17	0.1	48	0.28
13. Meghalaya	15	0.1	50	0.33
14. Nagaland	6	0.1	20	0.33
15. Orissa	1,000	0.1	1,600	0.16
16. Punjab	3,300	0.6	4,800	0.87
17. Rajasthan	1,000	0.1	2,600	0.26
18. Tamil Nadu	1,070	0.2	11,600	1.8
19. Tripura	20	0.1	120	0.6
20. Uttar Pradesh	6,000	0.3	7,000	0.35
21. West Bengal	1,800	0.2	5,000	0.56
22. Others			254	

Source: Mehrotra, 1989a.

The manufacture of pesticides started in India in 1952 with BHC (HCH). Excluding Japan, amongst South Asian and African countries, India has emerged as the largest manufacturer of basic pesticides. India is the largest producer of DDT in the world. It has been now banned in agriculture and being entirely used for public health purposes.

Indiscriminate use of pesticides

Although, pesticides have played a great role in agriculture and animal health, but indiscriminate and over use of pesticides have led to a number of problems including disruption of ecological balance. Many farmers apply pesticides in a routine manner particularly on vegetables when there is actually no pest in the field. Such a practice should be stopped as after some time the problem will be so grave that the pesticide will have no effect on the pest.

The cotton crop in Andhra Pradesh failed during 1987-88 season due to an unprecedented epidemic of American boll worm, *Heliothis armigera* Hub., particularly in Guntur and Prakasam districts. There were instances where cotton crop received 30 insecticidal sprayings including 8-10 synthetic pyrethroids at short intervals. Hence there should be a judiceous use of pesticides in the pest management programmes.

In India, only a few percentage of farmers exploit the use of pesticides while majority use very less or no pesticides. the average pesticide consumption in India is very less in comparison to other countries (Table 6).

Table 6. Pesticide consumption in different countries.

Country	Pesticide consumption g/ha
India	450
Mexico	750
Argentina	960
U.S.A.	3000
West Germany	3000
Japan	11000
Taiwan	17000

Mukerjee, 1985.

Pesticide hazards

1. Effect on soil, water and air

Most of the pesticides applied to crop ultimately reach the soil in the form of spray drift or run off and further move into the aquatic system and get accumulated in plant and animal. Water quality is affected by pesticides particularly by chlorinated hydrocarbon and their derivatives. Similarly, air is polluted while filling, loading, mixing, spraying or fogging the insecticides.

2. Toxic residues

Toxic residues are hazardous to health particularly those on fruits and vegetables are of great concern. In India, the pesticide residues detected in market samples is very high. Some of the pesticides remain in soil for a very long time. Brinjal, okra, cabbage, cauliflower, cowpea, spinach and radish in Delhi region have pesticide residues (DDT) above tolerance limit.

Toxic residues of DDT and HCH are found in stored wheat/flour, rice, maize, etc. and find their way to 'Chapatis' through flour. Residues of DDT and HCH have been reported in vegetable oil samples of coconut, cotton seed, groundnut, rapeseed and sesame in Punjab. Even 50 to 60% of baby milk powder and honey samples were found with the residues of several chlorinated hydrocarbon insecticides (Kalra and Chawla, 1983). Pesticide residues in market samples (mehrotra, 1989b) are given in Table 7, 8, 9, 10 and 11. Almost all the samples of meat, fish, eggs, milk, butter and ghee collected from different parts in India showed presence of DDT and BHC. Some of the pesticides are reported to alter the immune function of the body. DDT carrier women may show frequent abortions and complication during pregnancy. BHC is know to be a carcinogenic pesticide. Thus, one should be careful in selecting and using a pesticide.

3. Insecticide resistance in insects

Insecticide resistance in agricultural pest in India was first reported in singhara beetle in 1963 and since then many pests have developed resistance to different insecticides.

Table 7. Pesticide residues in market samples of crop commodities in India.

Commodity	Sampling sites	Number of samples	Pesticide detected	Incidence (%)	Rang of Residues (ppm)
Cereals and cereal products	A.P., Delhi, Haryana, Maharashtra, Punjab, U.P.	1670	DDT	53	*Tr. 12.0
			BHC	31	Tr.20.0
Pulses	A.P., Karnataka, Pubjab, U.P.	289	DDT	25	Tr.10.0
			BHC	20	Tr.10.0
Vegetables and fruits	A.P., Delhi, Haryana Karnataka, Maharashtra, Punjab	2244	DDT	27	Tr.35.0
			BHC	28	Tr.60.0
Vegetable oils	A.P., Delhi, Punjab, U.P.	83	DDT	95	Tr.25.7
			BHC	95	Tr.6.4
Animal feed and fodder	Punjab, U.P.	111	DDT	100	Tr.5.8
			BHC	100	Tr.6.6

*In Traces

Table 8. Pesticide residues in animal products in India.

Commodity	Sampling sites	Number of samples	Pesticide detected	Incidence (%)	Range of Residues (ppm)
Meat	A.P., Delhi, Punjab,	157	DDT	96	Tr.20.0
	U.P.		BHC	90	Tr.0.20
Fish	A.P., Punjab,	90	DDT	92	Tr.2.2
	West Bengal		BHC	85	0.02-1.1
Poultry	A.P., Delhi, U.P.,	110	DDT	60	Tr.8.0
eggs	Maharashtra, Punjab		BHC	20	Tr.2.5
Milk	A.P., Delhi, Gujarat,	1010	DDT	95	0.19-216.0
	Mahatrashtra, Punjab		BHC	90	0.12-040.0
Butter	A.P., Delhi, Gujarat,		DDT	100	1.72-9.81
	Haryana, Karnataka,		BHC	100	0.58-7.63
	Punjab, Rajasthan,				
	Tamil Nadu, U.P.,				
	Best Bengal				
Desi ghee	A.P., Punjab,	10	DDT	100	2.1-13.0
	Rajasthan, U.P.		BHC	100	2.0-9.8

Table 9. DDT and BHC residues in the human fat.

location	No: of samples analysed	No. of samples contaminated	Levels of residues (ppm) DDT Mean	BHC Mean
Agra	14	14	12.0	2.0
Ahmedabad	94	94	22.4	4.0
Bangalore	67	67	8.8	7.0
Bombay	34	34	6.2	1.6
Calcutta	45	45	6.2	1,6
Chandigarh	10	10	19.9	2.4
Ludhiana	51	51	9.1	4.3
Lucknow	-	-	1.8	2.3

Table 10. DDT and BHC residues in the milk of lactating females from India and other countries (ppm).

Country	DDT Complex	BHC-isomers
India	0.05	0.2
Canada	0.024	0.001
Guatemala	1.4	0.006
Japan	0.071	0.072
Sweden	0.078	0.007
Switzerland	0.075	0.0001
USA	0.22	0.02

Table 11. Average dietary intake of DDT and BHC residues in various countries (expressed as μg/person/day.

Country	DDT Complex	BHC-isomers
Australia	20.0	-
Canada	10.8	2.5
Czechoslovakia	99.0	53.0
West Germany	149.0	-
Italy	35.0	16.7
U.K.	12.0	-
USA	6.5	1.1
Yugoslavia	98.0	62.0
India (Vegetarian)	238.1 (428.4)	124.4 (313.3)
(Non vegetarian)	224.1 (396.2)	133.2 (320.8)

Figures in parentheses are the estimates of the maximum values of dietary intake.

It has been found that *Heliothis armigera* on cotton has become 150 to 300 fold resistant to cypermethrin in parts of Andhra Pradesh. In several part of the country, couliflower crop failed due to *Plutella xylostella* which has become resistant to a number of pesticides. Hence excessive use of insecticides especially synthetic pyrethroids should not be done. The pesticides should be judiciously used so as to prolong their useful life periods.

4. Mortality of parasites, predators and pollinators

The unwise use of pesticides kill the parasites and predators. The pollinators especially honey bees are killed by pesticides due to which cross pollinated crops are adversely affected that their yield is much reduced (Lal, 1973). It has been reported by Kalode et al. (1988) that spray with deltamethrin on rice killed the predator mirid. *Cyrtorhinus lividipennis* due to which Brown Plant Hopper *Nilaparvatha lugens* population was much increased and yield was reduced. Outbreaks of pyrilla (*Pyrilla perpusilla*) has now become a regular feature on sugarcane. Failure of the parasite, *Epiricania melanoleuca,* is one of the responsible factor for it.

5. Pest resurgence and outbreaks of secondary pests

Due to continuous and indiscriminate use of pesticides, many parasites and predators are killed and many pests have developed resistance to pesticides. This had led to resurgence of pests. Moreover, minor insects which were under check by parasites and predators, have become major pests.

In cotton, use of synthetic pyrethroids to control boll worm enhanced the population of white fly. In sugarcane, use of endrin to control borer increased the population of scale insects in 1956. In brinjal, use of acephate against fruit borer increased the population of mites and aphids, In chillies, use of cypermethrin, fenvalerate, permethrin and deltamethrin to control fruit borer increased the population of wheat mite. There are many more such examples which shows that only judicious use of pesticide should be done.

Impact of pesticidal hazards on quality of life

Pesticides may cause direct fatalities, damage the body surface, affect the internal organs and may induce malformation or carcinogenic growth. There may be other type of injury like dermatitis, asthma, urticaria, etc.

1. Effect of contaminated food and water

About 50% of the pesticides applied on crop foliage reaches the soil which further moves to aquatic system and get accumulated in plants and animals. Water quality is affected by chlorinated hydrocarbons, their derivatives, persistent weedicides and soil insecticides. Pesticides are also added to water to control disease vectors. Fresh water mussels, snails, prawns and many invertebrates collected from river Jamuna in Delhi showed high DDT residues ranging from 0.059 to 7.57 ppm. The pesticides contaminate fruits, vegetables, milk, fat, etc. affect our health as given in Table-12. The persons affected by pesticides may show symptoms of cough, tears, runny nose, blurred vision, vomitting, diarrhea, convulsion and salivation. It has been found that they affect the central nervous system, limbs, digestive systems, etc.

2. Pesticide accidents and poisoning

Although a number of pesticidal accident are reported in India and in the world, Some major accidents are given in Table 13.

Table 12. Some important pesticides, food they contaminate and nature of impact.

Pesticides	Food contaminated	Impact
Aldrin	Fish, milk, fat.	Accumulation in human tissues.
DDT	Fish, Milk, fat, meat.	Accumulation in human tissues, low toxicity, induction of liver enzymes.
Linadane	Fruits, cereals vegetables, fat, milk.	Accumulation in human tissues, toxic to bone marrow, probably carcinogenic.
Heptachlor	fish, milk, fat.	Accumulation in human tissues toxic to bone marrow, probably carcinogenic.
Malathion	Fruits	Acute toxicity, neurotoxic.
Parathion	Fruits, vegetables	Acute toxicity, neurotoxic.

Table 13. Pesticidal accidents in India.

Pesticide	Place	Year	Cause
Parathion	Kerala	1958	Food contaminated due to leakage.
	Bombay	1962	Inhalation in manufacturing plant.
HCH	-do-	1963	Rice contaminated
Endrin	-do-	1964	Food contaminated
DDT	Punjab	1965	'Chutney' contaminated
Diazinon	Pune	1968	Food contaminated
HCH	U.P. (Lakhimpur)	1976	Mixed with wheat
Endrin	Karnataka	1977	Contaminated crabs in rice field.
Aluminium phosphide	Rajasthan	1978	Contaminated food grains.
Methyl · isocynanate	Bhopal	1984	Storage tank leakage.

A number of workers were studied for poisoning by pesticides in different parts of the world. Maximum deaths are found due to organophosphate compounds and carbamate insecticides because thesse are more poisonous chemicals used in agriculture.

A survey on poisoning due to aluminium phosphide in northern India in 1988 revealed poisoning cases upto 115 in Rohtak (Haryana), 55 in U.P., 30 in M.P. and 28 in Rajasthan.

Table 14. Aluminium phosphide poisoning in north India.

State	No. of cases of poisoning
Haryana	114
U.P.	55
M.P.	30
Rajasthan	28

3. Pesticides and the Immune System

Pesticides viz., DDT, dieldrin, carbaryl, carbofuran, methyl parathion and endosulfan were reported to alter the immune function of body which is an important aspect of environmentally induced chemical toxicity.

4. Reproduction toxicity and teratogenicity

Several organochlorine pesticides have shown no evidences of teratogenicity e.g. DDT does not produce any teratogenic effect but DDT carrier women may show frequent abortions and complication during pregnancy. Recently 2,4,T and 2,4,-D have been reported having teratogenic effect on human and other animal species.

5. Mutagenicity and carcinogenicity

Some pesticides are known to induce heptic microsomal enzyme activity and thus affect the rate of steroid production and metabolism. DDT has been reported to cause hepatic cell tumour in rats. BHC is a known carcinogen. Recently murex has been shown to induce cancer in mice. Ethylene thiourea also produces thyroid tumour in animals and it is also a mutagenic. Carbon tetrachloride ($CC1_4$) as fumigant produced liver tumour in the mouse, hamster and rat by inhalation and oral ingestion.

Conclusion

We are using a lot of pesticides which is affecting our health and it should be stopped. We should remember that when DDt was discovered, it was thought that all the pest problems will be solved for ever. But this has not been the case. The pest problems are now much more realised than ever before. There is no doubt that the pesticides helped us to increase our agricultural production which paved the way for Green Revolution. But, the indiscriminate use of pesticides has led to many problems like environmental pollution, resistance to pesticides, adverse effects on non-target organisms, toxic residues, health hazards, etc.

In many parts of the country, cotton crop failed due to *H. armigera* and cauliflower due to *P. xylostella* in spite of heavy sprayings. Pesticide resistance was a major factor responsible for it.

The farmers particularly those near big cities are using highly toxic pesticides on vegetable crops, sometimes just before the harvesting. Even they dip the vegetables in pesticide solutions under the false impression that it will retain its fresh look for a longer time. Apple trees are being sprayed 10 to 12 times against scab disease which leaves toxic residues in fruits. Samples of meat, fish, egg, milk, ghee, butter, etc. from different parts of India showed presence of pesticide residues. So, almost every thing we are eating is polluted. Our air and water is also polluted with pesticides. Many people die due to pesticide poisoning every year. It gets accumulated in different organs of our body which may affect our immune system, nervous system and may cause abortion in pregnant women. Some pesticides are even known to be carcinogenic.

So, now it is high time for every one of us to think that the use of pesticides should be either stopped or it should be used to the barest minimum possible level. It should be used only under unavoidable situations and that too under a technical supervision. Other alternate methods of control should be encouraged so that a healthy and ecologically sound Integrated pest Management is maintained. If indiscriminate use of pesticides is allowed as such, a day will come when no pest will be controlled by any pesticide which will be disastrous.

References

Kalode, M.B., Pasalu, I.C., Krishnaiah, N.V. and Bentur, J.S. 1988. Changing insect pest complex in relation to crop and cropping systems-Rice, Natl. Seminar on changing Pest situtation in the Current Agricultural Scenerio of India. June 14-16, New Delhi.

Kalra, R.L. and Chawla, R.P. 1983. Final Technical Report of PL-480 project on "Studies on pesticide residues and monitoring of pesticide pollution", Punjab Agricultural University, Ludhiana, p. 230.

Kapadia, Z.J. and Mohla, D. 1980. Pesticides marketing in India. In Indian Pesticides Industry - Facts and figures. Ed. B.V. David, pp. 151-155.

Lal, O.P. 1973. Problems posed by pests, pesticides and pollinators in the seed production of cabbage. Proc. 60th Indian Sci. Cong. Part III, 545.

Lal, O.P. 1992. Integrated Pest Management for Sustained Crop Production. Proc. Natl. Symp. on Resource Management for Sustained Crop Production, Bikaner, p. 251-259.

Mehrotra, K.N. 1989a. The extent of pesticide use in India and its consequences - Pesticide Residues in Food, Consumer Education and Research Centre, Almedabad, 42-58.

Mehrotra, K.N. 1989b. Pesticide resistance in insect pests. Indian Scenerio. *Pesticide Res. J.* **1**: 95-103.

Mehrotra, K.N. 1991. Agriculture vs environment dealing with pests. The Hindu Survey of the environment 1991, 119-121.

Mukerjee, S.K. 1985. Hazards from pesticides on vegetables. Special lecture at the VIIIth workshop of All India Coordinated Vegetable Improvement Project at Sabour (Bihar) on 10th June, 1985.

Pooran Chand, 1992. Personal Communication.

Recent Advances in Indian Entomology
Editor: O.P. Lal
APC Publications Pvt. Ltd., New Delhi, 1996, pp. 140-165

Registration and Regulation of Pesticides in India

T. P. S. TEOTIA

Pesticides registration and regulation in India is governed by the Insecticides Act, 1968. The objective of this Act is to regulate the import, manufacture, sale, transport, distribution and use of insecticides with a view to prevent risk to human beings or animals, and for matters connected therewith. This Act became effective to the whole of India with the enforcement of the Insecticides Rules, 1971 on 30th of October 1971. The Insecticides Rules, 1971 have been subsequently amended through the promulagation of the Insecticides (Amendment) Rules, 1993 published *vide* GSR 533 (E) dated 6th August, 1993.

The implementation of the Insecicides Act is the responsibility of both Central and State Governments. The Centre looks after the policy making and grant of registration certificates for import and manufacture of pesticides, while the State Governments are entrusted with the issue of manufacturing and selling licences of pesticides.

Generally, the responsibility for monitoring of quality of pesticides and their proper usage rests with the State Goverments. In order to achieve these objectives each State has notified such functionaries as Appellate Authority, Licensing Officer, Insecticide Inspectors, Insecticide Analysts etc.

The Central government has constituted a "Central Insecticides Board" to advise the Central Government and State Governments on technical matters arising out of the administration of the Act, and a "Registration Committee" for registration of pesticides, besides other functions assigned to them by the Act and the Rules.

2. Registration of pesticides

2.1 Inclusion of new chemicals in the schedule to the Act :

All the insecticides used in agriculture and public health are required to be included in the Schedule to the Insecticide Act. Once the insecticide is included in the Schedule, only then one can apply for its import and manufacture in the country. On the recommendation of the Central Insecticides Board, the Central Government issues a notification for inclusion of the insecticides in the Schedule. Up-to-date, 514 chemicals have been included in the Schedule to the Act.

For including a new chemical in the Schedule, the applicants are required to submit the information in the following parameters :

1. Common name
2. Chemical name

3. Literature support on :

(i) Chemistry, (ii) Bio-efficacy, (iii) Toxicity, (iv) Registration status in other countries.

2.2 Import of sample quantity of insecticides not included in the schedule to the Act :

With a view to undertaking test and trial with a new pesticide in the country in order to generate data required for its provisional registration, a sample quantity of the product can be imported even during the pendency of its inclusion in the Schedule.

The Plant Protection Advisor to the Government of India is the competent authority for grating permission to the import of sample quantity of a pesticide not included in the Schedule to the Act. For this purpose applications are to be submitted to him on Proforma I (for Technical grade) and Proforma II (for formulation), appended herewith.

The following additional information is also required to be submitted alongwith the application :-

(i) Registration status in other countries.

(ii) Maximum Residue Limits (MRL) on crops/food commodities, if prescribed by the Codex Alimentarius Commission or country where the insecticide has been registered for use.

(iii) Literature support to arrive at a conclusion that the insecticide is not carcinogenic, mutagenic, tetragenic and has no effect on mammalian reproduction system.

(iv) Extracts of conclusions/appraisals by WHO/FAO Expert committee about the toxicity profile of the insecticide.

(v) Crops/pests on which the insecticide has been recommended for use in other countries with the mean dose level.

In case the information in respect of registration status, MRL and chronic status can not be given, the reason for non-availability of such vital information is to be provided.

2.3 Categories of registration of pesticides :

Any person desiring the import and manufacture of any pesticide can apply to the Registration Committee for registration of such pesticides under section 9 of the Act. The registrations are granted under three different sub-sections of Section 9. These are :

2.3.1 Provisional registration under section 9 (3B) : Provisional registration is granted to a new molecule of pesticide which is introduced for the first time in India for a period of two years for data generation under Indian conditions.

2.3.2 Regular registration under section 9(3) : Once the applicant submits the data as per requirement to the satisfaction of Registration Committee, registration is granted on regular basis under section 9 (3).

2.3.3 Repeat registation under section 9 (4) : Once a registration is granted under section 9(3) in favour of an aplicant, any other person can apply for registration under Section 9(4) by putting an application in the prescribed proforma along with the required fee and the registration is granted on the same condition on which the pesticide was orginally registered under section 9 (3).

Under the aforesaid three categories of registration, it is granted for (i) import of insecticides, (ii) manufacture of different formulations of the insecticide and (iii) indigenous manufacture of technical grade material for domestic use.

2.4. Data requirement for registration of pesticides :

The Registration Committee has prescribed data requirements on differen parameters in respect of :

i) Chemistry. ii) Bio-efficacy & Residues.

iii) Toxicology. iv) Packaging & Labelling.

These data requirements are not similar in all cases, but vary with the nature of pesticide to be registered, type of pesticide formulation, purpose of registration etc. Moreover, the Registration Committee has, from time to time, revised the guidelines. Hence while preparing the applications for registration of pesticides, the current guidelines should be followed.

2.4.1 Chemical pesticides : The parameters of data requirement presently prescribed for registration of chemical pesticides under different categories could be seen at Annexure I.

2.4.2 Botanical pesticide (Neem based) : The data requirements for registration of *neem* based botanical pesticides have been simplified by the Registration Committee to encourage entrepreneurs for easy availability of such pesticides for the benefit of farmers and the environment. These data requirements are given in Annexure 2.

2.4.3 Microbial pesticides (Bacillus thuringiensis and B. sphaericus) : The data requirements for registration of *Bacillus thuringiensis* and *Bacillus sphaericus* for pest control and vector control respectively are given in Annexure3.

2.4.4. Household pesticides : Household insecticides are defined as the insecticides which are used for direct domestic application in the form of different devices in order to kill/repel the insects and vectors of diseases. They are broadly classified into four categories depending on their physico-chemical properties, usage, packaging requirements, methods of application, and data requirements from safety angle :-

Category I - The household insecticides which are used as solid, e.g., baits and glue.

Category II - The household insecticides which are used as liquids, e.g., ready-to-use space and spot application, E C formulation and aerosols.

Category III - The household insecticides in technical form which emit vapours.gas/ smoke either as such or as a result of heating, burning or coming in contact with air/ moisture or with teh help of any other appliances/technology, e.g., mats, coils, cake etc.

Category IV- The products which do not come under the alone mentioned categories. The dta requirements for registration of these household insecticides are given in Annexture 4.

2.4.5 Combination products of insecticides : In the case of registration of combination products of insecticides, the toxicity data on the indivisdual components of formulated insecticides used in the combination product as well as on the end formulation (i.e. combination product itself) are required to be submitted.

2.5 Registrtion of pesticides for the purpose of export : The export of pesticides has not been covered in the Insecticides Act, 1968. However, in pursuance of the policy of the Government to encourage exports of various commodities including pesticides, the Registration Committee has created special category of applicants known as export category and has liberalised the data/information requirements for grant of registration to this category of applicants, without sacrificing the basic need to prevent the risks to human beings or animals, environment etc.

The data requirements for registration of pesticides for purpose of export only are given in Annexure 5.

2.6 Present status of registration of pesticides : Presently, 132 pesticides belonging to insecticides, fungicides, herbicides, fumigants, rodenticides etc. are registered on regular basis under section 9(3), and 19 pesticides have been registered on provisional basis under section 9 (3B). The Committee has refused registration for 18 products, banned the use of 12 pesticides, and restricted the use of 13 other pesticides in the country. Further, 17 more pesticides are being reviewed by the Government of India. The related information is listed in Annexures 6 (A) to 6 (F).

3. Grant of licences

Licences are required to be taken from the Licensing Officer appointed by the concerned State Government for carrying out manufacture, sale, storage and commercial pest control operations. Separate licences are required for each commecial pest control operations. Separate licences are required for each manufacturing, selling and storage place. Licences are valid for two years and subject to renewal. The Lincensing Officer can, after giving an opportunity to the licensee to show cause, suspend or revoke his licence on misrepresentation, failure to comply with the conditions of licence and contravention to the provisions of the Act and the Rules. The licence is also deemed to be cancelled when the corresponding Registration Certificate of the person has been cancelled. Provisions for transfer and succession of licences exist.

4. Inspection of manufacturing and selling units and analysis of insecticides

4.1 Appointment of insecticide inspector

The Insecticide Inspector can be appointed both by Central Government and the State Governments. The powers and functions of the Insecticide Inspector have been well defined under the Insecticides Act. Besides many other powers, the Insecticide Inspector can inspect and draw samples from both sale point and manufacturing units. The sample is drawn as per the prescribed procedure and sent to Insecticide Analyst for analysis.

4.2 Analysis of samples and reports thereof

The Central and the State Governments are competent to appoint Insecticide Analysts for any defined area and for a particular type of insecticide or class of insecticides. The qualifications of the Insecticide Analysts have been prescribed under the Act. The Insecticide Analysts analyse the samples received by them and send their reports to the Insecticide Inspectors.

On receipt of analysis report, the Insecticide Inspector has to deliver one copy thereof to the person from whom sample was taken. If that person intends to challenge the analysis report, he can, within 21 days, inform the Insecticide Inspector or the court, as the case may be. The court can direct reanalysis of the sample at Central Insecticides Laboratory, whose report is taken as final. No reanalysis is allowed if the sample ws analysed at Central Insecticides Laboratory itself.

The Central Government has established a Central Insecticides Laboratory at Faridabad and also two Regional Pesticides Testing Laboratories at Chandigarh and Kanpur. Besides, various State Governments have set up 40 Pesticides Testing Laboratories with a capacity to analyse over 45,000 samples per annum.

4.3. Launching of prosecution

The Insecticide Inspector is competent to launch prosecution witht the approval of the State Government. For initiating legal proceedings either an FIR could be lodged with the police or written complaint could be filed before the magistrate.

5. Packaging and labelling

Under the Insecticides Rules framed under the Act, no person can stock or exhibit for sale or distribute any insecticide unless it is packed and labelled in accordance with the provisions of these Rules. Every package containing an insecticide shall be of the type approved by the Registration Committee. It is also obligatory for the manufacturer to provide a label and leaflet approved by the Registration Committee along with each pesticide container.

The label provides particulars about the name of manufacturer, name of insecticide, chemical composition, registration number, batch number, date of manufacture, expiry date, antidote statement, net content, maximum retail price etc. Not less than 1/16 of the total area of the face of the label is occupied by a square, set at an angle of 45^0 (diamond shape), The square is divided into two equal triangles, the upper portion containing the symbol and signal word and the lower portion containing the colour, as specified under the Insecticides Rules, depending upon the classification of insecticides into four categories as given below :-

S.No.	Toxicity Category	Acute Oral LD 50 (mg/kg body wt.) of test animals	Acute Dermal LD 50 (mg/kg body wt.) ot test animals.
1.	2.	3.	4.
1.	Extremely toxic	1-50	1-200
2.	Highly toxic	51-500	201-2000
3.	Moderately toxic	501-5000	2001-20000
4.	Slightly toxic	More than 5000	More than 20000

Colour of identification band on the label	Symbol and signal word	Warning statements
5	6.	7.
Bright red	Skull & Cross bones and the word "POISON" in red.	Keep out of the reach of children. If swallowed or if symptoms of poison-ing occur call physician immediately.
Birght yellow	Word "POISON" in red.	Keep out of the reach of children.
Bright blue	Word "DANGER"	Keep out of the reach of children.
Bright green	Word "CAUTION"	-

The leaflet includes adequate directions concerning the use of the product, particulars regarding chemicals harmful to human beings, animals and wild life, warning and cautionary statements including the symptoms of poisoning, suitable and adequate safety measures and emergency first-aid treatment, cautions regarding storage and safe disposal of used containers etc. A statement showing the antidote for the poison and the irritant nature of the chemical, if any, is also included in the leaflet.

6. Transport and storage of insecticides in transit

Under the Insecticides Rules, packages containing insecticides, offered for transport by rail, are packed in accordance with the conditions specified in the Red Tariff, issued by the Ministry of Railways, and no insecticide can be transported or stored in such a way as to come into direct contact with food stuffs or animal feeds. The package containing insecticides are also required to be stored in separate rooms or premises.

7. Provisions regarding protective clothing, equipment and other facilities for workers engaged in handling of insecticides.

Under the Insecticides Rules it is obligatory that the person engaged in handling of insecticides during manufacture, formulation, transport, distribution or application, are periodically examined medically, adequately protected with appropriate clothing and other devices, and given first-aid treatment in case of poisoing. The manufacturers etc. are also required to keep sufficient quantities of antidotes and first-aid medicines, and to arrange for suitable training of workers in observing safety precautions. The Rules also require the packages to be broken and buried away from habitation and the safe disposal of surplus materials and washings so as to prevent environmental or water pollution.

8. Other measures to strengthen the regulatory requirements.

8.1 Standardisation of pesticides and quality control through certification

At the instance of the Central Insecticides Board, the Indian Standards Institution (now Bureau of Indian Standards) has formulated a large number of Indian Standards covering technical pesticides and their formulations, packaging of pesticides, methods of tests for pesticides, pesticide poisoning, pesticide residues, pesticides application equipment etc. The Certification Marks Scheme operated by B. I. S. makes it possible for Indian manufacturers to obtain ISI Licence and Mark on goods produced by them in conformity with the relevant Indian Standards. This scheme is governed by Indian Standards Institution (Certification Marks) Act 1952 and the Rules and Regulations framed thereunder. B. I. S. is thus making available quality pesticides to farmers through its quality certification scheme.

8.2. Enforcement of the Standards of Weights and Measures (Packaged Commodities) Rules 1977

These Rule have helped in safeguarding the consumer's interest particularly relating to the sale price and quantities of pesticide contents packed and other aspects of packing and labelling.

8.3 Enforecement of Insecticides (Price, Stock, Display and Submission of Reports) Order, 1986 under the Essential Commodities Act 1955 (10 of 1955)

The order directs the manufacturers and dealers to display stock and price list of insecticides and to give a cash or credit Memo to a purchaser of an insecticide and also to comply with certain requirements in regard to packing, and maintenance and submission of records etc.

Conclusion

It would be seen from the above that comprehensive legislative measures exist in India for the registration and regulation of pesticides. What is indeed important is the degree of their implementation both at the Central and State levels. With proper implementation, it is expected that the registration and availability of good implementation, it is expected that the registration and availability of good quality, safe and efficacious products would be ensured to the farmers and other consumers for combating insect and other pests, diseases and weeds in the country.

PROFORMA - I

Application for the import of sample quantity of Pesticides (Technical Grade) which are not included in the SCHEDULE to the Insecticides Act, 1968

1. Name of importer :
2. Name of overseas supplier :
3. Name of chemical :
   i) Common accepted name :
   ii) Chemical name :
4. Toxicity (LD 50 value) :
   i) Oral
   ii) Dermal
   iii) Inhalation
   } Laboratory animals
5. Main active group to which the chemical belogs and the %age of purity of the material
6. Whether being imported for the first time
7. Pest (s) against which the formulated material is proposed to be tested
8. Name of Institution(s) where it is proposed to get the formulations to be tested in the country with full address
9. Quantity asked for and the type of formulation (s) proposed to be made
10. Type of packing

11. Port of landing/entry

12. Whether being imported as free sample and its nominal value

(The importer shall be under an obligation to furnish any additional information called for by the Plant Portection Adviser to the Government of India).

In the event of the issuance of the Import Permit for the requisite quantity of Pesticide (Technical grade) I shall abide by the following conditions :-

1. The actual quantity of the Pesticide/Pesticides imported shall be intimated to the Plant Protection Adviser to the Government of India within 15 days of release of the consignment from the Customs Authorities.

2. The quantities of pesticide formulation(s) given to individual research institutions for evaluation will be intimated to the Plant Protection Adviser to the Govt. of India immediately.

3. The institutions where the evaluation of the pesticide is undertaken shall be supplied with the necessary technical information about the chemical composition, dosage recommendations, and safety precautions to be observed while conducting the evaluation.

4. The produce of the field where the pesticide has been evaluated shall be harvested separately.and shall not be released for human or animal consumption and the same shall be destroyed under the supervision of the responsible person of the Institution/organisation where the product is evaluated.

5. The Plant Protection Adviser to the Govt. of India shall be informed of the results of the trials conducted at various institutions. The said report shall be duly authenticated by the concerned authority of the institution/ organisation who have supervised the evaluation programme.

6. For any loss/damage to human life, animals, crop or the environment the total responsibility shall rest with us and the onus of any compensation towards any damage shall lie with us.

7. We shall abide by the decision of the Plant Protection Adviser to the Government of India to stop the evaluation of the imported pesticide at any stage, if there is any report to indicate that the use of such a pesticide is likely to cause risk to human beings or animals or environment.

8. In case of the second request for permission to import the same pesticide, the interim reports/data generated from the trials conducted with the formulation(s) of first sample of pesticide shall be submitted alongwith the application.

LACE : APPLICANT

ATE :

PROFORMA - II

Application for the import of sample quantity of Pesticides (in formulated form) which are not included in the SCHEDULE to the Insecticides Act, 1968

Note. 1 to 8 is same as given above in Proforma -I

9. i) Concentration of formulated material (in %) and its quantity proposed to be imported

   ii) Dosage per hectare of formulated material

Note. 10 to 12 is same as given above in Proforma -I

(The importer shall be under an obligation to furnish any additional information called for by the Plant Portection Adviser to the Government of India).

In the event of issuance of the Import Permit for the requisite quantity of pesticide in formulated form, applicant will have to abide by the conditions which are mentioned in Proforma-I.

Annexure 1

COMPERATIVE STATEMENT ON DATA REQUIREMENT FOR REGISTRATION OF CHEMICAL PESTICIDES UNDER VARIOUS CATEGORIES FOR HOME CONSUMPTION

Note : Please refer to Pesticide Directory as it is a very lengthy table.

Annexure 2

DATA REQUIREMENT FOR REGISTRATION OF BOTANICAL PESTICIDES (NEEM BASED)

Chemistry

i. Name of the plant (s) part to be used for extraction of active ingredient (s)/components (s).

ii. Method of experiment in detail clearly identifying the chemicals.

iii. Formulated neem extract should contain not less than 1500 ppm of azadirachtin a.i. in 'Kernel' based formulations and 300 ppm in 'neem oil' based formulations. When the insecticidal a.i. is other than azadirachtin, its (their) name (s) quality and quantity to be indicated.

iv. Chemical identity of the ingredient e.g. azadirachtin.

v. Physico - chemical properties.

vi. Specifications of the ingredient e.g. percentage or ppm.

vii. Method of analysis of azadirachtin/other insecticidal a.i. (s) other than azadirachtin.

viii. Analytical test report.

ix. Shelf - life

Bioefficacy

i. Bioeffectiveness.

ii. Phytotoxicity.

iii. Compatibility with other chemicals.

iv. Purpose of manufacture.

v. Direction concerning dosage.

vi. Time of application.

vii. Waiting period.

viii. Application equipment.

ix. Information regarding registration status in other countries, if any.

Toxicity

		Prov. Reg. U/S 9. (3B)	Regular Reg. U/S9 (3)
i.	Acute oral in rat and mice.	Yes	Yes
ii.	Acute dermal.	Yes	Yes
iii.	Primary skin irritation.	Yes	Yes
iv.	Irritation to mucous membrane	Yes	Yes
v.	Neuro - behavioural toxicity	No	Yes
vi.	Reproductive toxicity.	No	Yes
vii.	Carcinogenicity	No	Yes
viii.	Mutagenicity	No	Yes
ix.	Effect on spray operators (health records)	No	Yes

Packaging and labelling

i. Labels and leaflets as per the Insecticides Rules 1971 existing norms.

ii. Type of packaging.

iii. Manner of packaging-container content compatibility.

iv. Specification for primary package.

v. Specification for secondary package.

vi. Specification for transport package.

vii. Manner of labelling.

viii. Instruction for storage and use.

ix. Information regarding disposal of used package.

x. Process of manufacturing indicating material balance, generation of wastes.

Annexure 3

GUIDELINE/DATA REQUIREMENT FOR REGISTRATION OF *BACILLUS THURINGIENSIS* AND *BACILLUS SPHAERICUS*.

A. CHEMISTRY

Technical/Formulation

(i) Physico-chemical specifications and composition.

(ii) Systematic name and strain.

(iii) Common name.

(iv) Natural occurrence of the organism, its relationships to other species and history.

(*Note* : Information about the type of strains to ensure whether it is exotic one or it is different from indigenously available strains).

(v) Manufacturing process.

(vi) An appropriate test procedure and criteria used for identification, such as, morphology, biochemistry and/or serology/immunology.

(*Note* : 1. Morphology description : Particle size heat resistant spore count/ protoxins per mg of dry material (formulation).

2. Protein per mg dry weight of formulation. Protocols are required to be generated/developed for this assay preferably Lowry Method.

3. *Immunology assays* : Elisa test or any other sensitive starndard immunology test and rocket immunio assays are analytical methods. The Elisa test would have to be submitted by the Firm while rocket immunio assay will be treated as referal method for Insecticidal Crystal Protein (s) for Bacillus thuringiensis only.

4. *Routine Test* : Referee method could be the basic tool for surveillance to the occurence of the mutation, if any. Level of beta exotoxins to be identified if expressed. The methodology of determination of the toxin content is by Dot blot Elisa assay method.

vii) *Method of analysis :*

(a) The protein content per mg. - By Lowry's method.

(b) Viable spore counts.

(c) Determination of the toxin content by Dot blot Elisa assay method.

(d) Pladmid pattern if any.

(e) A technique for the separation and purification of the crystals for raising antiserum against pure creystals.

(viii) *Storage stability :*

1. Storage stability data has to be provided. Method of analysis will be bio-assay method. Experimental details and calculations, if any should also be provided.
2. The importer/manufacturer has to provide the sample of technical material for referal and storage purpose which could be sent to Indian Institute of Microbial Technology, Chandigarh under the Department of Bio-technology for storage and future references.

B. Bioefficacy

Technical

Laboratory tests : For Registration of microbial pesticides LC 95 or LD 95 values for each testing insect species under laboratory conditions should be generated. The data is required from two independent laboratories recognised by from two independent laboratories recognised by DST/ICAR/CSIR/ICMR and State Universities etc. for the purpose.

Formulation

(i) *Laboratory test* : For registration of microbial pesticides LC 95 or LD 95 values for each testing insect species under laboratory conditions should be generated. The data is required from two independent laboratories recognised by DST/ICAR/CSIR/ICMR and State Universities etc. for the purpose.

(ii) *Field test* : The data on bio-efficacy based on two seasons/two trials under two different agro-climatic conditions in the form of published/authentic data is required to be published. Information on phytotoxicity with the formulated material is also to be submitted alongwith bioefficacy report.

(iii) *Compatibility* : No data on compatability is required to be submitted. However, data on compatability will be required in case it is recommended to be used in mixture with some other pesticides. The data on interaction is desirable against more commonly used pesticides and fertilizers under field conditions.

(iv) *Time of application* : Information about time of application, equipment and the manner in which insecticides are used, is required to be submitted.

(v) *Data on non-targetted organisms* : Information on toxicity to two species of parasites and predators is required to be submitted.

C. *Toxicity*

(a) *Single exposure studies*

	Technical		Formulation	
	U/S 9(3B)	U/S 9(3)	U/S 9(3B)	U/S 9(3)
i) Oral toxicity/ pathogenicity.	Yes	Yes	Yes	Yes
ii) Dermal toxicity/ pathogenicity.	Yes	Yes	Yes	Yes
iii) Mucous membrane irritation.	Yes	Yes	Yes	Yes
iv) Primary skin irritation.	Yes	Yes	Yes	Yes
v) Inhalation toxicity/ pathogenicity.	Yes	Yes	Yes	Yes
vi) Allergy/sensitization/ immunio suppression.	Yes	Yes	Yes	Yes

Note : Allergy test has to be conducted as per standard protocol prescribed by BIS.

(b) *Repeated exposure studies*

i) Oral toxicity/ Pathogenicity-90 days.	No	Yes	No	Yes
ii) Dermal toxicity/ pathogexicity-20 days.	No	Yes	No	Yes
iii) Inhalation toxicity/ pathogenicity-14 days.	No	Yes	No	Yes

(c) Supplementary Toxicity/ pathogenicity

i. Mutagenicity	No	Yes	No	Yes
ii. Teratogenicity	No	Yes	No	Yes
iii. Carcinogenicity	No	Yes	No	Yes

(d) Eco-toxicity

	Technical		Formulation	
	U/S 9(3B)	U/S 9(3)	U/S 9(3B)	U/S 9(3)
i. Toxicity to birds	No	No	No	Yes
ii. Toxicity to fish	No	No	No	Yes
iii. Toxicity to honey bees.	No	No	No	Yes
iv. Toxicity to silk worm.	No	No	No	Yes

D. Processing, packaging and labelling

Technical/Formulation

(a) Manufacturing process/process of formulation

i. Raw material/material balance

ii. Plant and machinery.

iii. Process-Unit operation/unit process.

iv. Out-put (Finished product & generation of waste)

(b) Packaging :

i. Classification-solid, liquid or other types.

ii. Size-As per Indian Standards Specification.

iii. Compatibility - Glass bottles are not recommended

iv. Packaging specifications/systems of packaging.

(c) Labels and leaflets

As per Insecticides Rules 1971/as per existing norms indicating the common name, composition, antidote/storage statements etc.

(The packaging material should also be ensured freedom from contamination from handling storage and transportation).

Annexure -4

DATA REQUIREMENT FOR REGISTRATION OF HOUSEHOLD PESTICIDES

A. Chemistry

1. The common name on the labels and leaflets to be written as household pesticides containing 'X' --- % w/w and 'Z' ---- % w/w etc.
2. Source of supply of Techniccal material with minimum purity to be used in the formulation.

3(i) Chemical composition including the kind, name and percentage of all the ingredients used in the formulation.

(ii) Complete data on the identity and physico-chemical characters of various ingredients.

(iii) The number of active ingredients in the household combination product should not be more than two normally.

4. Specification.
5. Analytical Test Report.
6. Shelf-life data.
7. Method of analysis.

B. Bioefficacy and residues

1 Bio-efficacy claims to be given on the labels and leaflets in case of formulations of pesticides :

(i) A brief direction concerning the major usages of the product should be given on the labels. Detailed information indicating the names of insects, method of application, dosage, places of treatment, PP equipments to be used etc. should be given on the leaflets.

(ii) Whereever the Registration Committee has approved the product for restricted use, this fact should be indicated very clearly on labels and leaflets in capital letters.

(iii) Instructions regarding insecticide 'Not to be used on any food crop' to be given on labels and leaflets.

2. Bio-efficacy claims to be given on the labels and leaflets in case of technical grade material :

(i) The purpose of import/manufacture of technical grade material is required to be given on the labels and leaflets.

3. Data requirements on bio-efficacy and residues for formulations of pesticides :

(a) For provisional registration u/s 9 (3B) :

(i) Published/cited Indian data on bio-effectiveness in support of the claims. from national laboratories based on minimum repeated trials. supported by any published overseas data.

(ii) Information on secondary pest outbreaks particularly of ticks and mites where residual pyrethroids are being used.

(iii) Data on residues (persistence) of pesticides on different types of surfaces generated under foreign/Indian conditions from 2 laboratories.

(b) For regular registration u/s 9 (3) :

(i) Bio-efficacy test under Indian conditions minimum on two trials in each three national laboratories

(ii) As in 3(a) (ii) above

(iii) As in 3 a (iii) above from 3 national laboratories

4. Data requirements for the registration of new formulations of the approved pesticides :
Data on bio-effectiveness and persistence on different surfaces as required in case of 3 (b) above.

5. Data requirements for the registration of combination products :
Bio-efficacy data on combination products v/s individual products. Data on persistence on different types of surfaces should also be submitted. Data to be generted in the manner as indicated in case of 3 (b) above.

6. Methodology : Flying/crawling insects
Residual films of insecticides prepared by spraying insecticides on different types of surfaces, such as, glass, wood, mud and cemented surfaces. Insects to be exposed for 30 min. and then shifted to recovery chambers for 24 hrs after which the mortality count should be made and the satisfactory mortality would be more than 90%. The residual toxicity should also be studied at different intervals. Evaluation of space spray against flying insects should be conducted in Peet Grady Chamber as per ISI specification No. 1824. Mats/coils could also be evaluated inside the Peet Grady Chamber against caged mosquitoes and the knockdow effect is to be recorded at different intervals. Aerosols are to be evaluated inside a standard room. The tests are to be conducted as per WHO technical reports series No. 206.

C. Packaging and labelling

1. Proposed labels-as per Insecticides Rules 1971/Existing norms of the Registration Committee.
2. Proposed leaflets-as above.
3. Manner of Packaging :

   (i) In gerneral, the requirements for packaging of household pesticides are governed according to IS : 8190 (Part III) - 1979 and amendments thereof.

   (ii) Bottles of glass used as primary packaging material is not permitted for household pesticides.

   (iii) The following details shoud be submitted, if the packaging material does not conform to the specifications :

   Specification of primary packages

   Specification of Secondary packages.

   Specification of transporation packaging. History of conventional usage of packaging.

4. Compatibility :

   If the compatibility of container with pesticides is already established, the relevant IS No. may be quoted. In case compatibility is not established, detailed specifications for primary package, alongwith the analysis report on packaging and quality control as per test methodology approved by Registration Committee, are required to be furnished.

5. Instructions for storage and use including first-aid and precautionary measures proposed for labelling.
6. Information regarding disposal of used package, surplus material and washings of insecticides proposed for labelling.

D. Toxicity

Category I (solids)

i. Acute oral toxicity in rat and mice.
ii. Acute dermal toxicity in rabbit.
iii. Primary skin irritation.
iv. Irritation to mucous membrance.
v. Toxicity to fish.
vi. Toxicity to birds.
vii. Toxicity to beneficial insects.

Category II (Liquids)
Same as Category I plus Acute inhalation toxicity.

Category III (Vapours)
Same as category II plus Health monitoring study of the user by using the houshold pesticide in its actual use.

Annexure -5

GUIDELINES FOR REGISTRATION OF INSECTICIDES FOR EXPORT ONLY

The data/information requirements for registration of insecticides for export only have been liberalised by the Registration Committee with a view to increasing their exports expeditiously. Presently these requirements are as under :-

1. For pesticides already registered for use in the country :

   No additional data/information is required for export registration in respect of those pesticides already registered for use in the country, both technical grade and formulations as also their higher or low er concentrations.

   The barest minimum information required to be given for submitting such cases is given at Appendix-I.

2. For pesticides to be registered for the first time in the country :

   The data/information which are required to be sumitted along with such applications is given at Appendix-II.

Appendix-I

FOR PESTICIDES ALREADY REGISTERED FOR USE IN OUR COUNTRY

Application in Form-I with the following information :

Chemistry

(i) Name of the insecticide.
(ii) Chemical composition.
(iii) Physico-chemical properties of the product
(iv) Specifications.
(v) Method of analysis.
(vi) Source of import (in respect of technical grade to be imported for formulation and export).

Packaging :

(i) 7 copies of the proposed labels/leaflets (as per the existing requirements under the Insecticides Rules, 1971).

(ii) An undertaking to the effect that the insecticide shall be packed as per the requirements of the importing country.

Appendix-II

FOR PESTICIDES TO BE REGISTERED FOR THE FIRST TIME IN OUR COUNTRY

Data/information in repect of the following parameters should accompany the applications for registration of insecticides for *EXPORT* only :

Chemistry

(1) Source of import.

(2) Chemical composition.

(3) Physico-chemical properties.

(4) Specifications.

(5) Method of analysis.

Bioefficacy

Information on metabolism and persistence in soil and water in respect of technical grade insecticide only.

Toxicology

(1) Data on :

(i) Acute oral in rat & mice.

(ii) Acute dermal.

(iii) Acute inhalation

(iv) Primary skin irritation.

(v) Irritation to mucous membrane.

(vi) Sub-acute oral toxicity in rat/dog (Data/information).

(vii) Sub-acute dermal toxicity in rat/rabbit (Data/information).

(2) Information based on international organisations/Institutions on :

(i) Carcinogenicity.

(ii) Mutagenicity.

(iii) Health records of factory workers/spray operators.

(3) An undertaking to the effect that 'all the required information on the toxicity profile of the insecticide which is required to be manufactured shall be provided to the competent authrities at the State level implementing the Factories Act, 1948', be submitted by the applicant.

Packaging

An undertaking to the effect that the manner of packaging shall be as per the requirements of importing country, be submitted by the applicant.

Annexure -6 (A)

LIST OF PESTICIDES PROVISIONALLY REGISTERED FOR IMPORT/MANUFACTURE UNDER SECTION 9 (3B) FOR DATA GENERATION/LARGE SCALE TRIALS (AS ON 01.01.1994)

S. No. Name of Insecticides

1. Validamycin 2. Oxadiazinon 3. Lambdacyhalothrin 4. d-phenothrin
5. Tetramethrin 6. Trifluralin 7. MIPC 8. Fenpropathrin
9. Neem Triterpene emulsion containing Azadirachtin 0.15% w/w (1500 ppm) (Margocide CK 20 % EC).
10. Neem oil emulsion containing Azadirachtin 0.03% (300 EC). (Margocide OK 80 % EC).
11 Neem oil based WSP emulsion containing Azadirachtin 0.03% (300 ppm) (Godrej Achook).
12. Neem oil emulsion containing 300 ppm Azadirachtin (Nimbecidine)
13. *Bacillus thuringiensis* var. *Kurstaki* (Seroture 3a, 3b, SA-II WG (Deefin)
14. RD-9 Repellin Neem Oil 93 EC
15. *Bacillus sphaericus* 16. Iprodione 17. Isoprothiolane
18. Cyfluthrin 19. Methoprene.

Annexure-6 (B)

LIST OF PESTICIDES NOT APPROVED FOR USE (AS ON 01.01.1994)

S. No.	Name of pesticide	
1. Calcium arsenate	2. EPN	3. Azinphos-Methyl
4. Lead Arsenate	5. Mevinphos (Phosdrin)	6. 2, 4, 5-T
7. Carbophenothion	8. Vamidothion	9. Mephosfolan
10. Azinphos Ethyl	11. Binapacryl	12. Dicrotophos
13. Thiodemeton/Disulfoton	14. Fentin Acetate	15. Fentin hydroxide
16. Chinomethionate (Morestan)	17. Ammonium sulphamate	18. Leptophos (Phosuel)

Annexure-6 (C)

LIST OF INSECTICIDES REGISTERED ON REGULAR BASIS UNDER SECTION 9(3) OF THE INSECTICIDES ACT, 1968 (AS ON 01.01.1994)

S. No.	Name of Insecticide	S. No.	Name of Insecticide
1.	Acephate	41.	Dichloropropen and Dichloropropane mixture (DD Mixture)
2.	Alachlor		
3.	Aldicarb		
4.	Allethrin		
5.	Alpha naphthyl Acetic Acid	42.	Dicofol
6.	Aluminium phosphide#	43.	Diflubenzuron.
7.	Anilofos	44.	Dimethoate
8.	Atrazine	45.	Dinocap
9.	Alphacypermethrin	46.	Dithianon
10.	Aureofungin	47.	Diuron
11.	Barium Carbonate	48.	Dodine
12.	Benzene hexacholoride(BHC)#	49.	Diazinon
13.	Benthiocarb (thiobencarb)	50.	Dieldrin#
14.	Benomyl	51.	2, 4-Dichlorophenoxyacetic Acid (2, 4-D Sodium, Amine ester Salts).
15.	Bitertanol		
16.	Bromadiolone		
17.	Butachlor		
18.	Calcium cyanide	52.	Ethylene dibromide and carbon tetrachloride mixture (EDCT Mixture 3 :1).
19.	Captafol#		
20.	Captan		
21.	Cartap Hydrochloride		
22.	Carbaryl	53.	Edifenphos
23.	Carbendazim	54.	Endosulfan
24.	Carbofuran	55.	Ethofenprox
25.	Carboxin	56.	Ethephon
26.	Coumatetralyl	57.	Ethylene Dibromide (EDB)#
27.	Chlorfenvinphos	58.	Ethyl Mercury chloride (EMC)
28.	Chlormequat Chloride (CCS)	59.	Ethion
29.	Chlorbenzilate#	60.	Fenitrothion
30.	Chlorpyriphos	61.	Fenarimol
31.	Copper Oxychloride	62.	Fenthion
32.	Coumachlor	63.	Fenvalerate
33.	Cuprous Oxide	64.	Ferbam
34.	Cypermethrin	65.	Fluchloralin
35.	Chlorothalonil	66.	Formothion
36.	Copper Sulphate	67.	Fluvalinate
37.	Dalapon	68.	Fosetyl - Al
38.	Decamethrin (Deltamethrin)	69.	Glyphosate
39.	Dichlorodiphenyl Trichloroethane (DDT)#	70.	Gibberellic acid
		71.	Isoproturon
40.	Dichlorvos (DDVP)	72.	Kitazin

S. No.	Name of Insecticide	S. No.	Name of Insecticide
73.	Lindane (Gamma BHC)#	101.	Phenthoate
74.	Lime sulphur	102.	Phorate
75.	Malathion	103.	Phosalone
76.	Maleic hydrazide	104.	Phosphamidon
77.	Mancozeb.	105.	Pirimiphos methyl
78.	Methyl Chlorophenoxy acetic acid (MCPA)	106.	Phenyl mercury acetate (PMA)#*
79.	Metribuzin	107.	Prpanil
80.	Manazon	108.	Propetamphos
81.	Methodxyethyl mercury chloride (MEMC)	109.	Propoxur
82.	Metaldehyde	110.	Pyrethrum
83.	Methabenzthiazuron	111.	Quinalphos
84.	Methyl bromide#	112.	Sevidol (4 : 4 Carbaryl and Gamma BHC)
85.	Methyl parathion#	113.	Simazine
86.	Methyl bromide + Ethylene dibromide (1 : 1 & 3 : 1)	114.	Sirmate
87.	Metoxuron	115.	Sodium cyanide#
88.	Monocrotophos	116.	Streptocycline (Streptomycin sulphate + tetracycline hydrochloride, 9 : 1).
89.	Methomyl	117.	Sulphur
90.	Metalaxyl	118.	Temephos
91.	Mono Sodium methane arsonate (MSMA)	119.	Triadimefon
92.	Nickle choloride	120.	Tricyclazole
93.	Nocotine sulphate#*	121.	Triallate
94.	Oxydemeton methyl	122.	Trichloro Acetic Acid (TCA)
95.	Oxyfluorfen	123.	Trichlorphon
96.	Paris green (Copper acetoarsenite)	124.	Tridemorph
97.	Paraquat dichloride	125.	Thiometon
98.	Paradichlorobenzene	126.	Thiram
99.	Pendimethalin	127.	Thiophanate Methyl
100.	Permethrin	128.	Triazophos
		129.	Warfarin
		130.	Zinc phosphide.
		131.	Zineb
		132.	Ziram

* For Export only

\# Restricted in use.

Annexure-6 (D)

LIST OF BANNED PESTICIDES (AS ON 01.01.1994.)

S. No.	Name of pesticide	
1. Dibromochloropropane (DBCP)	2. Endrin	3. Pentachloronitrobenzene (PCNB)
4. Pentachlorophenol (PCP)	5. Toxaphen	6. Ethyl Parathion
7. Chlordane	8. Heptacholor	9. Aldrin
10. Paraquat dimethyl Sulphate	11. Nitrofen	12. Tetradifon.

Annexure-6 (E)

LIST OF PESTICIDES RESTRICTED FOR USE IN INDIA (AS ON 01.01.1994)

S. No.	Name of Pesticide	Type of Restriction (S) Imposed
1.	Aluminium Phosphide	It is to be sold only to Government, Government Undertakings/ Organisations and to be used under the strict supervision of Government Expert or Pest Control Operators whose expertise is approved by the Plant Protection Adviser to the Government of India.
2.	B. H. C.	The use of BHC on vegetables, fruits, oilseed crops and preservation of foodgrains is banned.
3.	Chlorbenzilate	Use in agriculture is banned. If required, it can be imported by Govt./Semi-Govt. Organisations and prepared folbex strips for making it available to bee keepers for controlling mites of honeybees.
4.	Captafol	Shall be used only as seed dresser. Its use as foliar spray is banned.
5.	D. D. T.	Use of DDT in agriculture is banned. In very special circustances warranting the use of Central Government may purchase it directly from M/s Hindustan Insecticide Ltd. to be used under Expert Government supervision.

		Use of DDT for the public health programme upto 10,000 MTs per annum, except in case of any major outbreak, is restricted.
6.	Dieldrin	Use of Dieldrin shall be restricted for locust control in desert areas by Plant Protection Adviser to the Government, of India
7.	Ethylene dibromide	Use of EDB shall be restricted as a Fumigant for Foodgrains through Central Government, State Governments, Government Undertakings, Government Orgainsations, such as Central Warehousing Corporations, Food Corporation of India and Pest Control Operators whose expertise is approved by Plant Protection Adviser to the Govt. of India.
8.	Methyl Bromide	Methyl Bromide be used as a fumigant in the manner in which Aluminium Phosphide is used.
9.	Sodium Cyanide	Use of Sodium Cyanide shall be restricted for fumigation of cotton bales by Plant Protection Adviser to the Govt. of India under Expert supervision.
10.	Phenyl Mercury Acetate (PMA)	Use of PMA is prohibited in India w. e. f. 1.1.1993 but it may continue to be produced in the country only for EXPORT purpose only and shall be subject to the modification in the Certificate of Regn. to this effect.
11.	Lindane	Use of Lindane formulation genertaing smoke for indoor use is prohibited in India, provided that Lindane may be continued to be produced in the country for use for the control of insect pests of field crops and subject to the modification in in the Certificate of registrtion to this effect.
12.	Methyl Parathion	Use of Methyl Parathion is permitted only on those crops where honey been are not acting as a pollinators and Methyl Parathion may be continued to be produced in the country for use in other crops and subject to the modification in the Certificate of regn. to this effect.

13.	Nicotine Sulphate	Use of Nicotine Sulphate is prohibited in India provided that may be continued to be produced in the country only for Export purpose and subject to the modification in the Certificate of Registration to this effect.

Annexure-6 (F)

LIST OF PESTICIDES WHICH ARE UNDER REVIEW BY GOVERNMENT OF INDIA (AS ON 01.01.1993.)

1. Alachlor
2. Benomyl
3. Copper acetoarsenite
4. Diuron
5. Ethyl Mercury Chloride
6. Fenarimol
7. Menazon
8. Methomyl
9. Oxyflourfen
10. Sodium Methane Arsenate
11. Calcium Cyanide
12. Phosphaidon
13. Thiometon
14. Triazophos
15. Tridemorph
16. Ziram
17. Monocrotophos.

Recent Advances in Indian Entomology
Editor: O.P. Lal
APC Publications Pvt. Ltd., New Delhi, 1996, pp. 166-173

Role of Economic Thresholds in IPM

V. K. SHARMA

The study of economic threshold probably the most often discussed issue in economic entomology and insect pest management today. The fundamental questions addressed are how many insects cause how much damage, and is the damage significant. A majority of the entomologists agree that a common decision rule or threshold should answer such questions and that such decision rules are the back bone of progressive pest control. The widespread and sometimes indiscriminate use of pesticides have disrupted the ecological balance. After introducing synthetic insecticides in to market, the concept of an economic threshold as the major criteron for persticide use had been essentially ignored. The literature reveals that economic significance of injury caused by a large number of pests, and the economic threshold of pest species is scanty. This has unquestionably led to frequent erroneous judgements and unnecessary control measures. In recent years the treatment level for a number of pests has taken a more definite meaning. Stern *et al* (1959) comprised the relationship of pest number to damage as an 'economic threshold' and defined as the density at which control measures should be determined to prevent an increasing pest population from reaching the economic injury level. The latter term was defined as 'the lowest population density that will cause economic damage'. Chant (1966) thought 'action threshold' as best decribed for this concept.

Shortwell (1935) was probably the first to establish a relationship between the density of a pest and potential crop damage. This occured during 1930's during grasshopper outbreak in the U. S. He designed a rating system of 5 catogories based on density i.e. natural, light, moderate, heavy and very heavy. However, shortwell (1935) mentioned that the rating system would change from one region to another depending on climatic conditions and whether the crops were irrigated or rainfed. Thus the rating system which has been very valuable and is still being used in grasshopper control does not represent damage to be expected under all climatic conditions. After this work and for a good number of years, there were little efforts for determining economic thresholds of pest species.

Today what is needed is more sophisticated knowledge? of pest control related to pest density and crop yield. This will help optimize the pesticides used and simultaneously deal with the injudicious use of insecticides resulting from our lack of knowledge of economic threshold of pest species.

The entire concept of pest control becomes even more complicated when economic variables are considered. Neverthless, in practical agriculture, a simple cost/benefit ratio is still the very best 'first approximation' of potential crop yield.

This is more easily understood by the growers. Obviously, the economic threshold concept is no so simple as orginally proposed by Stern *et al.* (1959). This concept is in a state of evaluation, and pest density treatment levels will surely become involved with the decision theory.

The fundamental principle in determining an economic threshold of pest species is to distinguish between mere presence in crop as opposed to a higher density that will cause reasonable loss in the quantity and/or quality of the harvest crop.

Ideally, the decision to apply a pesticide should be determined by the cost/ potential benefit ratio (i.e. cost of control meausre, balanced against the increased value of the crop that can be recovered (protected) which can be affected by market conditions, local grower, economics and investiment in the crop but not by the personal values of the people concerned.

Geier (1966) and smith and Van den Bosch (1961) have analysed many of the biological factors involved in pest control and generally concluded that economic control need not rely on killing the organisms but may be achieved in otherways to prevent the economic damage.

Economic damage and damage boundary

Economic damage was orginally defined as the 'amount of injury which will justify the cost of artificial control measure'. To understand this, there is a need to understand the difference between injury and damage. Injury is the effect of pest activities on host physiology that is usually deleterious. Damage is a measurable loss of host utility, most often including yield quantity, quality and aesthetics. As the concept is applied in pest management, economic damage begins to occur when money required for suppressing insect injury is equal to the potential monetary loss from a pest population. The term gain threshold has been used to express this beginning point of economic damage.

$$\text{Gain threshold} = \frac{\text{Management costs (Rs/acre)}}{\text{Market value (Rs./Bushels)}} = \text{Bushels/acre}$$

Another important damage level to consider is the damage boundary, defined as the lowest level of injury where damage can be measured. Specifically no injury level below the damage boundary merits control and injury estimated to result in economic damage does warrant action.

Economic injury level (EIL)

It is defined as the lowest number of insects that will cause economic damage or the minimum no. of insects that would reduce yield equal to the grain threshold. Although expressed as no. of insects per unit area as its name implies, the EIL is really a level of injury because injury is usually difficult to measure in a field situation. However, number of insects are used as index of that injury. In some instances particularly when several pest species causing similar injury are present, insect equivalents may be considered instead of insect numbers. An insect equivalent is the amount of injury that could be produced by one Pest through its complete life cycle. By understanding the feeding activity of various species and comparing these according to a Standard EILs can be developed and decisions made for managing the whole complex.

If the management action can be taken quickly and loss can be averted completely then the EIL can be expressed as follows :

$$V \times I \times P \times D = C$$

where : V = market value per unit of produce (Rs/acre)

I = injury units per insect per production unit (% defoliation/insect/acre)

P = Density or intensity of pest population (insects/acre).

D = Damage per unit injury (bushels lost/acre/% defoliation)

C = cost of management per area (Rs/acre)

then
$$P = \frac{C}{V \times I \times D}$$

and EIL = P

In some cases where some loss from the insect is unavoidable, for example, if injury can be reduced by 80 per cent then the relationship becomes :

$$P = \frac{C}{V \times I \times D \times K}$$

where K = Propotionable reduction of injury

In case of some insects (*viz.* piercing and sucking insects) it is a problem to differentiate between I and D variables. (For example I variable represent for plants photosynthate removed per insect and D variable would represent yield loss per unit of photosynthate removed). Since these variable are difficult to measure a coefficient 'b' representing the loss per insect is substituted. These 'b' coefficants are obtained from statistical regression analysis of data by using experimental populations and measuring the yield loss. The 'b' coefficient is obtained by following expression :

$$Y = a + bx$$

where y = Yield/area

a = constant

b = yield loss/insect

x = number of insects/area

and in such cases

$$EIL = \frac{C}{V \times b} \text{ or}$$

$$EIL = \frac{C}{V \times b \times K}$$

Economic threshold (ET)

The economic threshold (ET) is probably the best known term and most widely used index in making pest management decisions. The ET indicates the no. of insects (the density or intensity) when management action should be taken for this reason. It is some times could the 'action threshold'. Although expressed in insect numbers the ET is really a time parameter with pest numbers being used as an index for when to implement the management.

Once the relationship between ET and EIL is taken into consideration, normally the action (control) is taken when a population level exceeded the ET and the population is forced down before it could reach the EIL. 'No action' was taken at levels below ET.

Dynamics of EIL

Economic levels are very dynamic. They change with the changes in costs, values and production environments. An insect pest feeding on a crop at a given time can be expected to have a different EIL when feeding on the crop at another time in the same season or in other season. Variations in the levels may be significant at times or vary several folds at other times.

The major forces behind the change in economic decision levels as shown in the previous equation are (a) crop value (b) management costs (c) degree of injury per insect (4) crop susceptibility to injury (d).

Although the relationship among these factors are relatively straight forward, complexity is evident when attempting to predict the variability in the factors themselves. Primary factors are affected by complex secondary variables like the host damage, and insect-injury relationship. Moreover, these secondary variables are influenced by variables like weather, soil factors, biotic factors and the human social environment.

Crop value

Of the primary factors, crop value is one of the most variables and ii alone accounts for much change in the EIL. Crop market values are norovious perpetion and unpredictable. The relationship between EIL and market value is inverse. As market value increases EIL deseases and vice versea.

The quality of the commodity may be of much importance in determining market values in many instance. In situations where several specific grades exist, the value of the desired grade should be used in EIL. When quality is associated with appearance, the level of control should reflect the desired appearance. However, the variability in market values cann't be controlled by producers, hence, a marketing plan can be used to assist the estimation of EIL.

Factors that determine specific marketing plan are :

(a) Personal feelings about future trends

(b) Financial conditions unique to each situation

(c) Seasonal pattern in market prices

(d) Price outlook over several seasons.

Management costs

The cost of suppressing the pest populations must be estimated before profitability of an action can be assessed. As the management cost increases, net benefit of control decreases. Consequently EIL's must be caused to accomodate higher gain threshold. However, most of the uncertainity with this variable in EIL is caused by differences in cost between management options rather than between different seasons of crop grown. For this reason, management costs for calculting EIL's are probably the simplest of the primary factors to estimate.

Degree of injury per insect

The process of injury is a dual sided phenomenon governed both by insect and host populations. The insect aspect concerns a particular act or behaviors of the individuals that has a role causes impairment of a host abilities to survive, grow and reproduce. The host as the recipient of the behaviour plays a major role in determining the kind and degreee of injury.

For the purpose of calculation, the injury per insect usually has been assumed to have a linear relationship with insect density. However, crowding at high densities has been shown to reduce injury per insect in some insect species because of interfence between individuals and/or cummulative reduction in available food (Hill *et al.*, 1943)

Such observations show a density/injury relationship shown in the diagram relating relationship between EIL components and their variables. Although this phenomenon is of concern atleast for some species, the density/injury relationship is linear (i.e. crowains doesn't occur) at denisities upto and including more at the estimated EIL.

If the density/injury relationship is not known, evidence suggests that the relationship be considered linear until it is proven otherwise. Such an approach yields conservative (less listing) estimates in establishing operational EILs.

Crop susceptibility to injury

The relationship between injury and crop yield or utility is the most fundamental facts of the EIL. There are 5 major factors involved in the injury/plant response relationship. They are :

i. *Time of injury* : The time is a plants growth cycle when injury has a obvious influence on response to that injury.

ii. *Plant part injured* : Usually a destinction is made between injury to yield giving organs (direct injury) and injury to non yield promising organs (indirect indury).

iii. Type of injury : The common types of injury are already known and here it is important to note that the nature of injury is fundamental to know that how a plant responses to different intensities of a particular form.

iv. Intensity of injury : The relationship between the intensity or amount of injury in plant yield is the most important factor in injury of crop response relationship. To explain, it is of value to consider a generalised response curve of insect injury and plant yield.

v. Environmental effects : Factors in the enviroment often play an important role in determining plants response to injury within a given season. Enviromental factors may influence how long a plant remains susceptible to a specific type of injury. Similarly, between seasons, a plant's response to the same level of injury may be drastically changed. For example, as much as two folio difference in soybean yield loss per green cloverworm, *Platmypena scabra*, larva has been found between growing seasons, with different weather conditions. Such extreme variability emphasizes the need to calculate EILs over a large environmental conditions.

Calculation of economic decision levels

In developing economic indices to pest management decisions, the principal level to estimate is the EIL because it includes basic damage potential of the given population. Basically the steps required to calculate EIL are :

i. Estimate the loss per insect

ii. Ditermine the gain threshold as described previously.

$$\text{Then : EIL} = \frac{\text{Gain threshold}}{\text{Loss per insect}}$$

However, the first step is by far the most difficult. Estimate of losses are usually obtained from field observations and experimentation with various sized insect populations on a crop at specific times. Subsequently yields are measured and losses seen caused by the insects.

As far as using of economic levels in insect pest management programmes is concerned, the economic threshold is the practical operational level recommended to producers for making management decisions.

Implementation categories of economic threshold

The place and ultimate use of ETL in pest management programmes becomes clearer when state of development of insect is categorised. Accordingly they are placed in 4 categories mainly :

i. No thresholds

ii. Nominal thresholds

iii. Simple thresholds

iv. Comprehensive thresholds

No thresholds

It is a simple, identify and spray type of coventional practice, although it is having a limited use certain problems relating to following things are bound to arise.

- Pest sampling cann't be done economically
- Practical response to a problem cann't be implemented in a timely way.
- once detected the problem cann't be killed
- ET is immiserably low
- Poplulations are intense

Nominal threholds

This category respresents decision roles that are declared on the basis of managers experience. Although sometimes critisized for not beig based on vigorous research, nominal thresholds are in advance over using none at all because such decision roles are conservative and their use has resulted in reduced insecticidal application.

Simple thresholds

Here the levels are calculated on the basis of an average response of host to injury caused by an insect. The four major inputs of market value, management cost, tissue destroyed per insect and yield reduction are used to make the calculations.

Comprehensive thresholds

This category awaits a future development at the role of establishing rarely comprehensive thresholds for crops in an intimate understanding of the host plant and its reaction to combine stressess both biotic and abiotic.

Limitations of the EIL concept

Although the economic levels have been used on a practical basis for more than 25 years. They are not without drawbacks.

i. Fixing of decision levels for such pests namely insect vectors, medical pests and pathogens based on market values is practically not possible because the value of the human life cann't be quantified.

ii. In case of aesthetic pests, the quantification of injury is also practically impossible because the aesthetic sense of consumers cann't be measured.

iii. In case of forest pests, the calculations of a managment cost is a difficult thing by keeping in view the longest life spans of the forest trees. The predictions recorded may not be that much accurate as they are in case of crop plants.

iv. In case of non entomological pest the theory of no. of microorganisms will not work because of the fact the control measures for these are generally preventive rather than curative and virtually there is no use of developing EIL model based on the no. of microorganisms present on the plant.

Among the several limitations, the last one is considered as the greatest limitation in development of EIL concept which was originally developed with an objective of reducing insecticide use. Except from an assessment stand point, EILs are of limited use with preventive measures such as host plant resistance and most forms of ecological management.

Finally, the significant amount of research required and relative unsuitability when several diverse pests (those causing different kinds of injury) occur may present serious distreses in attempting to employ the EIL concept. But the injuries from different pests which produce the same host response and can be placed on a common basis, or the effects of different injuries are additive, the concept may find application for pest complexes.

The EIL concept has been one of the most important innovations in the development of insect pest management prograamme. Although not without drawback, EIL decision rules are the most widely accepted and practical tools for that purpose. However, these decision rules are yet to be established to many species and used for others. The EIL concept is possible in the form of comprehensive thresholds, which account for multiple pests as well as over environmental conditions.

References

Bordner, R. and Fletcher, K. E. 1974. Insect infections and their effects on the growth and yield of field crops : A review. *Bull, entomol. Res.* **64** : 141-160.

Chant, D. A., 1966. Integrated control systems. Scientific aspects of pest control Nat. Acad. *Sci., Nat. Res. Counc: Publ.* **1402** : 194-218.

Cieirer,. P. W. 1966. Management of insect pests. *Ann. Rev. entomol,* **11** : 471-490.

Hill. C. C., Udine, E. S. and Pinckaey, J. S. 1943. A method for estimating reduction on yield of wheat caused by Hessian fly infestations U. S. Dep. Agric. 663. 9 pp.

Recent Advances in Indian Entomology
Editor: O.P. Lal
APC Publications Pvt. Ltd., New Delhi, 1996, pp. 174-184

Post Harvest Grain Technology in India

G. K. GIRISH

A series of efforts have been made to initiate the work on post-harvest grain technology in India. The Asiatic Society of Bengal was the oldest which was established in 1785. It encouraged insect collection and their exhibition and its publication. Indian Museum was established at Calcutta in 1875. Bombay Natural History Society was established in 1883. Several studies were published on Biology, Bionomics etc. COATS (1888) published several papers on economic entomology from Calcutta. He gave emphasis on the biology of stored grain insect pests and their behaviours towards air-tight storage structures.

The first professional was appointed in the year 1905, known as Imperial Entomologist. Lefroy (1906) for the first time gave a tentative list of storage insects which was further supplemented by Fletcher in 1914. The problems of grain storage were discussed at the entomological meetings held at PUSA, Bihar, during the years 1917,1919,1923.

Fletcher and Ghosh (1919) gave some account of the occurrence of stored grain pest, the nature of attack and the amount of damage for different commodities and storage receptacles besides some chemical treatment. Except for some casual observations on the incidence of loss, research work on storage seems to have been in lull until 1938 when some research papers on entomology, taxonomy, biology, bionomics, incidence of storage pests, losses caused by them and various methods of control, were published by various workers.

Department of Food was established in 1942 by Government of India. The executive wing of the Food Department was responsible for the purchase, movement, transport and distribution of foodgrains.

The first step towards establishing an Institute in India was in the year 1950 when Central Food Technological Research Institute was established at Mysore with a mandate to train personnel and evolve technology relating to food processing and preservation.

The Central Warehousing Corporation was set up on 2nd March 1957 under the Agricultural Produce (Warehousing & Development) Act 1956. On 1st January 1965 Food Corporation of India was established.

In 1958, the Department of Food started a centre at Hapur known as Grain Storage Research & Training Centre with the main objective to train the personnel of State Government engaged in food procurement/storage and distribution. This centre was upgraded during 1968-1973 with the technical and financial assistance on UNDP and GSRTC was redesignated as Indian Grain Storage Institute. Two field stations at Ludhiana and Bapatla (later shifted to Hyderabad in March 1979) were established in 1968-69. Subsequently, three stations of IGSI at Jabalpur (October 1980), Udaipur (February 1981) and Jorhat May (1981) were also established. The objectives were:

The overall objective was to minimise the losses of foodgrains by development of modernised grain storage practices for the farmers, traders, coorperatives and largescale storage of foodgrains in the country and also evolve the methodology to be adopted by the Save Grain Campaign which has its headquarters in the Department of Food.

Food Corporation of India

History

The executive wing of the Food Department headed by the Director General of Food was responsible for the purchase, storage, movement, transport and distribution of foodgrains ever since the inception of the Department of Food in 1942. It was felt that the working of Government Department involved procedural delays and that operational decisions could not be taken with the speed and elasticity essential for marketing operations. A committee appointed by the Government recommended the setting up of a machinery similar to a business Corporation with its essential freedom and elasticity of operation. The suggestions were implemented in 1964 with the enactment of the Food Corporation Act (No. 37 of 1964). FCI, thus, came into being as an autonomous body under the administrative control of the Ministry. It started functioning from 1st January 1965 for fulfilling following Food Policies of the Department of Food.

(i) Effective price support operations for safeguarding farmers interest;

(ii) Distribution of foodgrains throughout the country at reasonable prices, and

(iii) To manage satisfactory level of operational and buffer stocks of foodgrains to ensure national Food Security.

Objectives of FCI

(i) To ensure that the primary producer obtained the minimum price that may be announced from time to time and to protect the consumer from the vagaries of speculative trade.

(ii) To undertake state trading in foodgrains in a commercial manner within the framework of government policy so that these operations would be in the interest of increased agricultrual production as well as in the interest of the common consumer.

(iii) To engage primarily in the purchase, storage, movement, distribution and sale foodgrains and other foodstuffs.

(iv) To ensure for itself a stratategic and commanding position in the foodgrain trade of the country and to build adequate buffer stocks.

Functions

The functions and activities which have been undertaken by the FCI for fulfilling the objectives are as follows

(a) Purchasing, storing, moving (transporting), distributing and selling goodgrains and other foodstuffs.

(b) Building and operating national buffer stock of foodgrains.

(c) Setting up or assisting in setting up modern rice mills, and other undertakings for processing foodgrains and other foodstruffs;

(d) Acting as a Price Support Organisation for foodgrains mainly-wheat/paddy, and price to the farmers;

(e) Importing foodgrains as per the policy of the Government of India from time to time, and

(f) Promoting, by such means as it thinks fit, the production of foodgrains and other foodstuffs.

The FCI in the course of time has also undertaken the following operations.

(i) Lifting of levy sugar from mills, moving, storing and distributing it in wholesale.

(ii) Handling of imported fertilizers at the ports (as 'Principals' on ownership basis), moving, storing and distributing it in the country.

Organizational Structure

It has a Board of Directors, comprising of a Chairman, a Managing Director, and six other Directors besides a Secretary. FCI network comprises of four zonal offices, 19 regional offices, 4 offices of port operations, more than 151 district offices/sub-offices and thousands storage and procurement centres.

Interests of both the farmers and of the consumers are taken care of while managing the National Food Security system.

Consumers' Interests

Subsidy is extended by the Government of India for the consumers particularly those belonging to the weaker sections. Uninterrupted supplies of foodgrains at subsidised prices is provided round the year. FCI is also ready to rush supplies within 24 hours in any part of the country affected by natural calamity caused by flood, drought or cyclone.

FCI's vast network of food storage depots is spreadover throughout the country for feeding the Public Distribution system to maintain the supply line on war footing.

Producers' Interests

Farmers are offered both market and price security. For increasinf foodgrains production farmers interests are well looked after. Over 99% market arrivals of wheat and 80% rice production is purchased by the procuring agencies in Punjab for instance. With the presence of FCI and other agencies no distress sale occur during the marketing season. FCI also reaches to the door steps of farmers for buying the grains through specially operated centres in rural areas. Minimum support prices are assured to the farmers resulting in their prosperity.

National food security

The essential requisites for building up of an effective national food security are adequate operational and buffer stocks, optimum dispersal of stocks in different parts of the country and an efficient delivery system including fair-price shops and attractive prices. Buffer stocks of foodgrains are maintained to tide over many a unforseen crisis on the food front. The food management in the drought areas bears ample testimony to the importance of buffer stocks.

Problems and issues

(a) *Buffer stocks of foodgrains:* By far the most important challenge is to build a large reserve of buffer stocks for ensuring the food security in the country. This would ensure not only the easy availability of foodgrains to the vulnerable sections of the society but also guard against speculative activities of the private trade and wide fluctuations in the open market prices of foodgrains. The Department of Food constantly reviews the question of the size of the buffer stock. According to the existing buffer stocking policy of the Government, F.C.I. is required to maintain a buffer stock of 10 million tonnes (5 million tonnes each of wheat and rice) besides operational stocks. On 1st April the size of the buffer stock and operational stocks should be to the tune of 16.5 million tonnes. Similarly, on 1st July, it should be to the tune of 21.4 million tonnes. On 1st October it should be 17.8 million tonnes and on 1st January it should be 20.1 million tonnes, each year. For such a vast country, to estimate the accurate size of the buffer and operational stocks in the different months is a cumbersome process.

(b) Procurement effort: The FCI frequently gears up its machinery not only for maximising the procurement but also for adopting improved and better methods of procurement so that the maximum results are achieved. This means establishment of a number of procurement centres, development of market yards/mandis, provision of adequate man and material machinery, so that the grain can be procured under a set of prescribed specifications. Provision of facilities for weighing, marking, bagging, transportation etc. are pre-requisities and are very complex one. Failure of one or the other will hamper the entire chain of procurement.

(c) *Larger and quicker movements:* This is very essential all over the country for catering to the PDS and for maintaining the buffer stocks at the various places. The maximum procurement of wheat and rice is being carried out mainly from Punjab, Haryana, and western U.P. while it is to be moved to the farthest end of the country for running the PDS efficiency and effectively. For this, jumbo rakes, special type of wagons, BCX and CRT have been developed by the railways. It would be necessary to adopt further improvements in this regard with particular reference to movement of grains in bulk.

(d) Scientific storage: For holding the buffer stock, scientific storage is a prequsite. The essential aspects are the location of the stocks, logistics of movement, optimum distribution of buffer stocks, development of integrated plans for locating the godowns in alignment with the rail/road movement capabilities etc.

Periodic inspection of the stocks, categorisation, liquidation of the stocks of low category and damaged stocks on priority basis and maintenance with regular curative and prophylactic treatment besides other hygenic factors are some of the essential components of the scientific storage of foodgrains.

(e) Improvement of preservation technology: Technology of preservation and storage, specifications/standards of quality are constantly reviewed for the upkeep of the health of the stored grain.

(f) Optimum utilisation of bulk storage structures: The requirement of storage capacity for bag, bulk and CAP storage is constantly reviewed so as to effect the optimum utilisation of the available capacity.

(g) Organizational and administrative challenges: On the organisational and administrative side, several emerging problems would have to be met in the future. These problems fall in the following areas

(i) Streamlining of methods and system in the various operations for ensuring better efficiency and effectiveness;

(ii) Adoption of better management practices for improving the management of the operations;

(iii) Intensification of the measures already intiated for the reduction in wastages, deterioration and losses both in transit and storage;

(iv) Improvements in the management information systems, and

(v) Improvements in the monitoring procedures for quicker decision making.

Recent achievements

(a) Installation of weigh bridges

Recently 93 new lorry weigh bridges were installed in godowns with capacity of 5,000 Mts. and above. With this, the total number of such depots having lorry weigh-bridges up from 65% to 85%.

(b) Electronic weigh bridges

Installation of four electronic wagon weigh-bridges so as to know precise weight of wagons in motion at the time of despatch and receipt are under way. Two of these weigh bridges will be in the despatching centres of north and 2 in the receiving centres of east zone.

(c) Computer programme

This is being developed to link the unconnected and missing wagons so that claims with the railways are finalised promptly.

Central Warehousing Corporation (CWC) and State Warehousing Corporations(SWCS)

History

The Central Warehousing Corporation was set up on 2nd March, 1957 under the Agricultural Produce (Warehousing and Development) Act. 1956, which was subsequently repealed and replaced by the Warehousing Corporations Act, 1962. The CWC acquires and builds warehouse at places and also operates them for storage of agricultural produce, fertilisers and certain other items both agricultural and industrial.

Under the warehousing Corporations Act, 1962, the general superintendence of the affairs and business of the CWC vests in a Board of Directors, who with the assistance of Executive Committee and a Managing Director exercise all powers and discharge all the functions which may be exercised or discharged by the Corporation.

Growth in Warehousing Capacity

The CWC started its operations in 1957 with a capacity of 7,000 tonnes in hired gowdowns at 7 centres. It has steadily increased its warehousing capacity and operating at 465 centres with a total capacity of 6.35 million tonnes as on 31.3.1989.

The Corporation owns as well as hires warehousing capacity. It addition, during the last few years, the Corporation has also constructed storage capacity under the World Bank assisted foodgrain storage project on behalf of FCI. Funds for construction of the storage capacity under the World Bank project were provided by FCI to CWC.

Norms for Selection of Godowns

The warehouses run by the CWC may be divided into three categories: (i) warehouses for storage of foodgrains (ii) warehouses for storage of commodities other than foodgrains and (iii) public bonded warehouses. The CWC has been constructing godwons for the exclusive use of the FCI on guarantee basis. The location and the capacity of the godwons is indicated by FCI. For construction of these godwons, the CWC is guided by the public interest and such construction are not taken as wholly commercial propositions. So far as the general warehouses for commodities other than foodgrains are concerned, the selection of the locations and the capacity to tbe carried out by the CWC. The public bonded warehouses are set up at locations as may be indicated by the trade and industry and approved by Central Board of Excise and Customs.

Perspective plan

The construction programme of the CWC is regulated within the framework of Five Year Plans. The target and outlay is determined with the approval of the Planning Commission for Five Year Plan as well as for Annual Plans. However, the corporation has entrusted a study to the Administrative Staff College, Hyderabad, to project the additional warehousing needs. This study will also keep in view the Corporation's need for long term perspective planning of warehousing infrastructure.

Storage losses

The overall storage losses in respect of foodgrain stocks in CWC warehouses were 0.41% during the year 1987-88 (14,749 tonnes).

State Warehousing Corporations

The CWC has 16 associates in the State Warehousing Corporations. The CWC is 50% shareholder in the equity capacity of the CWCs and the remaining 50% is subscribed by the respective State Governments.

National Cooperative Development Corporation

This was set up under the Agricultural Produce Corporations Act, 1956, later repealed and replaced by the National Cooperative Development Corporations Act, 1962. The main function of NCDC is to provide finance for organisation of rural and marketing cooperatives. The base of rural cooperative is put to three-fold use namely storage of inputs like seeds and manures, which are made available to the farmers; holding the agricultural produce procured from its members and also for storage of certain essential consumer items for distribution among its members godown, sizes ranging from 100 tonnes to 1000 tonnes.

Rural godowns

These godowns are small sized ranging from 100 tonnes to 500 tonnes except few godowns which may be as large as upto 3,000 tonnes. These godowns are located essentially in rural areas. Since this scheme was taken up in late 70's and the godowns are small sized, it has been possible to build up a capacity of about 1.5 million tonnes only over the period of last 10 years. Subsidy is given to the tune of 5% by the Central or State Government.

State Government and their agencies

Various State Governments and their agencies have gradually built up storage capacity to meet the requirement of public distribution system in order to supplement the efforts of FCI, CWC, SWCs and other agencies. According to the latest reports, a capacity of about 7.0 million tonnes is reported to be available with various State Governments and their agencies for foodgrain storage.

Important grain processing industries

Roller flour mills

The invention of Roller Flour mills in 1875 caused a revolution in grinding methods of wheat. In 1873 Wegmann, a miller of Naples in Italy invented flexible rollers which are in principle the same even today.

The flour milling industry in India has made a rapid development over the years and is the largest organized sector for utilization of wheat in the country. The industry has been closely associated with National Food policy of the Government which consists of procurement and distribution of wheat.

This industry is categorised into two sectors-the organized sector comprising of licensed RFMS involved in commercial milling operations and unorganised sector consisting of mainly Chakki flour mills.

In 1950, there were 54,000 Atta Chakki in India producing about 5 million tonnes of wheat products, today it is more than 2.8 lakh chakkies producing more than 3.5 million tonnes of wheat products.

The roller flour milling industry is one of the oldest established industries in the country and was started almost a century ago. About two decades earlier there were 211 mills with a capacity of 4.5 lakh tonnes to at present as on March 1989 the country has 557 RFMs and monthly milling capacity of about 8.7 lakh tonnes.

The wheat roller milling industry was governed by special regulations under the Industries (D&R) Act. Even the small scale wheat roller flour units were required to obtain two licences under this Act if they were employing 50 or more than 50 workers. But now the policy is liberalised. Since 1986-87 there are no restrictions on establishment of new RFMs or expansion in the capacity of existing units. No licence is required for manufacture of wheat products. There are no controls on price and distribution of wheat products either. RFMs falling under licenced sector are permitted to grind wheat upto 15% of the licenced capacity. Even though the mills are free to obtain their requirements of wheat from any source subject to availability of stocks at the prices fixed by the Government for open sale of wheat by the Corporation.

The industry needs to introduce improved milling technology and employ qualified personnel for ensuring quality of final products to meet the demand specifications of end users. In response of improved technology the mills are required to modernise themselves as early as possible.

In respect of having trained personnel an outstanding achievement is the establishment of International school of Milling Technology at CFTRI, Mysore, as a joint Indo-Swill Roller Flour Mills Projects. The Milling school has a 10 month course for flour milling besides a number of short courses. The Food Department through S&R Division is helping the admission of trainees in this school.

Rice milling industry

Modernisation of rice milling industry in the country was taken up in 1970 by suitably amending the Rice Milling Industry Act 1958, with the objective of increasing the outturn of quality rice and producing better quality bran and husk suitable for edible oil extraction and as a source of fuel respectively. In the first phase of modernisation, shellers, shellers-cum-hullers and multiple hullers were covered. In 1977, single hullers were also brought under the purview of modernisation. Modernisation started in the year 1970 i.e. in 1970 there was no modernised rice mill.

To propagate the programme on rice mills modernisation and better bye-product utilization nine regional extension service centres were set up in different paddy growing states under the Agricultural Universities/Research Institutioms etc. This network of nine regional extension service centres is established in the States of Andhra Pradesh, Karnataka, Madhya Pradesh, Maharashtra, Orissa, Tamilnadu, U.P. (two centres), and West Bengal.

By the end of 1988, out of 1,15000 rice mills in the coutry more than 35,000 rice mills had been modernised and now approximately 55% of the total rice produced in the country is milled through these modernised rice mills.

Besides, a scheme on modernisation of single hullers is also being implemented in 6 states viz., Andhra Pradesh, Maharashtra, Orissa, Tamilnadu, U.P., and West Bengal. In order to encourage modernisation of rice mills and production of mill machinery in the country. Three testing centres have been set up at Coimbatore, Kharagpur and Kanpur. These centres undertake testing of rice mill machinery according to the standards prescribed by Bureau of Indian Standards.

Pulse Milling

Among all the cereals, pulses are rich in proteins. They are consumed as 'Dal' and provide a source of nourishment. The process of manufacturing the raw grain into 'Dal' involves de-husking and splitting. The husk stronglyy adheres to the inner shell by natural gums and in traditional methods by rubbing/friction, we try to convert the whole Dal into splitted one. During this course, it causes approx. 10% loss which goes alongwith the husk (Chuni). In the villages this Chuni is being used as a blending material alongwith the milled products and serves as a rich source of protein for the poor. The modemisation of the pusle milling industry in the country is to minimise losses.

The modern pulse milling equipment consists of Roller grading machine, pearler ,grader, tampering unit, lump braker, conditions, splitter etc.

In India, the production of pulses in the yar 1988 was 14.5 million tonnes and targetted production for 1995 is likely to go high. The number of Dal Mills in the country is around 10,000 units.

Research development and training organisations for grain industry

A chain of Indian Grain Storage Institutes (IGSI), under Storage and Research Division of Deptt. of Food; Central Food Technological Research Institute at Mysore (CFTRI) under council of scientific and Industrial Research; Central Institute of Post Harvest Engg. & Technology at Bhopal (CIPHET) under Indian Council of Agricultural Research etc. are doing considerable work on research, development & training. Some State Agricultural Universities (SAU) under ICAR umbrella are solving regional problems.

Extension technology of scientific storage of foodgrains in the country

Nearly two-third of the total foodgrains produced in the country, are retained at farm level where considerable losses take place both in quantity and quality due to insect's, rodents and other factors as the storage conditions at farm level are not proper. For minimising the losses, the Department of Food have been implementing a rural based scheme of 'Save Grain Campaign'.

'Save Grain Campaign' has inbuilt system to identify the problems existing in the proper storage and preservation of foodgrains in different states, get the solutions worked out at the Indian Grain Storage Institutes of this Department and popularise them in selected areas, provide feedback needed for improving the codes of practices developed and also provide technical guidance to the State Governments and their agencies in extending the modern, Post Harvest practices to minimise on-farm lossess. It plays a catalytic role for the State Governments for giving due emphasis to this important programme.

(i) *Training*

The SGC teams organise two types of training courses

(a) Stipendiary courses of two weeks duration for the farmers; and

(b) Non-stipendiary courses of about one week's during for extension workers mandals, Balwaris, Anganwaris etc.). On successful completion of the stipendiary training each farmer trainee is awarded a certificate and a stipend @ Rs. 100/- per head.

(ii) Demonstration

Farmers in selected villages are given demonstrations in insect and rodent control methods and explained recommended practices. The demonstration programme include

(a) fumigation of foodgrains;

(b) rat burrow fumigation in the field;

(c) rat control measures in houses and

(d) prophylactic treatments; besides

(e) methods for improvements of storage structure.

(iii) Publicity

In order to make the farmers increasingly aware of the importance of saving the foodgrains during storage, multimedia publicity is arranged with emphasis on visual media.

Assistance to state governments

Under the scheme of Save Grain Campaign some loans and grants-in-aid have been provided

(a) For fabrication and sale of metal bins (improved storage structures) to farmers to 19 states and 1 union territory;

(b) For supply of various imputs required by farmers for improvement of their existing stroage structures to 16 State Governments;

(c) For making available storage pesticides to farmers to 14 State Governments; and

(d) For purchase of training and publicity material to equip the Farmers Training Centres to 15 State Governments.

Conclusion

The history of Post Harvest Grain Technology in India goes back as early as 1888 when COATS published several papers form Calcutta related to biology of stored grain insect pests. The problem of grain storage were discussed at Pusa. Bihar by entomologists in 1917. 1919 and 1923. Department of Food was established in 1942 by Government of India. In 1950, Central Food Technological Research Institute at Mysore was established to impart training and carry out research activities of food processing and preservation. In 1958, Department of Food, Government of India established Grain Storage Research and Training Centre at Hapur which was upgraded in 1968 as Indian Grain Storage Institute (I.G.S.I.). The prime objective of this institution was to carry out research and development activities on all aspects of storage of foodgrains with particular emphasis on farm level storage. Training and extension activities are also taken by this institute. At present IGSI Hapur is supported by its five field stations situated at Ludhiana, Hyderabad, Jorhat, Jabalpur and Udaipur.

In 1973 Department of Food, Government of India also launched Save Grain Campaign Programme at country level for training, motivation and persuasion of the farmenrs to adopt scientific technology of storage and preservation of foodgrains. The present paper deals with the History of Post Harvest Technology in India.

Recent Advances in Indian Entomology
Editor: O.P. Lal
APC Publications Pvt. Ltd., New Delhi, 1996, pp. 185-192

National Activities in Safe Storage of Grain in India

SONE LAL

Series of efforts have been made to initiate the work on post harvest grain technology in India. Lefroy, the 1st Imperial Entomologist appointed by the then British Government in India in 1905 had given a tentative list of storage insects in 1906. This list of storage insects was further supplemented by Fletcher in 1914. The porblems of grain storage were discussed at Pusa, Bihar, at the Entomological meetings held during the years 1917, 1919, and 1923.

Government of India had established Department of Food in 1942. Central Food Technological Research Institute at Mysore was established in 1950 with a mandate to train personnel and evolve technology relating to food processing and preservation. Central Warehousing Corporation was set up on 2nd March, 1957 under the Agricultural Produce (Warehousing & Development) Act. 1956.

The management of the food economy of the country is the primary responsibility of the Ministry of Food, Government of India. With a view to fulfilling this responsibility, various activities such as procurement of food grains, building up and maintenance of food stocks, their storage, movement of delivery to the distributing agencies are mainly undertaken through its one of the corporations viz. Food Corporation of India which was established in 1956.

Storage of foodgrain is inevitable both in times of deficit and surplus production. This single component in chain of post harvest handling of grains is responsible for economy owing to 65 to 70% of total foodgrains produced being stored at farm level. Besides quantitative loss, qualitative losses also take place that results in loss of nutritive value of foodgrains. Million tonnes of grain worth several hundred crores of rupees every year are either damaged or lost for want of knowledge of scientific methods of storage of foodgrains.

Strategy of Ministry of Food to minimise farm level foodgrains storage losses in India

For an agrarian country like India where more than seventy percent of the people are engaged in farming and allied vocations for their livelihood, agriculture assumes special significance in the over all scheme of national development. No wonder if agriculture was accorded a place of pride when the nation embarked on the task of forging its way towards the goal of self-reliance and ushering in an era of plenty and prosperity through planned development.

It has been as a result of cultivators rapidly taking to the modern methods of cultivation based on scientific research, with steady mechanization of farming, that the foodgrain production in the country has now gone up by almost more than three and half times as compared to what it was in the first year after the achievement of independence (52 million tonnes in 1947 to 180.5 million tonnes in 1993-94)

While the break-through in agriculture is a tremendous national achievement, brought about by adopting new strategy, it is discovered that the nation is losing more than ten percent of the total produce every year because of sheer ignorance of the farmers to save it from the natural enemies of foodgrains and lack of technical know-how to protect and preserve the foodgrains with adequate storage and grain handling facilities. Any reduction in losses between harvest and consumption will help to increase the availability of foodgrains. Therefore to derive the full benefit of increased production, it is essential to have proper storage facilities at the farmers as well as trade levels.

Foodgrain losses occur during harvesting, threshing, transportastion, loading, unloading, processing and storage. Keeping in view the fact that "A grain saved is a grain produced" and "A seed saved is a plant produced", the Ministry of Food, Government of India, has taken up number of programmes which may be broadly divided into :

I. Research and Development
II. Extension and Training

I. Research and development

The Govt. of India, Department, of Food, established Grain Storage Research and Training Centre at Hapur in 1958 to develop facilities of research and training in the field of storage and preservation of foodgrains. It was upgraded in 1968 as a national Institute and was renamed as the Indian Grain Storage Institute with assistance from the United Nations Development Programme. Subsequently, in 1980-81 five field stations were established at Ludhiana, Hyderabad, Jabalpur, Udaipur and Jorhat for tackling area-specific storage problems.The Hapur centre is recognised research and development unit for tackling post-harvest problems.

The main objectives of the Institute are :

To develop a code of practices for proper storage and handling of grain by recommending effective techniques for the control of insects, rodents and birds and checking of micro-organisms.

To investigate the nature and extent of losses due to these factors under different agro-climatic conditions.

To develop better types of storage structures, grain drying, handling, cleaning and grading equipment besides, improving the traditional stroage structures/ equipment using locally available materials

To evaluate the new methods before releasing for use throughout the country by developing suitable extension materials and semitechnical literature on grain storage and quality control.

To train people from various organisations in proper grain storage and handling practices.

The Institue at main station, Hapur has six technical divisions, namely, Engineering, Pest Control and Storage, Training, Extension, Agricultural Marketing and Economics & Coordination. As evolving the code of storage practices is one of the major activities, the R & D efforts are designed in such a way that 60 per cent of the programme is for the farm level and the balance for commercial storage.

The scientists of the institute are engaged in developing new and viable technology to be adopted at farm and commercial levels and also deliver this technology to the users either by organising Apex Level Training or through Save Grain Campaign teams of the Ministry of Food which are functioning in close collaboration with Agriculture and Extension Departments in each state.

Some of the important achievements of the Institute are mentioned as under:

Development of code of practices for safe storage of foodgrains for all the important cereals, pulses under different agro-climatic conditions and groundnut pods. Besides, code of practices for rodent control at farmers' level, warehouses, railway yards, flour mills.

Screening of new pesticides, standardisation of doses, period of exposure, frequency of use and recommendations for their use etc. at different levels.

(a) The pesticides being used in the country for the control of stored grain pests have been screened and recommended by this Institute. As an alternate of malathion, pirmiphosmethyl 50% E.C. @ 0.15 gm a.i/m^2 and Deltamethrin 2.5% W.P. and flow 25% @ 30 mg a.i./m^2 have been recommended and approved by the Govt. for use.

(b) Second generation rodenticides *viz.*, Bromadiolone, Brodifacoum etc. have been tested and found effective under lab. and field conditions.

- Assessment of the storage life of coarse grains.

- Estimation of storage loss/gain in wheat, rice and paddy due to variation in moisture content.

- Design and development of modern storage structures :

(a) Designed and developed more than 50 types of storage structures at various capacities with the use of locally available material for use at farm and community level storage. Developed technology for the improvement of locally used storage structures in various parts of the country. The metallic bins, pucca kothies and R.C.C. Ring bins developed by the Institute have become very popular amongst the farmers.

(b) Development of high moisture paddy storage bins, double walled bamboo bin for seed storage.

Development of grain dryers/cleaners:

The Institute has designed and developed farm level crop dryers using solar energy and agricultural waste for drying of different commodities of foodgrains and agricultural produce.

1. Solar grain dryer
2. Cabinet type solar crop dryer
3. Fuel operated crop dryer using agricultural waste.
4. Natural draught fuel operated crop dryer using agricultural waste.

- Substitute for wooden crates.
- Suitable material for fumigation covers are being tested and recommended for use.
- Sealing mechanism for sealing the fumigation covers which is one of the most important aspects for effective fumigation has been perfected and recommended for use in commercial godowns.
- The A.M.E. division, apart from studying the economic viability of the Institute's recommendations, regularly undertaking studies in various parts of the country, covering different aspects of post-harvest of foodgrains.

II. Extension and training at farm level

Realising the need for creating awareness of colossal storage losses and to extend the new techniques developed by IGSI, Agricultural Universities and other institutions for protection of foodgrains in storage at farm level, a pilot project entitled "Save Grain Campaign" was implemented by the Ministry of Food during the year 1965-66.

In 1969-70, a regular scheme of Save Grain Campaign was launched and two Save Grain Campaign offices at Patna and Bombay (Bombay office since shifted to Pune) and a small Central Unit at Department's Headquarter, New Delhi were established. Keeping in view the importance and the impact created by the initial teams of Save Grain Campaign, the Govt. of India have created by now in phases 14 regional teams functioning at Patna, Pune, Ghaziabad, Bhopal, Hyderabad, Ahmedabad and Guwahati and three sub-regional teams at Banaras, Raipur and Trivandrum and have covered entire country under the programme.

Aims and objectives

1. To disseminate the knowledge of scientific preservation of foodgrains and pest control techniques by arranging training, demonstration and publicity programmes so that the masses may get well acquainted with the latest technique of scientific preservation of foodgrains and thus eliminate huge losses.

II. Activities of Save Grain Campaign Teams

A. Training programmes

Two types of training programmes *viz.*, (i) Stipendiary and ((ii) Non-stipendiary Training

i Stipendiary Training Programmes

These programmes are mainly organised for cultivators and traders. The stipendiary courses are of two types :

(a) Stipendiary courses of two weeks duration : The duration of stipendiary training course has been changed from 3 weeks to two weeks only few years back. The number of persons to be trained in stipendiary training course will be 600 by each main team and 300 by each sub-team during 1994-95 by training 50 persons, in each course in 12 courses by each main team and 6 courses by each sub team.

(b) Preraks (Motivators) Training Courses : The number of persons to be trained in PRERAKS (Motivators) training courses will be 600 (in a maximum of 12 courses) by each main team and 300 in 6 courses by each sub-team during 1994-95, by training 50 persons in each course.

ii Non-stipendiay Training Programme

These courses are organised for the V.L.Ws/ State Govt. Officers/ personnels of organisations dealing with the foodgrains and volunteers.

These courses vary in duration from 3 to 7 days depending upon the time available at either end. Target fixed for non-stipendiary courses for VLWs/ volunteers, State Govt. functionaries etc. to be conducted during 1994-95 is 40 courses for the year by each main team and 20 courses for the year by each sub-team. Accordingly, a total number of 992 training courses (both stipendiary & non-stipendiary have to be organised by all the 17 SGC.

The main objective of arranging stipendiary and non-stipendiary training camp is to create a band of trained workers who may adopt scientific storage practices themselves and also act as volunteers to guide those who come in contact in improving their storage conditions.

B. Demonstration programmes

Save Grain Campaign Regional Teams have been conducting various demostration programmes in pest control which include prophylactic treatment with Malathion, grain fumigation with EDB fumigant, house rat control with anticoagulant and field rat control with aluminium phosphide and Zinc Phosphide.

During the 8th plan period (1992-97) 4000 villages are to be covered out of which 2000 villages would be nucleus villages. To achieve this objective, each year each main team is required to develop 26 nucleus villages out of 40 villages to be covered. Similarly, each sub-team is required to develop 12 nucleus villages each year out of 20 villages to be covered. The remaining villages have to be developed as medium contact villages.

In order to make the pesticides available to farmers, at the block level, Govt. of India, Ministry of Food had provided funds to State Governments as a grant-in-aid.

C. Publicity programme

With a view to create the desired impact, publicity is one of the major items. In this aspect, the regional teams have been distributing the literature to all the concerned free of cost and stencilling/painting walls of public importance with a comprehensive message of "Save Grain". Besides, the posters, hoarding, bus pannels and cinema slides have also been kept on display at the appropriate places.

D. Improvement of storage structures

An improved structure is the penacea of the ills of storage. The age old storage structures available with the cultivators and solely responsible for the considerable loss in the stored foodgrains.

Taking into account the basic need to introduce improved storage structures at the farmers level. the Govt. of India has sanctioned the interest-free loans to State Govts. for fabrication/distribution of metal bins. These bins are fabricated by Agro-Industries Development Corporation as per the design supplied by Indian Grain Storage Institute, Hapur and are available in 10 qtl., 7.5 qtl., 5 qtl. capacities at reasonably approved Govt. rates.

In order to make these structures more popular in the nook and corners of the State and for speeding up the distribution of bins, the State Govts. have involved various concerned mandi boards/departments like Agriculture and Panchayat etc.

The existing storage structures of the cultivators are also being suitably improved by supplying polythene sheet and shalikote T-10 free of cost making the structures reasonably air tight and storage worthy.

The small and marginal cultivators desious of constructing pucca structures for about two tonne of storage are helped in constructing such structures for which suitable guidance is given by the regional teams apart from the incentives like polythene for moisture proofing, inlets and outlets free of cost. The incentive is provided to the farmers of the Save Grain Campaign adopted village.

The achievements of Save Grain Campaign teams of last three years are given in Annexure-1. In addition, the following organisations are also functioning and contributing in minimising the post-harvest losses in foodgrains.

The Research Laboratories of the Council of Scientific and Industrial Research at Trivandrum, Hyderabad, Jorhat, Bhubaneswar and Jammu contributed on food and feed post-harvest technology,

The Ministry of Agriculture established the State Agricultural Universities (SAU) under ICAR Umbrella, results in establishment of Department/Sections, Agricultural processing/processing and food Engineering/ agricultural structures and process engineering in Agricultural Engineering, Deptt. of Entomology, Food Science/ Food Technology in the Faculty of Agriculture.

ICAR have launched All-India Coordinated Project on post-harvest technology, processing and storage of jaggery, oilseeds processing etc., ICAR's latest initiative is the establishment of Central Insitute of Post-Harvest Engineering and Technology (CIPHET). Though these programmes are not directly involved in safe storage of foodgrains but are very much concerned with minimisation of food losses.

Foodgrain storage organisation in the country

(1) Food Corporation of India (FCI)
(2) Central Warehousing Corporation (CWC)
(3) State Warehousing Corporations (SWC)
(4) National Cooperative Development Corporation (NCDC)
(5) Rural Godowns
(6) State Governments and their Agencies.

The above mentioned six organisations have got the technical by trained personnel who are reponsible for the management of foodgrains, hence the losses are minimum. The total storage capacity available with various agencies and the foodgrain capacity is as under:

Sr.No.	Agency	Total storage capacity Million Tonnes)	Foodgrains storage capacity (Million Tonnes).
1.	FCI	19.9	19.9
2.	CWC	6.4	3.2
3.	SWC	9.0	4.5
4.	Cooperatives	11.7	5.8
5.	Rural godowns	1.5	0.8
6.	State Govt. & agencies	7.0	7.0

While the capacity available with FCI is used mainly for storage of foodgrains that with CWC and SWCs is used for storage of foodgrains as also certain other items. The following table shows the position of covered storage/warehousing capacity available with the Public agencies during 1992 and 1993.

(In Lakh tonnes)

Agencies	As on 31.3.1992			As on 31.3.1993			As on 31.10.1993		
	Owned	Hired	Total	Owned	Hired	Total	Owned	Hired	Total
FCI	119.82	63.32	183.14	121.74	58.44	180.18	121.89	72.80	194.69
CWC	47.75	16.53	64.28	48.51	15.90	64.41	48.72	14.73	63.45
SWCs	67.46	22.01	89.47	69.38	21.36	90.74	69.82	29.60	99.42

The Government is fully aware of the colossal losses of foodgrains during its post harvest management at various stages and therefore making concerted efforts at all levels to minimise these losses both in quantity and quality. However, this could be done effectively if every citizen of this great nation realizes about this national problem and shares the responsibility.

Recent Advances in Indian Entomology
Editor: O.P. Lal
APC Publications Pvt. Ltd., New Delhi, 1996, pp. 193-196

Recent Advances in Safe Storage of Grains

T. D. YADAV

Indian food grain and seed storage system is blend of traditional and conventional practices. The system is dependent on utilization of local produce both for rural and urban level necessities. Seed upto 85 per cent is also stored at farm level. While food production has picked-up to the level of about 175 million tonnes, storage practices are predominated with usage of jute bags and a few other conventional structures.

A combination of most feasible methods both chemical and non chemical are invogue in the tropical countries. Low cost methods of grain storage such as use of mud, bamboo, stone, wood based structures along with ash, fine dust, frequent sun drying are practised. Yet during the recent years, use of metal bin, cement based structures jute bags, fumigation, or conventional insecticides have been increasing

Regional imbalance in food production and handling

The impact of green revolution was most quickly observed in Punjab followed by Haryana. The rise in production of wheat in sixties and rice in seventies was most glaring in these states. Due to surplus conditions, large scale movement of grain was expected to other states specially southern for wheat and rice in almost all regions. Similar condition was also found in case of bumper crops of rice from Thanjavur, in Tamil Nadu or some districts of A.P. Chickpea from Rajasthan and M.P. and Groundnut from Gujarat and A.P. are also moved out. The infestation and associated resistance in insects get distributed in this way. Besides storability of the food grain is affected due to prevalent humid conditions in the coastal and southern region.

Handling and storage practices

At national level, the present management pactices are based on use of malathion spray as fortnight prophylactic spray @150 mg/M2 and DDVP @2gm/ 100 M3 of space. Aluminium phosphide @ 6 gm/tonne of grain is also used upto 3-4 times annually. The practice has been followed for over 20-25 years. In the light of research inputs, developed under the Indian Institutions pertaining to grains storage, a large number of studies have been made. While these studies have a mixed relevance of changing scenario in world food grain storage under the direct impact of FAO, WHO, NRI, CSIRO, ACIAR, NAPHIRE etc. The relevant studies remained to be non feasible in the Indian conditions. This reminds one that storage is higly linked to ecological and socioeconomic conditions of a nation and subtropical situations needs to be more represented in planning safe storage. India has unique predominance of food grain productions linked to utilisation at farm level without much handling and associated processing at local level. However, use of edible oil like castor, mustard, groundnut are partially accepted in case of legume. The use of neem leaf, cowdung or wood ash and fine sand are also practised. Inspite of a large number of contributions relevant to effect of plant products & essential oils evaluated for ovicidal toxicity and hormonal imbalance nothing so far has been commercialised.

Use of activated dust and tricalcium phosphate to increase bulk density of the flour could not find application.

Application of radiation and principles of male sterile technique has remained elusive so far. Among the insect growth regulator methoprene appears to be useful in tobacco storage against *Lasioderma serricorne* Fab. Fluidised bed based on infra red super thermal heating of grain is expected of use in processed products.

Cryogenic storage of spices and extraction of oleoresins leading to later on blending of spices according to use is feasible. Air tight storage and controlled atomosphere have gained acceptability in developed countries. CO2 in dry ice form or exothermic production of nitrogen after removal of O2 are most feasible. However, under Indian systems the devices is being evaluated.

In situ metabolic use of O2 and accumulation of CO2 in dry grain mass stored in Pusa bin/Pusa kothar system has proved universally effective. The utility is obvious in dry season harvested crops. Yet use of drying has made it quite efficaceous in humid areas also. The performance of Pusa bin and Pusa kothar type of storage structures in the Indian situation and else where in the world are testimony to well developed grain storage principles and blend of technologically sound low cost devices. More recently the use of these structures in seed storage of wheat has shown benefit to rural community in terms of cheapest input of seed, higher yield due to better vigour, sustainability of high quality inputs and greater use of farm produce in human food.

Processing

Legumes are highly vulnerable to bruchid attack in whole form but converted into 'dhal' do not suffer damage by the primary insects. Other insects and micro-organisms related damage ensues in. Similarly paddy with husk is damaged by *Sitotroga cerealella* and *Rhizopertha dominica* do not remain active in flour and other products which attract *Tribolium castaneum* and *Corcyra cephalonica*.

Ecological concept

The climatic factors prevailing in a country influence the storage life of grains. In India, 4 distinct zones have been demarcated and safety to stored products have been assessed in a rated manner based on efficiency and deficieny of thermal, humidity value and other topographical features. The development of grain storage specially buffer stock can be planned to avoid serious losses in high risk areas. However, the present position of data on meteorological aspect is based on air and this is not accurate in case of storage facilities. Efforts in this direction are to be intensified.

Intensification of agricultural operation

The crop production system has undergone considerable changes during the last 3 decades. Various inputs such as newer varieties of seeds, fertilizers, pesticides, higher water application and larger area of crops have resulted into least time gap for post harvest operation and handling. The need for higher inputs in mechanisation of this phase of crop production is now witnessed in quick threshing due to larger number of diesel/electric based threshers or tractor mounted threshers. Even use of combine-harvester in wheat belt of Punjab and Haryana is now rising. In the storage system specially at rural level, the constraints of suitable storage practices have remained indicating lack of thrust in planning and development on this aspect. Thus the appreciable quantity of losses in food grains and seeds are remaining unchecked. This situation is not different in other parts of tropical and developing nations.

Innovative ideas

In Republic of China, rice husk is treated with a suitable insecticide and is mixed in rice during storage period. At time of consumption, the husk is removed by sieving. In Japan due to shortage of land space use of LDPE liners in storage of grain under sea/lake water has been tried. Bags of 0.5-1.0 tonne capacity were placed in water after introduction of CO2.

Long term storage

Several studies, on the long term storage of grain have been made with the objective of using low temperature near to zero, lowest O2 atmosphere near to 1.0 per cent, approximately 100 per cent N2 or CO2 above 60.0 per cent. Use of freon gas has also been attempted. Loss of germination, and alcoholic acidity are most important parameters to reflect deterioration.

Biological control

Parasitoids and predators have been studied in relation to different species of insects causing damage in storage. Insects from Braconidae, Trichogrammatidae, Eulophidae, Eupelmidae, Torymidae, Pteromalidae, Eurytomidae and mites have been found to be effective. However, exact role of bioagents in bringing down the population to accepted level of no loss status is not often discernable. Information on biology and ecology of damaging agent and beneficial agent in the actual condition of storage practices is yet preliminary.

Wet grain storage

In several countries of the world grain is harvested at moisture content range higher than 15 per cent which can not be stored in normal manner without immediate drying specially in tropics. Under this situation grain has been successfully stored by use of propionic acid, acetic acid, formic acid and sorbic acid @ 0.5 per cent level. The grain is used in cattle feeding. This type of storage in India has not been attempted much due to our climate and fodder requirement being different. However, the scope is only, in certain cool months in high hills.

Future scope-I

Improvement in processing facilities at rural level is expected

II) Large scale changes in houses based on cement is to favour change in insect complex

III) Use of solid fumigant and fabric treatment to be more acceptable

IV) Greater awareness to hygienic food is projected

V) Long period storage specially staple cereal is to be developed

VI) Quality seed at farm level would be more in demand

VII) Global needs of food and other commodities are likely to enhance.

References

Jayas, Digvir, S. ; White, Noel, D.G. ; William, E. Muir and Sinha, N.R. 1993 Controlled atmosphere storage of cereals and oilseeds. *J. Appl. Zool. Res.* **4** (1) : 1-12.

Mitsuda, H. ; Kawai, F. ; Kuga, M. and Yamamoto, A. 1973. Under water storage of cereal grains by CEM skin packaging technique. Abst. Int. Sym. Proceedings of wet harvested grain. 5-10 March, 1973, Paris.

Van Huis, A. 1991. Biological methods of bruchid control in the tropics : A review. *Insect Sci. applic.* 12 (1/2/3) pp 87-102.

Yadav, T.D. 1983. principles of safe storage of grain. pp. 387-403. In Agricultural Entomology, Vol. II AISWI, New Delhi. ed. P.D. Srivastava et. al.

Yadav, 1990. Development of future prophylactic for India. Fifth Int. working Conf. on stored prod. Protection. 9-14 Sept. 1990. Vol. III. pp. 1737-1746. Bordeaux, France.

Yadav, T.D. 1992. Status of storage and handling of legumes in India. Pages 187-193. Ed. J.N. Sachan, In New Frontiers in Pulses Research and Development : Proceedings of Nat. Sym. 10-12 Nov. 1989, Directorate of Pulses Research Kanpur.

Yadav, T.D. 1993. Sustaining Agricultural production : Present Scenario of food grain storage system in India. *J. Appl. Zool Res.* **4** (1) : 21-26.

Recent Advances in Indian Entomology
Editor: O.P. Lal
APC Publications Pvt. Ltd., New Delhi, 1996, pp. 197-203

Insecticide Resistant Insect Pest Management

K. N. MEHROTRA and AMRIT PHOKELA

Insecticide resistance is now recognized to be an important constraint in increasing agricultural productivity. The word "resistance" is used in instances where insecticide dosage that were formerly effective now meet with control failure. It is a dynamic, multidimensional phenomenon, dependent on biochemical, physiological, genetic and ecological factors. All these factors vary with the species of insect, its population and geopgraphical location. Resistant strains of pests develop through survivors and reproduction of individuals carrying one or more factors that allow survival after exposure to a given insecticide. The insecticidal stress sharply increases the frequency of resistant phenotypes within the exposed population. Resistance to insecticides develop mainly because of its intensive use in situations that would otherwise favour rapid pest build up. There are three phases in the development of insecticide resistance in a population : Firstly, when the frequency of resistant insects is very low (one in a million or more) and the insecticide gives a satisfactory control. Secondly, with the continuous use of the same or related insecticide the frequency of resistant individuals increases leading to occasion crop failures and thirdly, with the continuous pressure of insecticides there is a large increase in the numbers of resistant individuals in the population and the insecticide becomes ineffective. In India, insecticide resistance has now appeared to all major classes of insecticides and will go on to develop to future insectidies as well, including bio-pesticides.

The insecticide resistance in India was first detected in 1952 in mosquitoes (Pal *et al*., 1952) and, since then number of the insect pests developing insecticide resistance has increased (Mehrotra, 1993 a). In crop pests the first instance of insecticide resistance was reported in 1963 (Pradhan *et al*, 1963), but this event remained only of academic interest. It is only in crop season 1987-88 that insecticide resistance came into prominence when the American bolloworm or gram pod borer, *Helicoverpa armigera* was reported to have developed high degree of resistance to synthetic pyrethroids from cotton growing areas in Andhra Pradesh (Dhingra *et al*. 1988, McCaffery, *et al*. 1989). Now nearly 14 insect pests of public health importance, 7 of agricultural crops, and 7 of stored grains have become resistant to insecticides. The resistance reported is to nearly all major groups of insecticides, namely, chlorinated hydrocarbons, organophosphates, carbamates, synthetic prethroids and fumigants (See reviews Mehrotra, 1989; 1992; 1993 a and b). Few crop pests, for example, diamondback moth, *Plutella xylostella,* American bollworm, *Helicoverpa armigera* and, perhaps, tobacco caterpillar, *Spodoptera litura,* have developed multiple resistance. This increase in the incidence of insecticide resistance is important and requires structural changes in our current strategy of plant protection. Present article briefly discusses the problem and suggests what should be done to contain the problem of insecticide resistance.

Problem

Till to date insectidie resistance in insect pests was recognized only after repeated crop failures using the same or related insecticides. It is, therefore, essential that a continuous monitoring for insecticide resistance should be undertaken. This has not been done in this country except, perhaps, in case of vectors of various diseases, specially mosquitoes. Since there is a great difference in the physiology, ecology and in control strategies of disease vectors and phytophagous insects, the information from public health pests cannot be applied directly to agricultural pests. Because of this, here only the agricultural pests will be discussed.

Monitoring

Monitoring for insecticide resistance is essential to any insecticide resistance management program (IRM). The precision required in monitoring depends on what it is needed for. Some monitoring programmes are designed only to test whether an insecticide control failure was due to resistance rather than to the poor application of insecticides or other factors. Similarily, it can also be used to determine it resistance is a problem in a specific location to improve insecticide choice. This is usually done just to document the presence or absence of insecticide resistance in an area. This type of monitoring usually involve comparisions of LD_{50}s or LC_{50}s and slopes between field collected populations or laboratory strains or both. This method is adequate to document resistance that has reached a high level, but it is very inefficient for detecting an incipient resistance outbreak. The disadvantage of determining LD_{50} through complete dose-response lines (ldp lines) is that this methods is time consuming and insensitive to slight changes in resistance gene frequencies, specially at LD_{50} level. It has also been calculated that only a slight change in LD_{50} is observed when the frequency of resistant phenotypes have reached a 10 per cent or more (Roush and Miller, 1986). The advantage of determining the ldp lines is, that it *may* give some idea about the numbers of genes involved in imparting resistance, which *may* be useful in understanding the genetics of resistance.

A monitoring program to detect resistance before control failures occur requires a greater precision in estimating the frequency of resistant individual than the above mentioned documentation *program*. The precision of monitoring is important for two reasons. First, the resistance frequencies must be very low for formulating IRM strategies to be effective. Second, once an IRM strategy has been implemented, it needs to monitor for changes in resistance frequencies to ascertain if the program is effective or not.

If resistance monitoring is to be used in a practical way for IRM, it should be designed to detect resistant individuals at a frequency of 1 per cent or less of the population. A simple calculation using Hardy-Weinberg postulates would show that insecticides control can theoretically be lost, in as little as 1 to 6 generations depending on circumstances, insect species, whether the resistance is dominent or partially dominant etc., if the frequency of resistant phenotypes has reached 1 per cent. Although, it would be desirable to detect resistance at even lower frequencies to allow more time to develop an IRM program, 1 per cent may be a practical limit in most cases with the available techniques (Tabashnik and Croft, 1982).

Resistance to an insecticide could be due to less penetration in resistant insects, faster sequestration of insecticide, more hydrophobic storage of the insecticide, higher rate of metabolism, alteration in target site when compared to susceptible strains. Since, in a given strain more than one factor may be involved in imparting resistance, it can be safely presumed that resistance is polygenic in most cases and mongenic rarely. For monitoring for resistance a number of approaches have been used. These are :-

(a) bioassay using a diagnostic dose.

(b) biochemical approach

 i) *in vitro* metabolism,

 ii) serological techniques,

 iii) electrophoretic techniques,

(c) genetic approach

 i) cytogenetic techniques,

 ii) moleculare genetic techniques

Bioassay

Globally this is the most commonly used technique at present. It involves the determination of a diagnostic dose of an insecticide which would kill, ideally, all susceptible and spare the resistant insects. the technique has been used in public health, specially for determing resistance to DDT, HCH and malathion in malaria transmitting mosquitoes. Amongst crop pests, in India, only now are efforts being made to monitor resistance in *H. armigera* by using a diagnostic dose of 1 μg cypermethrin per third instar larva weighing between 30-50 mg. This diagnostic dose was computed from the data avaiable from ldp lines (Mehrotra and Phokela, 1992; Armes, *et al*, 1992). By using this diagnostic dose it appears that resistance is seen throughout India wherever cotton is grown. The resistance has also extended itself in *H. armigera* infesting chickpea and pigeon pea. When tested with this diagnostic dose, characteristics of populations occurring in north and south India are; a) resistance monitored with cypermethrin reflects not only resistance to cypermethrin but also to fenvalerate and deltamethrin, b) the degree of resistance is low in early crop season and there is a marked increase in the resistance in the late stage of the crop, c) various populations of the American bollworm show different degree of synergism with piperonyl butoxide (PBO), a known mixed function oxidase (MFO) inhibitor (Phokela and Mehrotra, 1993). But for these studies diagnostic doses for other pesticides against this pest have not been worked out as yet. There is an urgent need to determine the diagnostic dose for each and every insecticide used in the country against various pests.

Biochemical

Nearly 40 years of studies, all over the world, suggest that insecticide resistance could be correlated with the quantitative and/or qualitative changes in insecticide metabolizing enzymes or a change in target site or appearance of a protein in cuticle which retards the penetration of an insecticide of easy sequestration of insecticide by binding on a protein. Since in all these cases the insecticide resistance phenomenon is associated with a protein which may or may not have enzymatic properties, these proteins have been designated as insecticide Resistance Associated Proteins (RAPs). The insecticide metabolizing enzymes can be broadly classified as :

a) epoxidases, which act on aromatic (carbaryl), alicyclic (aldrin), heterocyclic (precocine I) and aliphatic (pyrethrins) moieties on an insecticides.

b) Hydroxylating enzymes, which affect aromatic (carbaryl), alicyclic (nicotine), heterocyclic (PBO), aliphatic (diazinon) moieties of the insecticides.

c) N-dealkylating enzyme (monocrotophos)

d) O-dealkylating enzyme (tetrachlorvinphos)

e) desulfurating enzyme (malathions,).

f) phosphoester hydrolyzing enzyme sometimes called phosphatases (OPs in general)

g) esterases cleaving the ester bond in a number of insecticides.

h) glutathione transferases, which metabolize large number of insecticides and cut across the broad chemical groups of insecticides.

These Resistance Associated Enzymes (RAEs) could easily assayed in laboratory. Efforts have been made to estimate the RAE's activity to surrogate substrates, usually naphtholic esters, whose products can be measured either in solution or on cellulose filter paper or on nitrocellulose membranes. Furthermore, these surrogate substrates can be used to probe the isozymes and their mobility variance (allozymes) following electrophoresis or electrofocussing. The esterase banding patterns have proved very successful in the diagnosis of insecticides resistance in aphids and jassids. This technique is straightforward to use when resistance is due to the over production of a detoxifying enzyme. In other words when the mechanism of resistance is due to gene amplification. The technique could be made more sensitive by the use of other confirmational macromolecules that are specifically related to RAEs. Specific antibodies or cDNA probes as reporter molecules, may give a high degree of diagnostic resolution for RAEs. Thus whole range of biotechnological methods can be used for developing a diagnostic methods for detecting resistance well in advance before the crop failures occur. A major disadvantage of these methods is that they either do not detect or underestimate the magnitude of resistance in multiresistant phenotypes.

Genetic

Resistance mechanisms arise through inheritable changes - mutations - in the genome of individiual insects. Mutations may include substitution in DNA base pairs, amplification of preexisting genes that confer defense mechanism, translocations, chrmosome inversion or other DNA rearrangements. Chromosome inversion specially in flies and mosquitoes have been used extensively but not in agricultural pests. Chromosome mapping using radioisotopes, fluorescence probes and immunologically activated chromophores can be used to analyze genes.

Gene amplification may be important in resistance based on sequestration of insecticides by binding to lipids or proteins or on a reduced rate of penetraton through cuticle. Base pair substitutions are also important in target site resistance, where the target macromolecules is so modified that it no longer binds with the insecticide. Both gene amplification and base pair substitution may be important in resistance that is due to increased metabolic detoxification. Aphids and jassids showing resistance to organophosphate insecticides are the best examples of gene amplification.

It may be mentioned here that, in India, but for bioassay, none of the above techniques have been used either to detect or forewarm the appearance of resistance in insect pest species. The sole exception to this is the work on carboxylesterases on three populations of *Helicoverpa armigera* which indicated, as early as 1986, that insecticide resistance is likely to appear in this pest species (Mehrotra and Phokela, 1986)

IRM in IPM

IRM in IPM should be an integral part of the strategy for the management of insect pests in any given cropping system. Since both IRM and IPM are interrelated and knowledge intensive the information on the following aspects should be generated:-

1. Biology of the pest, number of generations and the most susceptible and vulnerable stage of the pest.
2. Need based use of insecticides based on monitoring the pest populations in field.
3. Use of synergists, wherever feasible, which will enhance the toxicity of a given insecticide by inhbiting detoxification mechanism.
4. Alterartion of chemicals with unrelated mode of action.
5. Restrict the use a particular insecticide to not more than once or twice during the life cycle of an insect pest.

One or the other combination of the above five methods can be effectively utilized in combating the insecticide resistance once reported and will also delay its appearance. The key factor in this is the monitoring of pest pupulations as well as the resistance status of a pest to an insecticide Thus the IRM becomes an integral part of IPM. More information is needed to finely tune both IRM and IPM and these can be broadly divided into three categories.

1. Physiological, biochemical and genetical

This requires the knowledge of the rate of penetration of insecticide through cuticle, rate of excretion of unmetabolized insecticide, the rate of detoxification and the route of detoxification of insecticides in target species, the sensitivity of the target and the rate of appearance of insecticide resistant genome. An evaluation of these parameters by a systems analysis approach should give sufficient information to formulate an IRM strategy. Since insecticide resistance is either preadaptive or through gene amplification, it is essential to avoid the use of persistent insecticides and residual treatment of insect populations to reduce the selection pressure.

2. Biological

It is essential to determine the level of insect population which remains unaffected after the use of insecticides. It is better to use insecticide only at one stage of their life cycle and not subject the insect populations at every stage of their development and growth to insecticide selection. The use of ovicides, wherever possible should be encouraged because the eggs have very poorly developed detoxification mechanisms. It is also essential to continuously monitor for insecticide resistance in field populations. It is better to change the insecticide before the insecticide resistance actually appears.

3. Nature of insecticide

The insecticides used should be such that it leaves least amount of residues. The residues should be photodegradable and also preferably biodegradable. the chemical nature of insecticide should be such that the rate of appearance of resistance to them is slow and also they show least amount of cross resistance. Bioinsecticides and other biological control methods should also be incorporated in the IPM strategy. A knowledge of previous history of exposure of insect populations to insecticides is, therefore, important and can provide useful clues.

Conclusions

Although, insecticide resistance has been reported from India in a number of insect pest species, the information on mechanisms by which this resistance is imparted is scanty. In most of the cases the resistance was established only after the failure of an insecticide in controlling pests under field conditions. IPM or any other innovative strategy formulated for the control of pests and diseases, both in agriculture and public health should have a sound biological base. These strategies will differ from region to region and pest to pest. Whatever is true for public health may not be true for agriculture and vice versa. A technology for this will have to be generated in India and cannot be imported.

The era of IPM and IRM has unfortunately come after more than four decades of hard lessons, during which many of the important pest species have evolved resistance to a number of insecticides belonging to different chemical groups. During this period, there has been a tremendous loss of several agricultural resources, particuarly, a very valuable and probably nonrenewable susceptible genome of insect pests and effective insecticides. In Indian subcontinent fortunately, the magnitude of pest resistance is not as much as in the developed counries and we can learn a lot from their experience. This should go a long way in providing "Food for All, Health for All" and a better environment by 2000 A. D., provided a family planning program is also effectively pursued and the rate of development of population is reducted drastically.

References

Dhingra, S., Phokela, A. and Mehrotra, K. N. (1988). Cypermethrin resistance in the populations of *Heliothis armigera* Hubner. *Nat. Acad. Sci. Letters* **11**:123-125.

McCaffery, A. R., King, A.B.S., Walker, A.J. and El Nayir, H. (1989) Resistance to synthetic pyrethroids in the bollworm, *Heliothis armiqera* from Andhra Pradesh, India. *Pestic. Sci* **27** : 65-76.

Mehrotra, K. N. (1992) Pesticide resistance in insect pests : Indian scenario. In *Pest Management and Pesticides : Indian Scenario* ed. B. V. David, Namrutha Publications, Madras. pp.17-27.

Mehrotra, K. N. (1993 a) Status of insecticide resistance in insect pests. In *Pestcidides their ecological impact in developing countries*. eds. G. S. Dhaliwal and B. Singh. Commonwealth Publishers, New Delhi (India) pp 30-50.

Mehrotra, K. N. (1993 b) Pesticide resistance in India. *Chemistry & Industry* (London) 7 June, 1993, pp 399-403.

Mehrotra, K. N. and Phokela, A. (1986). Variations in the carboxylic ester hydrolase (EC 3.1.1.1) activity in *Heliothis armigera* hubner larvae. *Curr. Sci* **55** : 508-509.

Pal, R., Sharma, M. I. D. and Krishnamurty, B. S. (1952). Studies on the development of resistant strains of house flies and mosquitoes. *Indian J. Malariol.* **6** : 303-316.

Phokela, A. and Mehrotra, K. N. (1993). Pyrethroid resistance in *Helicoverpa armigera* Hubner. VI. Effect of PBO on toxicity of cypermethrin. *Pestic. Res. J.* **5** : 106-108.

Pradhan, S., Jotwani, M. G. and Sarup, P. (1963). Failure of BHC and DDT to control Singhara beetle, *Gallerucella birmanica* Jacoby. *Indian J. Entomol.* **34** : 176-179.

Roush, R. T. and Miller, G. L. (1986). Consideration of design of insecticide monitoring programs. *J. Econ. Entomol.* **79** : 293-298

Tabashnik, B. E. and Croft, B. A., (1982). Managing pesticide resistance in crop-arthropod complexes : interactions between biological and operational factors. *Environ. Entomol.* : **11** : 1137-1144.

Recent Advances in Indian Entomology
Editor: O.P. Lal
APC Publications Pvt. Ltd., New Delhi, 1996, pp. 204-215

Recent Advances in the Control of Desert Locust, Schistocerca gregaria *(Forskal)*

Y. N. SRIVASTAVA and O. P. LAL

Locusts are the oldest enemy of agriculture. They are capable of changing their social habits and behaviour under favourable conditions and multiply in large numbers, assume gregarious habit. Their hoppers move in bands and adults from swarm in search of food which migrate from one country to other. Thus, causing enormous amount of damage to agriculture and its produce.

Locust belongs to class 'orthoptera' in which grasshoppers come. There are about 19 species of locust (short horn grasshopper) found in the world. In India four species of locust are found viz.

1. Desert locust, *Schistocerca gregaria* (Forskal)
2. Migratory locust, *Locusta migratoria* L.
3. Bombay locust, *Patanaga succincta* Joh.
4. Tree locust, *Anacridium rubrispinum* Bei-Benko

Out of these, desert locust is most widely distributed affecting sixty one countries of old world and covering nearly 30 million sq. km. area i.e. from Mauritania to India and Tanzania to Turkey, which includes Northwest, East and North-east African countries, Israel, Arabia, Southern USSR, Iran, Iraq, Afghanistan, Pakistan, etc. This species alone is causing damage to agriculture equivalent to millions of rupees during swarming year in India.

Locust ecology

Locust being a polymorphic species occurs in three phases viz. Solitaria, Gregaria and Transiens phase in nature. These phases are density dependent and are reversible. The two extreme phases (Solitaria and Gregaria) differ morphologically and physiologically. The abiotic factors like food, temperature, rainfall and humidity etc. play important role in phase biology. Hussain and Ahmed (1936) reported no diapause in desert locust at any stage and observed 8 generations at temp. 40°C; 7 generations at 36°C and 5 generations at 30°C in a year.

(a) Soil effect

The locust normally breeds in loose moist sandy soil. However, it has been observed that locust lay eggs in almost all types of soil in Rajasthan which is composed of Grey-brown soil, red soil, red and black soil, mixed black and red soil, medium black soil and alluvial soil, etc. Locust has also bred in alluvial and deltaic alluvial soil of Gujarat; black and alluvial soil of U. P., Pedocal sierozene and alluvial origin soil of Haryana, etc. During 1968 a locust swarm laid eggs on the top of Aravalli hills near Alwar too.

(b) Food effect

Desert locust prefers to feed on green, succulent plants/leaves. Bhatia (1940), Bhatia and Sikka (1956) reported that it discarded even preferred plants of semi-dry conditions when offered fresh and succulent plants. Hussain et al. (1949) listed 198 species of plants into three categories i.e. plants not eaten at all (9), plants eaten with great reluctance (29) and plants readily eaten by locust (160). An additional list of 19 plants of Solitaria locust was reported by Bhanotar *et al.* (1973).

Desert locust adults in association with desert aromatic shrub *Commiphora myrrhae* and *Boswellia elegans* which possess high percentage of terpenoids, get quicker maturity Carlisle et al. (1965). Ibrahim et al. (1972) studied the effect of different plants like castor, bean, maize, clover, *Panicum* and *Casurina* on longevity, preoviposition, sexual maturity and fecundity in females.

(c) Temperature effect

Temperature plays an important role on lucust ecology. The daily activity of locust hopper and adult start after sunrise. When the surface temperature slightly warms, hoppers go for basking and start marching in groups between 24°C to 36°C. Again at the time of sunset hoppers go for roosting. Adults start flight activity when atmospheric temperture reaches 14°C and activity decreases when air temperature increases beyond 40°C. The optimum temperature for locust breeding ranges from 27°C to 31°C. Above 45°C temperature may prove fatal to locust. Gregarious hoppers reared at 24°C to develop an unusual deep extensive black pattern, at 33°C the pattern is reduced and at 44°C it is almost absent (Hussain and Ahmed 1936). The author observed that in adults reared at 27°C and above possess seven eye stripes while at 20°C and below possess six eye stripes under laboratory conditions.

(d) Moisture effect

The soil moisture 10 to 25 per cent good both for egg laying and development in desert locust. In soil moisture below 10% either eggs do not hatch or very few eggs hatch out. In excess of soil moisture over 50% due to heavy rainfall eggs get killed or start rotting/decay.

(e) Air-wind effect

The wind has not much role on lucust development but strong wind current can divert locust migration to such a place like sea mountain and areas of high humidity where they die or perish. The classic work of Rainey (1951) on locust plague dynamics in relation to meteorology solved the mistery related with arrival of swarms along with rain in affected countries. According to Rainey (1963) locust swarm moves towards the zones of convergent winds. One such Zone is called 'Inter Convergence Tropical Zone (I.C.T.Z.).' During the northern summer, swarm moves into I.C.T.Z., which at that time lies across Africa from Senegal to Sudan, over the gulf of Aden and across west and northern India.

Role of biotic factors

Locust plagues are governed by periodicity known as locust cycles. It consists of swarming periods and non-swarming periods. Roonwal (1953) recorded ten locust cycles since 1860 to 1963 in India. Each cycle or an average lasted for 11 years which included 7.1 year as swarming period and 3.9 years as recession period

Pradhan (1965) profounded a "new biotic theory" to explain the periodicity of locust cycles in India and stated "Although the biotic factors are themselves modified by abiotic factors, they play quite a decisive role in determining the periodicity of locust cycles and delimitation of areas in which successful locust breeding can take place. Pradhan (1961) stated that "during the period corresponding to the maxima of the sunspot cycle, the weather is mild and therefore, the density of the natural enemies of the locust is high enough even in semi-desert areas not to allow the gregarious breeding of the desert locust. On the other hand during the period of sunspot minima the weather becomes much drier and hotter thereby creating adverse conditions resulting in annihilation of the natural enemies. After reduction of enemy population when sunspot cycle reverts to maxima period, the weather becomes mild and the first showers creating green vegetation tempts locusts to arrive and concentrate in favourable patches or niches, and locust breeding start in an environment of reduced biotic resistance leading within few generations to initiate phase transformation. Thus, locust population explosion continues till the enemy population also gets build up again, either by rehabilitation from the periphery or by local breeding thereby again exerting their check on vigorous locust activity.

There are large number of desert locust natural enemies (Singh, 1970) and majority of them are only as record. Very little work on natural enemies utilization has been made in India or abroad against desert locust. Srivastava and Bhanotar (1983a) worked out predation rate against locust and biology of preying mantid *Mantis religiosa* L. under laboratory conditions. *Nosema locustae* Canning - a protozoan pathogen was for the first time recorded from field population in 1978 of desert locust (Mohangarh : Jaisalmer) (Srivastava and Bhanotar 1983b). The pathogenecity was tested against hoppers and adults (Srivastava and Bhanotar, 1986). Recent status and management in arid region of Rajasthan was published by Bhanotar and Srivastava (1981).

Behaviour of solitarious locusts

Rao (1940) was the first Indian who experimentally proved that it is not only the gregarious swarms which migrated during day time but solitarious individuals also migrated but only at night time due to recent man-made changes brought in Rajasthan desert. It has now been observed that during winter period solitarious population which used to migrate back to West have now started taking shelter in *Khadins,* a man-made catchment area constructed near foot hills where rain water gets accumulated in the rainy season and where some crop is raised (Bhanotar and

Srivastava, 1972). Such area provides suitable ecological nitches for solitarious locusts to exist. The Khadins have now mushroomed up in several places of arid region in Barmer and Jaisalmer. Solitaries are usually found within open grass patches or between bushes when temperature ranges from 25°C to 37°C. In the morning they bask in sun and during noon when surface temperature goes above 37°C roost in bushes to avoid solar glare. Again when temperature comes down below 25°C they become active and continue to bask till they retire for the night in some bushes.

Population density of desert locust

During the non-swarming years adult population density never exceeds 1000-2000/sq. mile in permanent locust breeding ground. For swarming years, Roonwal (1953) gave the figures as 30,000/sq. mile. Various population behaviours are as follows :-

(i) In non-swarming period - 1000/sq. mile
(ii) Inaugurating year of new cycle-20,000/sq. mile.
(iii) In swarming period - 30,000/sq. mile.
(iv) For hopper density - in swarming period figures are :
  (a) In first instar - 25,000/sq. metre.
  (b) In 5th instar - 1,050/sq. metre.

Life history of desert locust

The study of life history of desert locust is important to know for effective and economic control. The life cycle of phase *gregaria* and *solitaria* individuals are identical in respect of mode of copulation, egg laying, oviposition, number of instars, however, there is physiological and colour differentiation. Below is given in brief the life cycle of gregarious individuals as this phase is more important from economical point of view.

There are three stages in locust life cycle

(1) The egg stage : This is the period between depositing the eggs by females into the soil and their hatching. This is commonly known as incubation period.

(2) The hopper stage : The stage between emergence of hoppers and their fledgling into adults. This stage is also known as nymphal stage.

(3) The adult stage : The males are smaller than females. After a fortnight the average weight of male is about 2.2 gm and that of female 3.2 gm. Each adult locust feeds daily equal to its body weight. In a small locust swarm minimum ten millions (1,00,00,000) locust are found who would consume 2,00,000 tonnes of green food daily. Females have two pairs of black tipped ovipositor values at the end of abdomen and are used for digging holes in the soil for egg laying.

Colouration

Immature locusts are pink in colour and later on attain maturity, it turns yellow. Males are much brighter yellow than females.

Copulation

The male crawls upto the female and by a single jump rides on its back while tightening it holds over females by its legs. Polyandry and polygammy are quite common. Copulating pairs are very sluggish and provide good target during control operations. The copulation may lose upto 14 hours. Female lays eggs three times in her life-time and then die with natural death.

Egg laying

Female selects open space with very little vegetation as laying site. She attempts several trial bores with abdomen before finally laying and digs 6 to 10 cm deep hole in moist soil and deposit generally 60 to 80 eggs in a pod and later plugs the pod with frothy substance and withdraw her abdomen. *Solitaria* phase female lays 100 to 120 eggs in a pod. The eggs are rice-shapped and are 7-8 mm in length. Egg laying takes one to two hours. When the soil is dry and hard; they drop eggs on soil surface or on vegetation because fully mature eggs can be retained in the body upto three days only. Eggs can be laid parthenogenetically.

Locust situation in India during 1968-1994

The locust situation for the years 1968-94 is given in Table 1. This clearly indicates that locusts have always remained with us in some intensity and the entomologists are to remain ever alert and should seek new ecological approach to control this pest.

Table 1. Locust situation in India (1968-1994).

Year	Swarms		Remarks
	Indigenous	Exotic	
1968	-	six	Breeding occurred at several localities.
1969	-	-	No locust activity
1970	-	Two (pink)	Entered Jaisalmer from West, hopper breeding in several localities were controlled
1971	-	-	No locust activity
1972	-	Twc	Entered Barmer and Jaisalmer through West. Due to strong wind and drought swarm got scattered and thus no gregarious breeding.
1973	-	six	Entered Jaisalmer from West in August to December and were controlled.
1974	-	Six	138 swarm movement reported during June, July in several district of Rajasthan. With vigorous control operations locusts got controlled.
1975	-	Nineteen	Entered Jaisalmer and Barmer from

Contd......

Year	Swarms		Remarks
	Indigenous	Exotic	
			West in Sept. and November. Adults- and hoppers got controlled.
1976	-	Two	Entered Jaisalmer from West. Breeding in 22 villages got controlled.
1977	-	-	No locust activity.
1978	-	Fifty	Twenty swarms entered India directly from West crossing Arabian sea and others from Pakistan between 8th June to 30th July, 1978 invading Districts of Gujarat (8), Rajasthan (8), Punjab (3), and Haryana (2). Later on 30 swarms entered India between 20th Sept. to 5th Nov. 78. Massive control operations were undertaken and no swarm was allowed to escape.
1979	-	-	No locust activity
1981	-	-	-
1982	Countless population built up	-	Local gregarisation in locust at several places at Jaisalmer and Barmer districts. All population got controlled
1983	-	Thirty	Entered from West during Aug. to Oct. 83. Hopper breeding several localities and were got controlled.
1984	-	-	No locust activity.
1985	-	-	- "
1986	-	-	- "
1987	Few locust Swarms reported	-	Controlled.
1988 to 1992	-	-	No locust activity.
1993	97 exotic swarms	-	Entered from Pakistan in June-August, invaded 640 localities in Rajasthan and Gujarat, which were successfully controlled.
1994	-	-	No locust activity.

International cooperation in anti-locust control

After Uvarov's phase theory (1921, 1928), a series of work was started all over the world to control desert locust. Uvarov suggested that countries having different political views to act on a single platform in International conference to fight this menace, and work unitedly. In initial years such International conferences were held in Italy (1931), France (1932), U. K. (1934), Belgium (1938), Egypt (1936, 1945, 1951), Iran (1942, 1944), India (1943, 1950, 1962) and Pakistan (1949) to exchange and discuss strategies to fight locusts. Anti-locust Research Centre (now C.P.P.R.), London under Uvarov became a prominent centre for locusts. In 1951, FAO recognised the importance of international cooperation in desert locust control and since then has actively monitored the locust information all over the globe and has helped several effected countries with finance for purchase of insecticides, equipments and training of field staff. Field courses were organised (1960-1966) in several countries for the benefits of technical staff. The Govt. of India has agreed since 1955 to participate in International Campaign against the desert locust in foreign lands primarily with the intension to reduce the locust swarm incursion into India. Under UNDP project, India received equipment, books, vehicles and research guidance for its major locust laboratory called " Field station for Investigations on locusts (F.S.I.L.) at Bikaner".

Desert locust management

Male invididuals are smaller in size than female. The surface area occupied by males and females *gregaria* individuals (in q. cms) was recorded by Weisfogh (1952) as 11.9-13.9 (Mean 12.9) and 14.6-16.6 (Mean 15.7) respectively. This surface area could be helpful while finalising the insecticidal doses required for a maximum kill. Desert locust management has now been perfected from what it was in practice in earlier decades. The days of flame throwers, trenching and inorganic chemicals are now over. Locust is now being fought both from air and at ground levels. Since eggs are laid in the soil and the young hopper bands move over the ground surface and the resultant adults are good flier and migratory in habits, so the control operations has to cover all the stages of locusts.

A. Chemical control

1. Baiting : Before the use of DDT and BHC in 1945, inorganic chemicals like sodium fluosilicate and sod. arsenate baits prepared in wheat bran were used. Baiting gave good results under certain favourable conditions only. In India baiting has not been recommended and found unfavourable as it involves lot of labour, water and transportation, which is difficult to get in thinly populated desert regions.

2. Dusting : BHC has proved very successful against locust hoppers and adults. In 1949 BHC dust was first used against locust in India. In 1962 alone 9000 tons BHC 10% was used against locust adults and hoppers at the rate of 20-25 kg/ha. BHC 2.5% (i.e. 0.325 gamma isomer) is effective against first instar, 2nd instar and fledglings. BHC 5% (i.e. 0.6 gamma isomer) is effective against 3rd instar hoppers and BHC 10% (i.e. 1.3 gamma isomer) is effective against 4th, 5th instar hoppers and adults. Dusting is done generally in the evening or early morning when the temperature remains low and locust remains sluggish on bushes (roosting period) or forming groups (Basking period) with the help of mannual or power driven dusters.

3. Spraying : In 1951 aldrin was first sprayed against desert locust in India. Till 1958 conventional ground operation and spraying through aircraft were used and large quantity of water was required. Sayer (1959) evolved exhaust nozzle sprayer where minimum water is needed. The aldrin, dieldrin, lindane and malathion are more popularly used for locust control. Their dose rates are given in Table 2. One gallon of insecticide is sufficient to cover over 10 hectare against hoppers by exhaust nozzle sprayer, which is fixed on petrol driven jeep. It is so designed that 7 litres of sprays liquid per kilometre is sprayed when vehicle runs at 7 km/hour in Ist gear. If the width of sprayed area is between 100-250 metres, average deposit is 0.5 lit/ha with average droplet size being 70 μ. For control of eggs in egg-laid fields by spraying of higher percentage (2.5%) neem kernel solution would reduce the percentage of hopper emergence due to ovicidal action (Bhanotar and Srivastava, 1984). This will minimise the environmental pollution.

4. Aerial spraying : The arial spraying is resorted where the infested area is vast and vehicular approach is not possible and the time is short. The aircraft usually flies 30 metres high with speed of 8-10 km/hour and spray droplet size of 100-150 μ. Two types of sprayers are used in aircraft (i) Boom and nozzle-used for normal crop protection, (ii) micronair rotary automiser—mainly used for locust control and occasionally for crop protection. In a day an aircraft could spray 1000 hectare with boom and nozzle and 1500 hectares by micronair rotary automizer. Many new techniques such as barrier spraying, blanket spraying, target spraying etc. have been evolved to economise on locust control.

The air-craft used, under Directorate of Agricultural Aviation, are Beavers and Basant and is supplemented by private operators fleet.

Earlier desert locusts were controlled by mechanical and chemical methods. With the invention of exhaust nozzle sprayer (Sayer 1959) for ground operation and micro-automizer (Micronair), for aerial operations have revolutionized the effectiveness.

Table 2. Dose rate of insecticides used in locust control.

Name of insecticide	Dose rate	Remarks
Aldrin	70-280 gr (1 to 4 Oz.)/ha	Cheap, effective toxic and persistent
BHC	150-560 (1 to 4 Oz.)/ha	Relatively cheap, effective less persistent than dieldrin.
Dieldrin	20-150 (0.3-2 Oz.)/ha	Highly toxic to locusts, highly persistent. Mammalian toxicity more than Aldrin.
Malathion	280-1400	Quick acting than BHC, less persistent but more expensive.

better type of control and minimize the insecticidal wastage. Sayer (1959) used ULV (Ultra low volume) drift spraying technique against desert locust as lower dose as 0.30 lit/hectare of dieldrin in the form of fine spray. The average droplet size being 70 μ to the sparse vegetation and thus could get effective control of hoppers under various weather conditions both in desert and bush. With the advent of micronair, it becomes possible to reduce the droplet size of liquid insecticide upto 50 micron, a swath range of 144-160 metres and a deposit range of 5.5 grams to 14 grams of technical dieldrin per acre is obtained. Still it appears that more scientific methods of control is to be developed in view of Uvarov (1928) statement.

B. Biological and ecological management

Several workers in recent years suggested biological and ecological control strategies to tackle the desert locust problem (Bhanotar and Srivastava, 1983).

Gunn (1960) has stated that "not one has yet been deliberately brought under control by ecological methods". At present all effective and economical control depends on chemical insecticides and it is unlikely that ecological control will soon out chemical control. Pruthi (1967) stated that role of predators and parasites is not important, only weather is important factor in locust ecology. He stresses on the need of efficient forecasting system and suggested that planned sampling of soil of dormant eggs in the autumn be undertaken to assess the maximum population possible after hatching in the following spring.

Pradhan (1961, 1965) in his "biotic theory" stressed mainly in reptile predators found in locust breeding zones of Thar deserts. The quantitative estimation of predation rate of single *Uromastix hardwickii* in 24 hours against various instars of desert locust was worked out in laboratory at IARI, Delhi (Bhanotar *et al.* 1980). Pradhan (1969) suggested an integrated control method for locust control, where he suggested that first try to check locust population explosion in locust out-break areas with the help of natural enemies. If out-break occurred, then spray 0.1% aq. solution of neem kernel, an antifeedant on the crops before locust invasion. It remains effective for 15 days (Pradhan *et al.*, 1962 and 1963). Lastly, for wiping out settled locust swarm on sprayed crop, use insecticides. Singh (1974) has found an ornamental herb "Sukhdarshan *(Crinium bulbispermum)* possessing antifeeding property against locust.

Exploitation of the weak links in the life cycle of desert locust

For economical and effective control attention on the weak links found in life cycle should be exploited in desert locust (Srivastava and Bhanotar, 1981). Weak links will give equal opportunity to other control methods and will significantly reduce the excessive use of insecticide which cause problem of health hazard and environmental pollution. Weak links in locust life are :

(1) Period of copulation,

(2) Period of egg laying,

(3) Incubation period and hatching period.

(4) Controlling egg fields,

(5) Roosting period,

(6) Basking period,

(7) Moulting period and

(8) Movement and grouping behaviour of hoppers.

Keeping the above weak links in view, the controlling cost could be easily reduced if proper time and dose of insecticide is applied i.e. at weak link points.

A squadron consisting of 3-4 squads may only manage settled swarm/mating swarm/ovipositing swarm from location to location, other specific squad for controlling egg fields, similarly other for noon, dusk and early morning operations and finally few squads for assisting aerial operation. Thus specific squad should perform some specific functions only for undertaking specific control based on scientific methodology.

Management of desert locust—the future problem

Srivastava and Bhanotar (1981) discussed the impact of new Rajasthan Canal "Indira Gandhi Canal" on the locust ecology. New agricultural development now being planned in arid region of western Rajasthan because of coming up of 'Indira Gandhi Canal", would increase the incidence of pests complex manifold. The ecological features also reveal that any means of increasing local water supply in the desert tends to produce conditions favourable for the desert locust. The irrigation scheme on the Red Sea coast have already proved stable habitats to locusts. Uvarov (1960) Cited several examples from USSR, Australia and North America where acridid population have shifted from grasses to cultivated crops.

Similarly in India, it is apprehended that instead of sustaining the locust problem, it may rather increase the locust activity. Viewed against locust requirement and seasonal migration in solitarious population, it is visualised that favourable eco-nitches may mushroom up in the desert favouring locust multiplication, leading to phase transformation and swarm formation, thereby causing serious damage to new and coming up cultivation.

Future plan for preventive technology

1. Locust surveys should be based on more scientific lines and technical staff should be made fully aware of the population dynamics of locusts which provides sufficient hint for the future shape of population likely to occur.

2. An effort to rehabilitate the predatory fauna controlling locusts be taken up at right earnest. This may also help in checking population explosion.

3. Technique to deploy specific pathogen affecting locusts be developed and sprayed in large scale on green patches at least during summer monsoon periods. When newly emerged hoppers will feed on such vegetation they will be contaminated and will die slowly.

4. Serious efforts to plant foliage trees like 'neem' be initiated on large scale in the desertic region along Rajasthan canal (Srivastava and Bhanotar,

1981). These plants after establishing will stabilise sand soil, provide shady umbrella and increase the nesting habitat of predatory birds. This may further help in checking the build up of solitarious population by using neem kernel suspension as insecticide.

5. More light traps on a permanent basis to catch locusts at night be installed at suitable nitches particularly during monsoon and post monsoon period. This would also help in eliminating locust concentration.

References

Bhanotar, R. K., and Srivastava, Y. N. 1972. Behaviour of solitary locust in relation to physical environment in Thar Desert. *Ent. Newslet.* **2** (6) : 40.

Bhanotar, R. K. and Srivastava, Y. N., 1981. Status and management of desert locust - the future problem. *Trans. Isdt. A. Ucds.* **6** (4) : 20-22.

Bhanotar, R. K. and Srivastava, Y. N. 1983. 'Desert Locust' in : *Agricultural Entomology.* Vol. II Pub. All India Scientific Writers' Society, New Delhi, pp. 438.

Bhanotar, R. K., Srivastava Y. N. and Bhatnagar, R. K. 1973. Additional list of food plants of solitaries in the desert region of Rajasthan, *Ent. Newslet.* **3** (9) : 59-60.

Bhanotar, R. K. and Srivastava, Y. N. 1984. Effect of neem kernel suspension in the development of eggs of desert locust *Schistocerca gregaria* (Forskal). *Neem Newsletter,* **1** (3) : p. 30.

Bhanotar, R. K. Bhatnagar, R. K., Srivastava, Y. N. and Mahto, Y. 1980. Bio-ecological studies on spring tailed lizard "*Uromoastix hardwickii* Gray, Part III. Quantitative Analysis and consumption of various instars of desert locust hoppers by *Uromastix* sp. *J. Zool. Soc. India,* **30** (1+2): 21-35.

Bhatia, D. R. 1940. Observations on the biology of desert locust (*Schistocerca gregaria* Forsk.) in Sind-Rajputana desert area. I. the preferred food plants of the locust. *Indian J. Ent.* **2** (2) : 187-192.

Bhatia, D. R. and Sikka, H. L. 1956. Some striking cases of food preference by desert locust *Schistocerca gregaria* (Forsk.) *Indian J. Ent.* **18** (3) : 205-210.

Carliste, D. B., Ellis, P. E. and Betts, E. 1965. The influence of aromatic shrubs on sexual maturations in the desert locust, *Schistocerca gregaria. J. Ins. Physiol.* **11**: 1541-1558.

Gunn, D. L. 1960. The biological control background of locust control. *Ann. Rev. Ent.,* **5** : 279-300.

Hussain, M. A. and Ahmed, T. 1936. Studies on *Schistocerca gregaria* Forskal. VI. Influence of Temperature on the intensity and extent of black pattern in the desert locust hoppers bred crowed. *Indian J. Agri. Sci.* **6** (3) : 624-664.

Hussain, M. A., Mathur, C. B. and Roonwal, M. L. 1949. Studies on *Schistocerca gregaria* (Forsk.) XIII. Food and feeding habit of Desert Locust *Indian J. Ent.,* **8** (2) : 141-163.

Ibrahim, S., Tawfik, M. F. S. and Abd El Hadi, N. H. 1972. The effect of different food plants on the adult stage of *Schistocerca gregaria* Forsk. Part. II. UNDP/DLP, No. AGPDL/TS/**11** : 18-27.

Pradhan, S. 1961. Probable role of biotic factors in the periodicity of locust cycles. *Indian J. Ent.* **23** (1) : 1-6.

Pradhan, S. 1965. A new biotic theory of the periodicity of locust cycles. *Indian J. Ent.* **27** (1) : 93-101.

Pradhan, S. 1969. Integrated pest control for developing countries. *World Sci. News.* **6** (2) : 10-12 and 22.

Pradhan, S. Jotwani, M. G. and Rai, B. K. 1962. Neem seed deterrent to locusts. *Indian Farming,* **12** (8) : 7, 11.

Pradhan, S., Jotwani, M. G., and Rai, B. K. 1963. the repellent properties of some neem products (to locust). *Bull. Reg. Res. Lab. Jammu,* **1** (2) : 149-151.

Pruthi, H. S. 1967. "Agricultural Entomology" Pub. Indian Council of Agricultural Research, New Delhi, pp. 284-300.

Rainey, R. C., 1951. Weather and movements of locust swarms, a new hypothesis. *Nature, London,* **68** : 1057-1060.

Rainey, R. C. 1963. Meteorology and the migration of desert locust : applications of synoptic meteorology in locust-control *World Met. Org. Tech.* **54** :117.

Rao, Y. R. 1940. The desert locust in India. Pub. Indian Council of Agricultural Research, New Delhi, pp. XIX + 721.

Roonwal, M. L.1953. Food preference experiments on the desert locust, *Schistocerca gregaria* (Forskal) in its permanent breeding grounds in Mekran. *J. Zool. Soc. India, Calcutta,* **5** (1) : 44-58.

Sayer, H. J. 1959. *Bull. ent. Res.* **50** : 371-386.

Singh, K. M. 1970. Bibliography on the natural enemies of locusts and other aricdids, Division of Entomology, IARI, New Delhi, pp. 27.

Srivastava, Y. N. and Bhanotar, R. K. 1981. Suggestion for planting Neem *Azadirachta indica* (Syn : *Melia indica)* along with Rajasthan Canal. *Trans. Insdt.* and *Ucds.,* **7** (1) : 56-69.

Srivastava, Y. N. and Bhanotar R. K. 1983a. Studies on predation rate of preying mantid *Mantis religiosa* L against desert locust and its biology. *Indian J. Ent.* **45** (3) : 348-352.

Srivastava, Y. N. and Bhanotar R. K. 1983b. A new record of a protozoan pathogen "*Nosema locustae* Canning infecting desert locust from Rajasthan. *Indian J. Ent.,* **45** (4) : 500-501.

Srivastava, Y. N. and Bhanotar, R. K. 1986. Laboratory studies on effect of protozoan pathogen *Nosema locustae* Canning against eggs and hoppers of desert locust. *Indian J. Ent.* **48** (4) : 420-427.

Uvarov, B. P. 1921 "Phase theory in desert locust. *J. Bombay Nat Hist. Soc.* **27**: 803-812

Uvarov, B. P. 1928. "Locusts and Grasshoppers" A hand book for their study and control. Imp. Bur. Ent. London, pp. XVII + 352.

Uvarov B. P. 1960. UNSF Prog. Rep. UNSF/DL/TC/2 : 3-8.

Recent Advances in Indian Entomology
Editor: O.P. Lal
APC Publications Pvt. Ltd., New Delhi, 1996, pp. 216-227

Host Plant Resistance Against Insect Pest in India

O. P. LAL and Y. N. SRIVASTAVA

Insect pests cause heavy losses each year due to which about 25 to 30 percent crop yields are damaged in fields and stores by different insect pests. The crop loss which was estimated to be about Rs.6000 crores a year in 1973-74 has been now estimated to be around Rs.20,000 crores a year (Lal, 1992) With the advancement in production technology, we have now a number of high yielding varieties, hybrids, chemical fertilizers, increased irrigation facilities and good agronomical techniques. The high yielding varieties and hybrids getting high dose of fertilizers and irrigation have also good plant growth which is quite suitable for pests and diseases (Lal, 1992). On such plants, insect pests develop fast and their fecundity rate is also high. As a result they become very vulnerable to insect pests and are badly damaged.

There are a number of insecticides which may be used to control insect pests. But, the indiscriminate use of pesticides has led to many problems like adverse effects on parasites, predators, pollinators, toxic residues causing health hazards, resurgence of insect pests, development of resistance in insects to insecticides and pollution of the environment. It has been also observed that in the areas where more pesticides are used, are getting more pests problems. The states of Andhra Pradesh, Tamil Nadu, Punjab and Haryana, which are using more pesticides have more problem of insect pests and also more pest out break are recorded there.

The best control method of insect pests is, therefore, to grow a resistant variety, if available. Even a partially resistant variety will be of much importance as it will require less insecticidal sprayings. It also fits well in the Integrated Pest Management programmes without any extra cost to farmers. Hence it provides the best alternative control of insect pests.

Resistance

It is the inheritable trait that enables a plant to inhibit the growth of insect population or to recover from injury caused by populations that were not inhibited to grow. Painter (1951) defined resistance as the relative amount of heritable qualities possessed by the plant which influence the ultimate degree of damage done by the insect. According to him, there are three main categories of resistance:-

1. *Non-preference* : It is the inherited feature of a host plant which discourages the feeding, colonisation or oviposition of an insect pest.
2. *Antibiosis* : It is the adverse effect on development and reproduction of insect.

3. *Tolerance* : It enables a host plant to withstand an attack by insect without suffering severe damage.

Other terms used for resistance are :-

(i) Non-acceptance

(ii) Antixenosis

(iii) Pest avoidance

Pest avoidance is expressed as tendency of the plant to escape infestation. Kogan and Ortman (1971) have proposed the word "Antixenosis" for non-preference.

There are six main types of resistance (Table 1) as proposed by Jones (1979).

Developing insect resistance programmes

In order to develop a programme for insect resistance, one may need several informations essential for taking up the job. This will include:

1. the nature of the crop
2. insect pests involved
3. initial informations and data regarding:-
   (a) possible sources of resistance from varieties and strains locally available.
   (b) possibility of acquiring new germplasm that may carry resistance.
   (c) determination of some of the basic properties of the plant responsible for resistance.
   (d) hybridization to combine gene for resistance with desirable agronomical characters.
   (e) study of resistance in advance generations.
   (f) study of genetics of plant resistance to the insect.
   (g) study of resistance of released varieties in plots to evaluate resistance on an insect control method.

Genetic nature of resistance

Knowledge of genetic nature of resistance of host plant can be of great value in breeding programme. The mode of inheritance has been used to distinguish three main categories of resistance viz., monogenic, oligogenic and polygenic in which resistance is controlled by one, few or many genes. Oligogenic resistance is normally dominant and relatively stable from environmental influence. But, the majority of monogenic resistance is biotype specific.

Some of the above said basic aspects have been utilised in defining the mechanism of resistance and its inheritance in crop plants. Several techniques are applied to use the available resistant genes.

(a) *Sequential Release of varieties* - It is to incorporate the genes into several improved varietal background and releasing them sequentially enabling the farmers to choose the variety depending on the biotype identified.

(b) *Accumulation of Major Genes* - It arised by combining two or more major genes into an improved variety.

Table 1. Different level of resistance in relation to insect pests and host plants.

		Cause of resistance	Reaction of pest to host	Effect on host	Multiplication
(1)	Immunity	Unattractive unpalatable Unsuitable toxic	Ignored or quitted	Undamaged	None
(2)	Susceptibility	Attractive palatable nutritions non-toxic	found fed-upon	heavily damaged	to limit
(3)	Hypersensitivity	toxic saliva or excreta	found	damaged	decreased or
(4)	Resistance	Less attractive	foundless readily fed	less heavily	decreased
	Non - preference Antibiosis	less pallatable less nutritious some what toxic	upon less raidly non-preferred	damaged	
(5)	Tolerance	as (2), (3) or (6)	As (2), (3), or (6)	replaces damage parts or list neighbours	(As (2), (3), or (6)
(6)	Evasion	Life-cycle of host and pest not fully synchronised	As (2), (3), (4) of (5) of	escape some damage	decreased

Source : Jones, 1979.

(c) *Developing multiline Varieties* - It involves incorporation of several major genes into an isogenic background and mixing the seeds of these lines to form a multiline variety.

(d) *Accumulation of Minor Resistance Genes* - The minor resistance genes accumulated through a process of recurrent selection and the intensity of resistance is increased.

Field screening and gradation of varieties

While screening the varieties under field conditions, population of insect pests are recorded and according to the pest populations or the intensity of infestation. The varities are classified into different grades. These may be:

1. Immune
2. Highly Resistant
3. Resistant
4. Moderately Resistant
5. Moderately Susceptible
6. Susceptible
7. Highly Susceptible

The population of insect in each category may vary according to pest infestation level or the type of insect. For example, if it is aphid on mustard, it may be classified differently but if it is shoot and fruit borer in brinjal it may be a different type of classification.

Varietal resistance in different crops

Rice

Several varieties with resistance to major pests are grown in the country. Gal midge (*Orseolia oryzae*) whcih is a serious pest in Warangal distrtict of A.P. caused 100% losses in late sixties and early seventies. But, with the introduction of resistant varieties like Surekha and Phalguna, it is of no consequence these day (Balasubramanian, 1986). Similarly, in Raigad district of Maharashtra, stem borer tolerant varieties are being widely used. The resistance in rice as reported againt different insect pests is given below:-

(a) *Stem borer (Scirpophaga incertulus)*

The variety TKM-6 has a wide base of resistance to insect pests. It gave rise to the high yielding and resistant variety, Ratna. Many accessions like IET 3093, 5265, 5561 and 8284 were identified as resistant. IR-20, 26, 36 and 42 possessed resistance to stem borer (Balasubramanian, 1986). It has been reported by Pathak (1990) that there is no systematic inheritance of resistance in rice. Misra et al., 1990, reported that low nitrogen and high silica is responsible for resistance.

(b) Gall midge (Orseolia oryzae)

The varieties, Phalguna, Surekha, Usha, Vikram, Shakti, Kakatiya, BPT 1235, Divya and PTB series have resistance to gall midge (Balasubramanian, 1986). Antibiosis was found to be the chief mechanism of resistance to this pest (Joshi and

Venugopal, 1985; Paroda and Malik, 1990). The resistance is governed by a single dominant gene. But, gall midge biotypes have also developed due to severe selection pressure on the pest.

(c) *Brown Plant Hopper (Nilaparvata lugens)*

ASD-7, 9, ARC-6650, 14766, PTB-18, 21, 33, Manoharsali, Leb Mury, Nahug, Karivennel are reported as donors of resistance to BPH (Paroda and Malik, 1990). PTB-33, IR-64, IET-7575, 6315, MTU-4870, 5293 are also resistant to BPH.

(d) *Green Leaf Hoppers (N. virescens, N. nigropictus)*

Jogannath, IR-60, 62, 40, 30, 32,PTB-18 are resistant to leaf-hoppers and Rice tungro virus are not associated (Narasimhan et al., 1990).

(e) *Multiple Resistance*

The lines which posses resistance to 5 or more pests like BPH, gall-midge, stem borer, white backed plant hopper, leaf-folder, green leaf hopper and thrips are:

CR 294-54-8, OR 131-6-154, AD 14-1-39, RP 2068-18-3-5, RP 2068-18-3-5, RP 1579-58,VRS 163-2-2-2-3, TNAULFR 831311, RP 1579-1633

Sorghum

It is attacked by many pests from germination till harvesting. Singh and Rana (1986) reported the varieties CSV-5, 6,7R, 8R and hybrids CSH-7R and 8R as resistant to shootfly, *Atlherigona soccata*. Glossy leaf surface and trichomes are associated with resistance and are governed by monogenic recessive genes (Gibson and Maiti, 1983). The varieties, CSV-2,3,4,5,6,7,8R, 9R, 10, 12 Swati and hybrids, CSH-7R, 8R, 9R, 10, 12, are reported tolerant to the stem borer, Chilo *partellus*. Recently P-311 and SPV-1015 (PGS-1) have shown resistance to the stem borer, (Prem Kishore, 1987, 1991). The sorghum midge *(Contarinia sorghicola)* has assumed major importance with the introduction of high yielding varieties of sorghum. However, ICRISAT variety, ICSV-197 is found resistant to this pest. JAM-2566 and IS 12666C also possess resistance. Short and compact glumes, and tannin content of grains conferred resistance to this pest (Singh, 1987). However, all midge resistant sources are highly susceptible of shootfly. Early sowing and high seed rate give less shootfly damage.

Maize

Stem borer, *Chilo partellus*, is a serious pest of maize. Panwar and Sarup (1980) identified source of resistance in Antigua Gr1, EH-4030, 4027, JML 305 etc. Deccan Hybrid, Ganga-5, Kundan and Chanda are tolerant to *C. partellus* and antibiosis is the chief mechanism of resistance composite-217 is tolerant to stem borer and shootfly (Siddiqui et al., 1987). Harsh and Ashwini are tolerant to stalk borer.

Pearl millet

Thakur and Sharma (1990) reported that the lines ICMPES-28, 29 and 38 have resistance to the armyworm, *Mythimna separata*. Pearl millet varieties viz. MP-9, MP-15, MP-60, MP-95 and PSB-8 showed multiple resistance to shootfly, stem borer and grey weevil (Prem Kishore 1991).

Wheat and barley

Brown wheat mite is an important pest of wheat against which the variety C-306 is found tolerant. In barley, DL-117 or DL-200 is resistant against the aphid, *Rhopalosiphum maidis*. Vidya Sagar Singh et al., (1980), reported that hull less barley varieties are more susceptible to attack by the the weevil, *Sitophilus oryzae* than hulled varieties.

Sugarcane

It is attacked by a number of insect pests. Some of the varieties showing resistance/tolerance against different pests in different regions are mentioned below in Table-2 (David, 1985).

Pulses

In pigeon pea, PDA-881E, PDA-882E, PDA-883E are resistant/tolerant to pod-fly *(Melanagromyza obtusa)*, while JAT-10, ICPL-6, UPAS-120, ICP-903E, BDNT, ICP-1903E are resistant/tolerant to pod borer *(Heliothis armigera)* (Anonymous, 1988-89). In green gram, ML-337, 428, 4221 are tolerant to whitefly, jassid and yellow mosaic virus, ADT-3 is tolerant to yellow mosaic virus and stemfly.

Oilseeds

In groundnut, NCA-17840, NFG-79 and EC-21898 are reported resistant to tobacco caterpillar, *Spodoptera litura* (Rajagopal et al., 1989). Kundu and Misra (1985) found that soybean variety, PK-625, was resistant to stem fly *(M.sojae)*, root knot nematode and YMV. Laha-101 of mustad is found resistant to the aphid, *Lipaphis erysimi* (Teotia and Lal, 1970). Bhaketia (1991) has given some morphological and biochemical characters associated with aphid resistance (Table 3 and 4)

Table 2. Insect pest resistance in sugarcane varieties.

Pest	Region	Resistant/tolerant varieties
Top borer	Punjab, Haryana & Western U.P.	Co J. 65, Co 859, Co 1158, Co. 7224.
Internode borer	Tamil Nadu & Andhra Pradesh	Co 975, Co 62175 Co 6806, Co C 77-1.
Scale insect	Andhra Pradesh, Tamil Nadu, Maharashtra, UP, Haryana	CoC 671, Co 62174 CoC 6907, CoS 611, Co 1132,
White grub	Tamil Nadu	Co 6304.
	U.P.	Co 1158, CoS 510.

Table 3. Some morphological characters for aphid resistance in oilseeds.

S.No.	Plant character	Plant species	Aphid reaction
1.	Smaller and hardy inflorescence with loosely packed buds	*B. Juncea*	Resistant
2.	Darker leaves	*B. juncea, B. campestris* var. brown sarson	"
3.	Non-waxy plants	*B. rapa*	High resistance
4.	Hairiness of plant	*Eruca sativa*	Non-preference
5.	Tender inflorescence and thickly packed pattern of buds	*B. campestris*	Susceptible
6.	Yellowish and green leaves	*Brassica rapa*	More attraction

Table 4. Some biochemicals responsible for aphid resistance in oilseeds.

S.No.	Biochemical constituents	Plant species	Reaction/response to aphid
1.	Total sugar, phenolic compounds, potassium, sulphur and iorn.	*S. sativa*	High concentration imparted resistance
2.	Total ash, nitrogen and phosphorus	*B. napus*	low in resistant varieties
3.	Glucosinolates, especially sinigrin	*Brassica* spp. *B. juncea* *B. cannata* *B. oleraceae*	High conc. responsible
4.	Allylisothioxyanate (1x10-4 mg/min as threshold level of activity)	*Brassica* spp.	Higher release rate as repellent and lower rate as stimulant

Fibre crops

Cotton is attacked by 130 insects and many are quite injurious to the crop. Bikaneri Nerma is moderately hairy and resistant to the jassid, *Amrasca biguttula biguttula*. Surekha PS-141 and Kanchan are resistant to whitefly (*Bemisia tabaci*). A-1318, Bes-9-7 and FBRN-3-7 are tolerant to boll worms, *Earias vitella and H. armigera*.

Vegetables

In tomato, Parker, Bonus and VFN-8 are report to possess resistance against the borer, *H. armigera* (Lal, 1985). Kakar et al., 1990, reported the genotype 122775 and the cross Rick X Solan Gola resistant to *H. armigera.*

In okra, AE-30 has resistance against the leaf-hopper, *A biguttula. biguttula.* Other cultivars with tolerance to the leafhopper are Punjab-7, IIHR-4, AC-3375, Reshmi, Sangli-2, Okra Red, Sel. 8 and Sel. 6 (Lal, 1991). It has been also reported that all the crosses involving Ghana species had very high jassid population but visibly they had no curling, cupling or hopper burn symptoms (Lal, 1991). At Rahuri, AE-79 was found tolerant to the shoot and fruit borer, *E.vitella*. AE-22, 52, No. 5325, 6327, 6701, AE-71, Narnaul Special and Perkins Long Green are also moderately resistant to *E.vitella.* Kalloo (1985) have reported the pressure of sticky hairs as a mechanism of resistance to green house white fly and aphids on tomato. Mahal and Singh (1982) reported that inheritance of resistance to the leaf hopper in okra was polygenic and was controlled mainly by additive gene effects.

In brinjal, Sm. 202, Sm 17-4, Pant Samrat and Pusa Purple Cluster are resistant to the shoot and fruit borer, *Leucinodes orbonalis* Guen (Lal, 1991 a). The varieties BI-7 and 435-20 are having resistance to the leaf hopper, *A. biguttula biguttula.*

In cabbage, Red Rock Mammoth, Red Drum Head, Glory, Early Queen, Express Mail, Rainy Princess, All season and Red Pickling are moderatily resistant to the aphid, *L. erysimi* (Lal et al., 1987). Many of the red cultivars were resistant to the cabbage caterpillar, *Pieris brassicae* (Lall, 1991b).

In califlower, the cultivars resistant to the head borer, *Hellula undalis* are ES-97, E.S.-96, Kartiki (J.B.) K.W.-.5, KW-8, K.W.-10, Kuwari (R.B.), Kathmandu Local, Early Patna, EMS-30 and P.S.K.-16.

In chilli, Pusa Jawla, is resistant to mosaic, leaf curl, aphids, jassids and nematodes. Among cucurbits, Arka Suryamukhi of Muskmelon, Improved Long Green of cucumber, Pilibhit Padmini of spongegourd, Soft Green of bitter gourd, Smaller Sugar of pumpkin, Akra Tinda of round gourd and Sultans Long Green of bottle gourd are reported to be resistant (Lall, 1977).

Fruit trees

There has been very less work on the varietal resistance in fruit trees. In ber, the cultivars Tikadi, Illaychi, Katha, Umran, Tasbataso and Desi Alwar have been reported comparatively resistant against fruit fly, *Carpomyia vesuviana* Costa. (Sharma et al., 1993). In mango, the varieties with relatively soft flesh sush as Alfonso, Bangalora, Neelum etc. are more susceptible to the mango store weevil, *Sternochaetus mangiferae.*

In guava, Seedless Guava *and* Apple Guava are more attached by bark eating caterpillar, *Indarbella* spp.

Insect nutrition and its impact on insect pests

The plants vary in food quality and its nutrients which affects the development, fecundity and mortality of insect pests. Concentrations of nutrients in plants may be affected by its varieties, age, growing conditions and ecological factors. Fertilizers and water change the proportion of nutrients in crop plants which may result into increased pest population. All the three major nutrients viz., nitrogens phosphorus and Potash are required for a good crop production. But, in many cases, farmer's generally use only the nitrogen fertilizers. This increases the amino-nitrogen concentration of plants which is condusive for quick development and fertility of insect pests. However, Potash can counteract the ill effect of nitrogen and provide a general resistance to plants. Enhanced supply of potash reduces the amino nitrogen and increases the tissue silica concentration which hardens the sclerenchymatous tissues and the cell wall. This makes the leaves unpalatable for pests and they also find it difficult to bore through the firm outer tissues.

Carbohydrates are the most widely occurring substances in the plants. Sugar content of plants may act as a feeding stimulant to pests (Table-4). Low level of soluble sugar in plants affects the fertility of aphids. Higher level of total carbohydrates and lower total nitrogen and amino acids are found responsible for leaf hopper resistance in cotton. Lower concentration of amino acids in pea and maize make them resistant to aphid and corn earworm, respectively. Lower fertility rate of pea aphid is found in plants deficient in C, Mg, N, P, and K. Lack of thiamin, nicotinic acid, folic acid or chaline reduce the fecundity of the fruitfly, *Dacus dorsais*. Ascorbic acid deficient corn lines are resistant to European corn borer. Similarly plants that do not provide enough fat to insects have a detrimental effect on insects body.

Role of allelochemicals in plant resistance

Chemicals of this group may be put in different categories:

1. Isoprenoids.
2. Acetogenins
3. Shikimic Acid and Acetate Derivatives
4. Alkaloides
5. Protease Inhibitors
6. Glycosides.

Isoprenoids include full range of terpenes which act as allomones in plants. This may affect insect behaviour, sensory, physiology, metabolism and endocrinology. Some plants contain isoprenoidal hormones affecting metamorphosis, fecundity, development and longevity of insects.

DIMBOA and other benzoxazolinones from maize are important acetogenous molecules which impart resistance to certain insects.

It has been reported that in young corn plants DIMBOA and related compounds are more which deter the larvae of European corn borer from feeding and their development and survival is also reduced. Presence of DIMBOA in wheat seedling gives resistance against aphids. Gossypol, a yellow polyphenolic pigment contained in the subepidermal glands of cotton plants is a source of resistance to the boll-worm *Heliothis sp.*

Similarly, tannin, charysanthemin and anthocyanin contents of plants affect the insect population of plants.

Alkaloides which are basically nitrogen containing compounds affect the nervous system of insects. The true alkaloid nicotine in *Nicotiana* functions as allemone against many insects and hence used successfully as an insecticide. Solanine acts as a deterrent against *Leptinotarsa decemlineata* in potato but tomatine may be an attractant for *Epilachna* beetles. Gramine is found to give resistance in barley against aphids. The mustard oil glucoside e.g. sinigrin and sinalbin are responsible for attraction and feeding to the larvae of *Plutilla xylostella*. Arcelin, a major seed protein in french bean has toxic effect on bruchid pests. Oxalic acid in rice varieties is a chemical factor governing their resistance. Barley resistant cultivars possess high concentrate of benzyl alcohol which is a resistance factor against green bug, *Schisophis graminum*.

Thus, it is clear from above discussions that the chemically based resistance is a major component of plants total defence armament against insect pests.

Problems associated with breeding for resistance to insects

Although resistant varieties are the best ways to tackle the pest problems and to save the crops from tremendous losses but there are certain reasons that we are not able to get a break through in this line of pest control. Some of these are:-

1. Failure of entomologists and plant breeders to complete their task of developing resistant cultivars with desirable agronomic attributes after identifying the insect resistant germplasm.
2. Inadequate knowledge of the biology and feeding habits of the insect and the mechanism of resistance.
3. Inadequate knowledge of inheritance of resistance.
4. Development of insect biotypes.
5. Problem of incorporating the multiple resistance.
6. Lack of efficient screening techniques.
7. The tendency to separate crop production and protection.

Summary

The insect pests are causing tremendous losses in fields and stores. Chemical control measures pose problems of toxic residues, adverse effects on parasites, predators, and pollinators, resurgence of insect pests, insecticides resistance in insects, health hazards, environmental pollution etc. Hence the best method to control the insect pest is to develop a resistant or a tolerant vartiety. It fits well in the integrated pest management programmes. There has been some work on varietal resistance in rice sorghum, maize, sugarcane, cotton, oil seeds, pulses, vegetables etc. But, the work done so far is not enough and needs more concerted effort on this aspect. It requires more knowledge on insect biotypes, efficient screening techniques, mechanism of resistance and incorporation to multiple resistance. The knowledge of genetic engineering should be used to increase the vigour, yield and pest resistance in plants. Hence there is a need of better inter-disciplinary approach to develop pest resistant varieties.

Some of the varieties/hybrids reported with different degrees of resistance to insect pests are : Surekha, Phalguna, Ratna of rice; CSV-5, CSV-6, CSV-7, CSV-7R, ICSV-197, JAM-2566 of sorghum; Antigua Gr 1, JML-305, Deccan Hybrid, Ganga-5, Kundan, Chandan of maize; DL-117, DL-200 of barley; C-306 of wheat; PDA-881E; PDA-882E, JAT-10, ICPL-6, UPAS-120 of pigeonpea; ADT-3 of green gram; PK-625 of soybean; Bikaneri Nerma of cotton; VFN-8 of tomato; AE-30, 11HR-4 of Orka' Sm 202, Pant Samrat, Sm 17-4 and B1-7 of brinjal; Red Rock Mammoth, Glory, Red Drum Head of Cabbage; Pusa Jwala of chilli etc.

References

Anonymous 1988-89. Annual report. Department of Agricutural Research and Education, Ministry of Agriculture, Govt. of India, New Delhi.

Balasubramanian, M. 1986. Host resistance in Integrated Pest Management in rice. *Plant Prot. Bull.,* **38** : 31-34.

Bhaketia, D.R.C. 1991. *Insect Pests. in Oilseed Brassica.* In Indian Agriculture (Ed. Chopra, V.L. and Shyam Prakash). Har. Anand Publisher, New Delhi, 315 p.

David, H. and Alexander K.C. 1986. Integrated pest and diseases management in sugarcane. *Plant Prot. Bull:* **38** (1-4):37-49.

Gibson, P.T. and Maith, R.K. 1983. Trichomes in segregating generations of sorghum matings. 1. Inheritance of presence and density. *Crop. Sci.* **23** : 73-75.

Jones, F.G.W. 1979. Proc 1979 British crop-Prot. Conference-Pests and diseases, **3**:741 - 752.

Joshi, R.C. and Venugopal, M.S. 1985 Gall length and its relation to plant height in rice varieties susceptible to gall midge. *Oryza* **22** : 132-133.

Kalloo, G. 1985. Vegetable Breeding, Vol 11 CRC Press 96-135.p.

Kogan and Ortman 1971. Antexenosis - a new term for antibiosis. *Bull. of Ent. Soc. of Amer* **17**:250.

Kundu, G.G. and Misra, S.D. 1985. Field evaluation of some germplasm for resistance to stemfly *Melanogromyza sojae*, root knot nematode, *Meloidogynae incognita* and yellow mosaic virus. *Ind.J.Ent.,* **47** : 230-232.

Lal, O.P. 1985. Field resistance of some tomato cultivars against the fruit worm, *Heliothis armigera (Hb) in Tripoli. Bull. Ent.* **26**:96-97.

Lal, O.P. 1991. Pests of Okra and their management. A decade of research of Horticultural crops (1980-90). In : Group discussion of Entomologists working in the coordinate projects of horticultural crops 12th and 13th August 1991. Central Instt. Hort. for Northern Plains, Lucknow. SIV 53-59.

Lal, O.P. 1991a. Vareital resistance in the egg plant, *Solanum melongena* against the shoot and fruit borer, *Leucinodes orbonalis. J. Pl. Dis. and Prot.,* **98**:405-410.

Lal, O.P. 1991b. Studies on field resistance of different cultivars and crosses of cabbage against cabbage caterpillar, *Pieris brassica* L. (Lepidoptera:pieridae). *J.Pl.Dis. and Prot.* **98**:40.

Lal, O.P. 1992. Toxic hazards of Pesticides in Indian Agriculture. *Proc. Workshop on safety hazards in Process Industries*. Organised by Varanasi Regional Centre, Indian Institute of chemical Engineers Nov. 14,1992, pp. 17-28.

Lal, O.P. 1992. Pest management in Vegetable crops. *Farmer and Parliament* **27**: (7) : 17-24.

Lal, O.P., and Gill, H.S., Ram Singh 1987. Varietal resistance in cabbage against mustard aphid (*Lipaphis* erysimi). *Cruciferae Newsletter*, U.K. 12:83.

Lall, B.S., 1977. Pests in tropical crops. True Files, *Cacus spp*. In. Diseases, Pests and weeds in Tropical crop. (eds.) Kranz, J., H. Schmutter and W. Koch (1977), Verlag Paul Parey, Berlin, pp. 524-26.

Misra, B.K., Sontakke, B.K. and Mohapatra, H. 1990. Antibiosis mechanisms of resistance in rice varieties to yellow stem borer - *Scirpophaga incertulas. Ind. J. Plant. Prot.* **18**:81-83.

Mahal, M.S. and Balraj Singh 1982. Inheritance in okra to the cotton jassic. F.Field studies. *Indian j. ent.* **44**:1-12.

Narasimhan, Y., Sivaraj, K. Kareem, A.A., Khush, G.S., and Saxena, R.C. 1990. Registance of selected breeding lines to GLH and rice tungro virus. *Int. Rice Res. N.L.* **15**:10.

Painter, R.H. 1951. Insect Resistant in crop plants Mcmillan, NY. 520.

Panwar, V.P.S. and Sarup, P. (1980). Differential development of chilo partellus (Swinboe) in various maize varieties. *J.Ent. Res.* **4**:28-23.

Paroda, R.S. and Malik, S.S. 1990. Rice genetic resources, its conservation and use in India. *Oryza.* **27**:361-369.

Pathak, M.D. (1990). Genetics of sorghum, maize, Rice and Sugarcane resistance to the cereal stem borer, *Chilo spp. Insect, Sci. Applic,* **11** : 689-699.

Prem Kishore 1987. P-311-A new sorghum variety to stem borer, Chilo partellus (Swinhoe). *J.ent.Res.* **11**(1) : 115-118.

Prem Kishore 1991. Sources of resistance amongst world pearl millet *Pennisetum typoides* (Burm.), germplasms to important insect pests. *J. ent. Res.***15**.

Prem Kishore 1992. SPV-1015 (PGS-1)-A new sorghum variety endowed with multiple resistance to shootfly, *Altherigona soccata* Rondani and stem borer, *Chilo partellus* (Swinhoe) *J.ent.Res.,* **16** (4):319-321.

Rajagopal, K., Nandagopal, N., Bhagat, M.R. (1989). Preliminary screening of virginia ground nut genotypes against tobacoo caterpillar *S. litura. S. oilseeds Res.* **5**:162-165.

Siddiqui, K.H., Marwaha K.K. and Sarup, P. (1985). Performance of early maturing composites to locate sources of multiple pest resistance in mansoon (Kharif) and Spring sown maize. *J.Entomol. Res.* **12** (1):1-3.

Singh, B.U. (1987). Varietal resistance in sorghum to midge, *Contarinia sorgicola* coquillet (Diptera: cecidomycidae). *Insect Sci. Applic* **8** : 129-144.

Singh, B.U., and Rana, B.S. (1986). Varietal resistance spotted stem norer *chilo partellus* (Swinh). *Insect Sci. Applic.* **10**:3-27.

Teotia, T.P.S. & Lal, O.P. (1970). Differential response of different varietie and strains of oleiferous brassicae to aphid - *Lipaphis erysiimi Labdev Journal of Science and Technology,* **88**:219-226.

Thakur, R.P. and Sharma, M.C. (1990). Identification of pearl millet lines with multiple resistance to erogt, smut and oriental army worm.*Ind.J.Pl. Dis. and Prot.* **18**(1) : 47-52.

Singh, Vidya Sagar, Bhatia, S.K. and Murthy, B.N. (1980). Effect of hull on the resistance of barley varieties to the rice weevil *Sitophilus oryzae* (Linn) infestation. *Indian J. Ent.* **42**: 576-581.

Recent Advances in Indian Entomology
Editor: O.P. Lal
APC Publications Pvt. Ltd., New Delhi, 1996, pp. 228-234

Role of Appliances and Application Technique in IPM

DINESH CHANDRA and RAJESH KUMAR

Crop production in our country is not increasing for the last 3-4 years. It seems, it has reached at peak. In major crops both of rabi & kharif, yield estimates are not able to reach near the target. It is a matter of great concern. While mentioning about production in agricultural field, one has to think about protection also. Here comes the need of *Pesticide Umbrella*, which Dr. Pradhan, a renowned Entomologist has visualised and suggested over 2½ decades ago. "Protection Research more needed than Production Research", (Pradhan, 1973). Crop protection is in the limelight for two reasons : *first* EC countries are currently formulating a unified crop protection policy as part of "Europe 1992", *second*, the use of pesticides is under review by several national governments, owing to increasing environmental concerns, (Jansma *et. al.*, 1993).

The damage due to pests in India is estimated to be nearly 5500 crores annually (Rajagopal, 1993). There is a great need to characterize both quantitatively and qualitatively, the population dynamics and mobility of our major crop-pests. For both the pests and their natural enemies, much basic information is required, (Glass, 1975).

Long term life table studies are perhaps generally prohibitive, but short-term population dynamic studies in which the causes and extent of mortality are tabulated in a life-table will often prove useful relative to insect pests. Particularly essential in basic information on the potential of natural enemies to fully or adequately control various actual or potential pests in the absence of disruptive effects of agricultural practices (especially pesticides, but may also include, for example, the manner of crop refuse destruction). These studies will require extensive, detailed and often frequent monitoring and sampling of population, (Glass, 1975). Most of the work in plant protection in India is concentrated on studying biology of the pests and bio-efficacy of the pesticides. Recommendations are made on intelligent usage strategies to minimise side effects or retard resistance (Narasimhamurthy, 1994).

Control of pest species and conservation of non-target species is the new strategy to be adopted in the IPM (Rajagopal, 1993). For IPM we require minimal use of pesticides applied so as to give an optimal effect in controlling the specific target pest population, while conserving natural enemies. Complete elimination of target pest will indirectly render the environment sterile for support of natural enemies, (Matthews, 1973). The judicious use of chemical pesticides is a critical part of population management. The important consideration is that pesticides to be used in a manner that is harmonious with other elements of agro-eco-system and; if possible, to augment other control agents. A highly specific environmentally sound microbial material that gives a lower initial kill may be better than a conventional pesticide that kills 100% of a target population, (Matthews, 1973).

As such the use of pesticides has been widely practiced to get the effective control, the maximum kill to have immediate suppression of pest during natural outbreaks like the one of recent locusts which has covered larger areas of Rajasthan, part of Gujarat, Haryana, Punjab and and M. P. In the present situation, increasing amounts of pesticides are being used throughout India, mostly with manually operated equipment, (Matthews, 1993).

Proper and Rationale use of these chemicals with suitable appliances will not only serve the purpose but also help in reducing the pollution in the environment. Pollution to either air, food, fodder, fibre or water finally becomes hazardous to animals and human beings. The chemical, thus used should hit the target either directly or indirectly, (Chandra, 1982). Unfortunately, crop protection is often poor as a result of incorrect timing (aimed at large visible larvae rather than young instars hidden inside bracts) poor choice of pesticide and inaccurate application (Matthews, 1993). Target may be insect itself or leaf or stem (since insects are harbouring and damaging). It is possible only if a proper appliance is used. (Chandra, 1982).

Pesticides are applied with various type of appliances. For the selection of appliance it is essential to have a thorough knowledge of *Pest* - situation, susceptible stage along with mobility, *Pesticide* - formulation, concentration and equipment available (Chandra, 1982). *Spraying equipment* often selected without consideration of objectives of rational pesticide management viz., correct dose, controlled placement and accurate timing of application (Narasimhamurthy, 1994). So it is very important to know the droplet spectra from the nozzle/disc/orifice in all machines/equipments, as this gives the drift potential and the likelihood of inhalation and dermal absorption of chemicals (Thornhill, 1991). The spray volume per second, the volume of spray, the concentration in the spray tank, the pump pressure, the throttle speed, the length of the lance or discharge tube, the height of the nozzle/disc/orifice, and the number of nozzles are all important features, that greatly influence the safety of the spray operator, (Lee, 1993). Farmers should avoid making cocktails of insecticides as such inseciticide is likely to be applied at sublethal dose resulting in inadequate control and there are chances of developing resistance, (Deshmukh, 1994). *The main appliances are :*

Dusters, Granule applicators, Sprayers, Fogging machines,

Smoke & Vapour generators, Agricultural helicopters/aircrafts.

Dusters : useful, less expensive, less toxic and easy to operate with little maintenance. Four types: bellow type, plunger, rotary and mist blower-duster.

Common principle - to utilize air blast that spreads the particles to form a cloud. Dust present in the chamber is constantly agitated with the help of brushes, scrappers, plates or by air itself. Pollution hazards and the inability of the dust to stick properly with the foliage are the draw-backs. Wet dusting is often considered better.

Granule applicator : Granules are the dry formulations of a particular size ranging between 250-2000 microns (μ). Granules are applied in different ways depending on the type of application, crops etc. Insecticide granules or even urea may be easily applied by the broad-casters. Granules are normally applied in the seed furrows or in the leaf whorls (in case of maize, sorghum, sugarcane etc.), hand held applicators are also used.

Spraying equipments are used for application of chemicals mainly PESTICIDES

1. *Insecticides* against crop pests, storage & godowns, houses, kitchen etc.
2. *Acaricides* against mites
3. *Bactericides & Fungicides* to control diseases
4. *Herbicides* to control weeds
5. *Hormones* to increase fruit set or to prevent dropping of premature fruits.
6. *Nutrients* plant nutrients as foliar spray.

Sprayers working on *pneumatic energy*

1. Hand Sprayer 2. Hand held compression sprayers
3. Knapsack compression sprayer

Sprayers working on *hydraulic energy*

1. Syringe 2. Hand sprayer 3. Bucket pump - stirrup pump
4. Knapsack sprayer 5. Rocking sprayer 6. Foot sprayer
7. Wheel barrow sprayer.

Energy/Nozzle : appliance

Hydraulic - flood jet, flat fan, cone.

Gaseous - pneumatic - mist blower, spray gun, mist sprayer

Centrifugal - rotary

Thermal or Hot tube - swingfog

Hydraulic - energy nozzel - small chamber with a device inside it to atomise the liquid before it comes out of the orifice.

Types :-

1. Jet or solid stream nozzle - simple exit hole or a tube designed for producing jet.
2. Impact or flooding or defector nozzle - a jet of liquid strikes on a smooth solid surface at a high angle of incidence to have a deflector of stream.
3. Flat fan or fish-tail nozzle - It has low atomising power.
4. Cone or swirl nozzle - The spray liquid rotates owing to its passage through a

spiral groove. It forms a hollow cone or ring shapped spray pattern.

5. Solid cone nozzle - Forms the disc pattern spray.
6. Triple action nozzle
7. Low volume nozzle

Gaseous - energy nozzle

1. Shear or blast nozzle - It feeds the spray fluid into the air stream with in the nozzle (eg. motorized knapsack). There is one another where water pipe comes at right angle to air current (eg. spray gun).

2. Annular nozzle - (eg. exhaust pump).

Centrifugal energy nozzle - spinning discs are capable of producing U.L.V. spectrum.

Kinetic energy nozzle or vibrajet nozzle used for herbicidal application. *Thermal energy nozzle* or hot tube nozzle used for making fog. Fluid is injected into a stream of hot gases to shear the liquid into finar droplets which are immediately vaporised & fog is formed.

G.S. Sandhu at Ludhiana University introduced a copy of the Turbair battery-driven spinning disc sprayer known as the *Fogair* and both ASPEE and Thomson Motoronics, Ahmadabad (Helispray) have marketed a similar machine. (Matthews, 1993)

Coverage

1. Contact pesticide - Full coverage, coarse droplets

2. Fungicide - Full coverage, coarse droplets

3. Herbicide - Course to the need coarse droplets, not to drift.

4. Systemic pesticide - Fine droplets - which are abosrbed by plant surface. No drift problem, dosage is exact.

5. Granular pesticide - Specific coverage, dosage exact, placement at the target.

Coarse droplets -runs off lossess, placement of chemical only on one side.

Finer droplets - Drift losses and pollution hazards, placement of chemical on both sides due to Gyrating movements.

Vol.	*Energy*	*Appliance*
H. V.	Hydraulic	Knapsack, hand compression sprayer
C. L. V.	Hydraulic	foot sprayer

L. V.	Gaseous	Motarised knapsack mist blower
U. L. V	Centrifugal	Mist blowers or C. D. A., electrodyne
Fog	Thermal	Fog machine
Smoke	Thermal	Smoke generator
Vapour	Thermal	Vapour generator

Droplet size classification

(Vmd)	>400 u	Coarse spray, high volume 300-500 litre/ha.
	201 - 200 μ	Medium spray 200-350 litre/ha.
	101 - 200 μ	Fine spray 5-100 litre/ha.
	51 - 100 μ	Mist spray 1-5 litre/ha.
	1 - 50 μ	Aerosol/Fog 1 litre/ha (less than 1 litre)
	.001 - 1 μ	Smoke
	<.001 μ	Vapour

Vmd = Volume median diameter

μ = micron, 1mm = 1000 microns

Droplet size	*Situation & nature of pest insects*
10 - 30μ (Aerosol)	Flying mosquito & small insects
30 - 60μ (Aerosol)	Locust hoppers & forest pests
70 - 150μ	ULV crop spraying
150 - 220μ	LV crop spraying
200 - 240μ	CLV (conventional low volume)
240 - 360μ	HIV (High vol. crop spraying)
250 - 400μ	Highly toxic chemical spraying
400 - 1000μ	Herbicide spraying

Distance from which spectrum of droplets is to be taken

45 cms (1.5') - High vol. to conventional low vol.

90-120 cms (3'-4') - L. V.

4.5m-6m (15'-20') - U. L. V.

Mist Blower - The discharge rate is adjustable, between 0.6 to 2.5 litres/ minute (speed of engine 5500 to 6000 r.p.m.) volume of air discharged will be 1000 times the volume of discharged liquid. As well as 4 times higher concentrate chemical as compared to high volume spray may be used. Nozzle should not be held closer than 2-3 metres from the target otherwise deposition will not be proper.

Thermal or hot tube nozzle - Fog composed of very fine droplets which can be produced by condensing a pesticide which has been injected into a stream of hot gas to shear the liquid into droplets which are immediately vaporised. These vapours condense into very fine droplets which can float in the air. Thus, coverage is excellent, suitable for green houses or in closed vicinity i.e. godowns etc. more suited for flying or hidden pests.

Finer droplets will be more effective than the bigger ones provided the right concentration of the chemical has been utilized and it reaches the target before drying or evaporation. It can reach the target more easily with the slightest wind current while bigger ones rolls off. Summarising our knowledge, we may say that study of application technique is undoubtedly necessary but study of pest, its biology and ecology is equally important. A proper appliance has to be selected for desired results before selecting an appliance, we must keep in our mind -

i) What type of discharge required

ii) What type of nozzle will be suitable

iii) How much distance to be maintained between nozzle and target to get effective spectrum

iv) What should be the speed of the operator and

v) What time the operation be undertaken.

By slight modification, the equipment may be made more useful for wider application purposes. So, This study will help us :-

1. To enhance the efficacy of the (sprayer) appliance to suit the particular situation
2. To make it available for wider uses by slight modification
3. To modify it avoiding maximum drift, hazards and overflow of material.
4. For this we have to be familiar with the different parts of the appliance/machine and their role on the machine in detail.

References

Agarwal, R. A., Gupta, G. P., Kishore, P. and Chandra, D. (Eds) 1982. Principal and Concepts of Integrated Pest Management. ICAR, New Delhi. 212 pp.

Bals, E. J. 1975. The importance of controlled droplet application (CDA) in pesticide application. *Proc. 8th Br. Insectic. Fungic. Conf.,* : 153 - 160.

Bals, E. J. 1977. A new Rotary stacked disc nozzle. *British Crop Prot. Conf. Pest and Diseases.* : 523 - 525.

Bhatia, G. N. 1971. Recent developments in Pesticide Application Technology. *Pesticides Annual.* 1971. 43 - 44.

Bindra, O. S. and Singh, H. 1971. Pesticide Application Equipment. Oxford and IBH, India.

Chandra, D. 1982. role of appliances in pest management. *In* Principals and Concepts of Integrated Pest Management. ICAR : 114-118.

Deshmukh, M. G. 1994. Safety in storage, transportation and field application of pesticides. *Pestology* **18** (1) : 5-8.

Glass, E. H. (Coordinator) 1975. Integrated Pest Management : Rationale, Potential, needs and implementation E. S. A. special publ. 75-2. *Entomol. Soc. Am.*, College Park, MD.

Groner, H. 1979. Application Technology, Spl. Edition, 26 pp BASF, Germany.

Jansma, J. E., van Keulen, H and Zadoks, J. C. 1993. Crop protection in the year 2000 : a comparison of current policies towards agrochemical usage in four European countries. *Crop Protection* **12** (7) : 483-489.

Johnstone, D. R. 1973. Insecticide concentration for Ultra - low volume crop spray application. *Pest. Sci.,* **4** : 77-82.

Lee, S. A. 1993. Reducing the exposures of spray operators to pesticides in Malaysia. *Pesticides Information* **21** (2) ; 25-34.

Matthews, G. A. 1973. Nozzles for Pesticide Application in the Tropics. *PANS,* **19** (4) : 583-600.

Matthews, G. A., and Clayphon, J. E. 1973. Safety precautions for Pesticide Application in Tropics. *PANS,* **19** (1): 1-12.

Matthews, G. A. 1979. Pesticide Application Methods. Longman, London & New York.

Matthews, G. A. 1993. Pesticide Registration, Formulations and Application in India. *Pesticides Information* **17** (2) : 1-4.

Mehrotra, K.N. 1989. The extent of pesticide use in India and its consequences In Pesticide Residues in Food, Ahmadabad : Consumer Education and Research Centre, : 42-58.

Narasimhamurthy, R. L. 1994. Application - The most neglected aspect of plant protection. *Pestology* **18** (1) : 13-14.

Patel, S. L. 1975. Recent Advances in Application Techniques. *Agril. Mech. in Asia* : 64-71 & 82.

Patel, S. L. 1978. Plant Protection Equipment for small and marginal farmers. *Agril. Mech. in Asia* : 24-30.

Rajagopal, D. 1993. Integrated Pest Management and Judicious Use of Pesticides. *Pest Management* Nov. 1993.

Pradhan, S. 1973. In Tropics - Protection Research more needed than Production Research. *Pesticides Annual,* Bombay.

Thornhill, E. 1991. Measurement of spray droplets. Lecture *6th Course on Pesticide Application Technology* organised by UPM and MAPPS (notes compiled by Dzolkhifli Omar and Matthiew Vanhaeke) 11-15 Nov. Serdang Selangor.

Turnbull, G. J., Sanderson, D. M. and Crome, S. J. 1985. Exposure to pesticides during application. *In* Occupational Hazards of Pesticide use. :35-49. Taylor & Francis, London & Philadelphia.

Recent Advances in Indian Entomology
Editor: O.P. Lal
APC Publications Pvt. Ltd., New Delhi, 1996, pp. 235-244

Recent Advances in Biological Control

ATMA RAM

Biological control which is the ingenious and resourceful use of parasitoids, predators and pathogens to suppress pest population is a strong weapon in our pest control armoury. The value of naturally occurring parasitoids and predators in regulation of pest density is well established. Out of 30 lakh species of insects only 10,000 are considered as pest of crops, man and domestic animals (Sailer 1969). It is variously estimated that from less than 1 to perhaps 2 percent phytophagous insects that are potential pest ever achieve the status of even minor pests (DeBach 1974). This method has proved effective and economical in dealing with pest problems (Rosen and DeBach, 1992). The natural enemies are either introduced from one area to another or the performance of those already present in area is improved by various manipulations. In this paper I will discuss some of the major developments in the recent years.

Classical biological control

Importation and establishment of exotic natural enemies is known as classical biological control. Ever since the spectacular success of *Rodolia cardinalis* (Muls.) into California in 1888 to control *Icerya purchasi* (Mask.), the classical biological control remain the most promising approach for forseeable future. This method of control offers effective, permanent, economical and hazard free solution to serious pest problems. Worldwide attempts at classical biological control have been made against 416 species of insect pests belonging to eight orders and significant degree of success has been achieved against two fifth of them - 75 species have been brought under complete control, 74 under substantial control and 15 under partial control in atleast one country. Many of these successes have been repeated in other countries, so that altogether there have been 384 successful projects worldwide, of these 156 have been rated as complete success, 164 as substantial and 64 as partial control (DeBach and Rosen, 1991, Rosen and DeBach, 1992).

Besides its effectiveness the classical biological control is economical. The benefit cost ratio of classical biological control has been calculated as 30 : 1 as compared with 3 : 1 ratio for chemical control (Pimentel et al. 1981, DeBach and Rosen 1992).

The rate of establishment is also impressive, out of 4226 natural enemies importation in 153 countries, about 30 per cent have established, the fate of an additional 22 per cent is unknown (DeBach and Rosen, 1991, Rosen and DeBach, 1992). This is an impressive record considering that by comparison only one in the 15,000 or 20,000 candidate chemicals ever makes it to the marketing stage as a pesticide (Rosen and DeBach. 1992). Thus the importance of exotic natural enemies remains to be the leading force in controlling the insect pest.

Mass production of parasitoids and predators

The potential use of parasitoids and predators for mass release programme will depend on our ability to produce large numbers of quality adults at a reasonable cost. The mass rearing of parasitoid on host is an expensive and difficult proposition. with a few exception such as production of *Trichogramma* (Morrison and King. 1977). However. numerous centres throughout the world are engaged in mass production and field releases of parasitoids and predators for controlling a variety of pest species. Large scale rearing programmes have used factitious laboratory host for the production of parasitoid and predators. Morrison and King (1977) described techniques for mass multiplication of parasitoids and predators of some important insect pests using factitious hosts.

Trichgramma spp. are mass reared on the industrial scale in countries like China. USA. USSR and a few others. A mass production system used in USA for rearing *T. pretiosum* on eggs of *Sitotroga cerealella* yielded on an average 13.465,000 parasitoids daily for field releases (Morrison, 1985).

In china about 500 million *Trichogramma* adults per day are being produced (Hsiao and Kang, 1981). In canada a system was developed to produce large numbers (30 x 10^6 per week) of *T minutum* using eggs of *Sitotroga cerealella.*

Development of artificial diet for parasitoid is a major development in the mass production technology. The initial attempts to rear parasitoid, Trichogramma was with limited success (Guan etal. 1978; Hardin and Hoffman, 1981; Xie etal. 1986, 1988). The microbial contamination was major problem associated with the rearing technique. Several antibiotics with concentration as high as 0.5 % were found non-toxic to parasitoid and inhibiting microbial growth, permitting successful rearing of parasitoids. *Trichogramma* spp. viz., *T. dendrolimi, T. chilonis, T. nubilale, T. chilotraeae, T. Japonicum* and the Eupelmid, *Anastatus* sp. were mass reased in vitro on artificial diet. The medium consist of 2-3 ml oak silkworm (*Antheraea* pernyi) pupal haemolymph, 0.5 ml of yolk, 1.5 ml of 10% malt solution and 0.5 ml Neisenheimer's salt solution (Li etal. 1988). Vinson etal. (1988) also reported successful rearing of *T. pretiosum* in Vitro on artificial diet using egg yolk, powdered milk suspension, distilled water and haemolymph of sphingid, *Manduka sexta.* Liu etal. (1988) reported a technique to produce successfully the artificial egg cards for *Trichogramma chilonis* and *Anastatus* sp., using a micro-computer-control machine. The machine automatically complete five processes, setting up the synthetic membrane, forming the egg shell, sealing the double - layered membrane and separating into egg cards. The machine could produce 1200 cards per hour which could be used to rear 6-7 million individuals of *Trichogramma.* The antiseptics were used to prevent the egg cards from becoming contaminated with fungi or bacteria. The antiseptics were effective during continuous rearing of *T. chilonis* for 40 generations. The adults of *T. chilonis* produced from artificial eggs are qualitatively as good as those reared on the eggs of Oak silk worm, *Antheraea pernyi.*

Mass production of *Trichogramma* spp. and *Anastatus* in vitro on artificial diet and manufacture of artificial host egg cards has also been reported by Li (1990). The artificial diet consisting of the medium haemolymph (44.4%), wheat germ+milk powder (33.3%), egg yolk (11.1%) and Nielson salt (11.1%), was also developed (Xing and Li 1990).

The braconid, *Bracon millitor* and the pteromalid, *Catlaccus grandis* ectoparasitoids of *Anthonomus grandis* were reared in vitro for the first time on artificial diet devoid of insect component. When these parasitoids were reared in vitro, on larvae of *A. grandis,* on semiartificial diet, the duration of life cycle from egg to adult were ca. 15.5, 16.5.and 15.0 days, respectively for *B. mellitor* and 16.0, 17.5 and 15.5 days, respectively for *C. grandis*. The percentage adult emergence was 58, 25 and 65, respectively for *B. mellitor* and 50, 28 and 60, respectively for *C. grandis*. Male Female ratios were ca. 45 : 55 for both the species.

Augmentation of natural enemies

The importance of entomophgous arthropods (parasitoids and predators) in suppressing pest population has been well established. However, sufficient numbers may not always present to provide adequate control. Biological control by augmentation of natural enemies that are mass produced in the insectaries or collected from areas of abundance and transferred to desired area, provide means of insuring the presence of number that are adequate to maintain pest population at desired level. Augmentation may involve (1) inoculative releases where the control of the pest is expected from the progeny produced by the released natural enemy or (2) inundative releases where the natural enemies are released in sufficiently large numbers so as to have direct and immidiate impact on the pest population.

Biological control by augmentation of natural enemies against a variety of pest worldwide has been reviewed by various workers (Ridgway and Vinson, 1977). However, efforts made in India are not covered there in.

The natural enemies utilized in USA include, 3 billion adults of *Trichogramma* spp. against lepidopterous pests on wide range of crops, 203 million adults of *Aphytis melinus* against california red scale, 26 million *Cryptolaemus montrouzieri,* a predator of mealybug, 18 million *Chrysoperla carnea* a general predator of soft bodied insects (Starler and Ridgway, 1977).

In USSR, 50 billion adults of *Trichogramma* are produced per season and used for the control of many species of lepidopterous pests that damage grain, deciduous fruits, citrus, mulberry, sugarbeet and vegetable crops in glass houses and fields (Beglyarav and Smetrick, 1977).

In China *Trichogramma* species are produced in large numbers and released extensively against European corn borer, pine caterpillar, rice leaf roller, sugarcane borer and other lepidopterous pests infesting cotton. On cotton alone, the parasitoids have been released in over 680,000 hectare annually (Huffaker, 1977). During 1985-87, *T. chilonis* produced on artificial eggs were inundatively released in 400 ha. of sugarcane field for control of sugarcane borers. The rate of parasitism was 70-86% which was similar to that obtained when the parasitoids were reared in the eggs of Oak silkworm *Antheraea pernyi* (Liu et al. 1988).

Besides these 170 million *Encarsia sp. (Hym., Encyrtidae)* annually was released for control of whiteflies in glass houses in Britain and the Netherlands, 32 million *Opius sp.* (Hym., Braconidae) against olive flies in Sicily and over 4 million *Lixophaga* (Dip., Tachinidae) against sugarcane borers in Cuba (Starler and Ridgway, 1977).

Effective control with *Trichogramma* has also been demonstrated in Europe where the European corn borer is controlled by *T. evanescens* on approximately 20,000 ha. of corn (Ferguson, 1992).

In Canada releases of *T. minutum* at rate of 8-23x10^6 female/ha against *Choristoneura fumiferana* resulted in 42-83% reduction in the larval population of the following year while release of 23 x 10^6 female/ha and one application of Btk following spring reduced budworm population by 93% (Smith etal. 1990).

Augmentation of natural enemies in India

(a) Sugarcane borers

In India, the results of field releases of *Trichogramma* against sugarcane borers carried out between 1930 and 1950 were claim of both success and failure. According to Shankaran and Manjunath (1988) the results of early experiment may not be wholly reliable due to unsatisfactory experimentation and presentation of data. The results of the experiments carried out during 1970s and later indicate that the inundative releases of *Trichogramma* may be useful. In the Tamil Nadu weekly release of 40,000 adult of *T. chilonis* per ha from the 4th to the 11th month of the crop have been found to reduce damage by *C. sacchariphagus indicus* (Sithanantham etal. 1973). Large sclae field trial carried out in different sugarcane factory area of Tamil Nadu also showed usefulness of this parasitoid (Varadharajan, 1976; Sithananthan, 1980; Solayappam, 1980). In Andhra Pradesh periodic releases of *T. chilnis* in continuous area of sugarcane field throughout the year can give effective control (Kamalakara, Rao, 1980). In Punjab the parasitoid is reported to be effective against *Acigona steiniellus* Hamp., when releases were made at the rate of 1,25,000 parasitoids/ha. against each brood (Varma etal. 1979). In north Bihar weekly releases of 50,000 adults/ha were reported to give encouraging results in control of *C. infuscatellus* and *C. auricilius* (Mishra etal., 1986; Pawer, 1987)

(b) Sugarcane leaf hopper, Pyrilla perpusilla walker

Banerji (1974) observed lower incidence of *Pyrilla* in the eastern Uttar Pradesh and attributed this to greater conservation of its natural enemies, *Epiricania melanoleuca* and *Tetrastichus pyrillae.* This has led to the use of *E. melanoleuca* for the control of *Pyrilla* in the area of scarcity. The population of *E. melanoleuca* is fairly high in eastern Uttar Pradesh. The cocoons and egg masses of the parasitoid are collected and transported to areas of scarcity within the state and to other states for field releases. The parasitoid has been colonized in states of Andhra Pradesh, Gujarat, Karnataka, Kerala, Madhya Pradesh, Maharashtra, Orissa,

Rajasthan, Uttar Pradesh and West Bengal (Mishra and Pawar, 1981, 1982, 1984, Pawar etal. 1980; Jagdish Prasad etal. 1981). Generally 4000 to 5000 cocoons or 4 to 5 lakh eggs per ha. provide good control of the pest (Pawar 1983). The number of cocoons to be released per hectare varies with population density of host (Madan, 1985).

(c) Cotton bollworms

Periodic releases of egg parasitoid *Trichgramma chilonis* and predator *Chrysoperla carnea* have been found to result in substantial control of *Helicoverpa armigera* and *Earias vittela* infesting cotton in Tamil Nadu, Gujarat and Karnataka (Manjunath 1992).

(d) Spodoptera litura

Releases of the egg parasitoid *Telenomus remus* (Hym., Scelionidae) introduced from Papua, New Guinea @ 50,000 parasitoids 0.2 ha./ week gave parasitisation up to 60% of *S. litura* infesting cauliflower at Anand (Gujarat). However, mass releases of *T. remus* and *Chelonus heliopae* an egg larval parasitoid against *S. litura* on tobacco proved ineffective (Patel et al. 1979).

(e) Scale insects

The predatory lady beetle *chilocorus nigritus* and *Pharoscymnus horni* were multiplied in the laboratory at Nidadavole (Andhra Pradesh) and released in sugarcane field for control of sugarcane scale *Melanaspis glomerata*. The total number of adult predators released during 1980-81, 1981-82 and 1982-83 were, respectively 13,300, 49,000 and 50,100, *C. nigritus* and 3,000, 41,200 and 45,700 *P. horni*. The predators are reported to have become well established (Muktilingam et al. 1983).

(f) Chilo partellus

Periodic releases of a high fecundity strain of *Trichogramma exiguum* gave complete control of *C. partellus* on sorghum at Delhi (Anonymous, 1982).

Genetic improvement of natural enemies

Genetic improvement of natural enemies could be achieved by employing artificial selection. The natural enemies may be selected for climatic tolerance, improved sex ratio; host finding etc. The selective breeding resulted in the improvement of different biological characteristics (see table 1). Hybridization is the other way to bring about genetic improvement. Hybridization has been attempted for improving vigour or fitness through heterosis. Hoy (1975) found that a hybrid strain of gypsy moth parasitoid *Apanteles melanocelus* was easier to mass rear than the three strains from which it was derived.

Semiochemicals in the management of parasitoids

The behavioural chamicals involved in the habitat and host location offer great potential for the manipulation of parasitoids. It is now well established that the host plant provide cue to host habitat. Elzen etal. (1984) identified several terpenes from cotton that would attract *Campoteles sonorensis* an ichneumonid attacking

Helicoverpa on cotton from a distance. These chemicals or synomones (a chemical release by one organism that effect the behaviour of a second for the benefit of both) differ from one variety of plant to another and some cultivars were unattractive to parasitoids and lack terpenoids (Elzen, William and Vinson, 1984). This suggest that certain unattractive plant could be made attractive by application of these chemicals to plants by spraying. Thus Altieri etal.(1981) applied extract of *Amaranthus sp.* (a weed plant) to soybean and tomato and increased parasitism of *Helicoverpa* eggs by *Trichogramma* sp. *Aphidius ervi, Praon* sp., *Aphelinus flavus, Lysiphlebus fabarum* and *Aphidius rhopalosiphi parasitoids of Rhopalosiphum padi* responded positively to the odour of the plant. *A. rhopalosiphi* showed strong positive response to 3 wheat volatiles (Cis - 3 - hexenyl acetate, cis - 3 - hexen-1-01 and trans-2-hexanal) as well as to indol-3-acetaldehyde (a break down product of tryptophan in honeydew to its host *R.* padi). *A. rhopalosiphi* distinguish and prefer the variety of wheat on which their development has occurred compared to other wheat varieties (Wickremasinghe and Van 1992). The pattern of attact by the tachinid parasitoid *Cyzenis albicans* on its host *Operophtera brumata* indicated the

Table 1. Results of selective breeding.

	Species	Traits	References
1.	*Dahlbominus fuscipennis* (Zett.) *Aphytis lingnanesis Compere* *Trichogramma chilonis Ishii*	Improved climatic tolerance	Wilkes (1942) White etal. (1970) Abraham & Pradhan (1976)
2.	*Dahlbominus fuscipennis* (Zett.) *Aenoplex carpocapsae* (Cushman)	Sex ratio	Wilkes (1947) Simmonds (1947)
3.	*Trichogramma minutum* Riley	Host finding ability	Landaluze (1950) Urquijo (1946)
4.	*Horogenes molestae* (Uchida) *Paratheresia claripalpis* (Wilkes)	Host preference	Allen (1954) Box (1956)
5.	*Macrocentrus ancylivorus* Rohwer *Trioxys pallidus Holiday* *Aphytis holoxanthus DeBach* *Metaseiulus occidentalis* (Nebitt)	Insecticide resistance	Pielou & Glasser (1956) Hoy etal. (1990) Robertson (1957) Hoy (1986)
6.	*Trichogramma fasciatum*(Perkins) *Dahlbominus fuligenosus* (Nees) *Dahlbominus fuscipennis* (Zett)	Increased fecundity	Atma Ram and Sharma (1977) Szmidt (1972) Wilkes (1947)

use by parasitoid of a volatile cue present in the foliage of host plant. (Oak) (Roland and Denford 1992).

Annadurai etal. (1992) showed that *Trichogramma chilonis* and *Chrysoperla scelestes* exhibited behavioural response to host plant odour from different cotton cultivars. The efficiency of both *T. chilonis* and *C. scelestes* varies within different parts of the cotton plant . This may be attributed to various volatiles profile as well as to the allelo chemicals complex of the host genotype.

Once the habitat is located the parasitoid must find host in the habitat. The three basic behaviour associated with these chemicals or kairomones (chemical released by one species that affect the behaviour of second species to the detriment of the releaser). The kairomones may be attractant which orient and bring the parasitoid from one place to another, searching stimulant which aggregate the parasitoid in the local area and increase turning and movement and the host examination, ovipositor probing and ovipositor drilling stimulant that hold the parasitoid in a very local spot examining or probing the host. This can be referred to as attractant kairomones, searching kairomones and decision kairomones, respectively. The first two have potential for manipulation of parasitoids for increased biological control.

The attractant kairomones could be used to bring parasitoid to particular location and act similar to synomones. The searching kairomone which stimulate parasitoid to search and retain parasitoid in the treated area have already shown potential in biological control. Lewis and his co-workers have applied searching kairomone to crop which increased the parasitisation by retaining parasitoid in the treated area.

References

Abraham. C.C. and Pradhan, S. 1976. Studies on developing races of *Trichogramma australicum* Girault suitable for high temperature - low humidity conditions. *Madras Agric. J.* **63** : 550-556.

Allen. H.W. 1954. Propagation of *Horogenes molestae* an Asiatic parasite of the Oriental fruit moth on potato tuberworm, *J. Econ. Entomol.* **42** : 278-281.

Altiere. M.A., Lewis, W.J., Nordlund, D.A. Gueldner, R.C. and Todd, J.W. 1981. Chemical interaction between plants and *Trichogramma* wasps in Georgia soybean field. *Prot. Ecol.* **3** : 259-263.

Annadurai. R.S. Murugesan. S., Senrayan, R., Gurusubramanian, G. and Anantha Krishnan, T.N. 1992. Tritrophic interactions in *Heliothis armigera* Hubner and its natural enemy systems : A chemical ecological approach (Noctuidae : Lepidoptera). In Emerging Trends in Biological control of Phytophagous Insects. T.N. Anantha Krishnan (Ed.). Oxfort and IBM Publ. Co. PVT. Ltd. New Delhi. pp. 83-101.

Anonymous. 1982. Use of parasitoid to control a serious pest of maize and sorghum. IARI Newsletter. **4** (2) : 5.

Atma Ram and Sharma. S.K. 1977. Selective breeding for improving the fecundity and sex-ration of *Trichogramma fasciatum* (Perkins) (Trichogrammatidae: Hymenoptera) an egg parasite of lepidopterous hosts. *Entomon.* **2** : 133-137.

Beglyarav. G.V. and Smetnik. A.I. 1977. Seasonal colonization of entomophages in the USSR. *In* Biological Control by Augmentation of Natural Enemies. R.L. Ridgway and S.B. Vinson (Eds.). New York. Plenum Press. pp. 283-328.

Box. H.E. 1956. Battle against Venezuala's cane borer I. Preliminary investigations and the launching of general compaign. Sugar, 25-27 : 30-45.

Banerji, S.N. 1974. Epidemic of sugarcane *Pyrilla* in India. *Pl. Prot. Gleanings* **3** : 1-3.

DeBach, P. 1974. Biological control by natural enemies. Cambridge Univ. Press. Cambridge, 323 pp.

DeBach, P. and Rosen, D. 1991. Biological control by natural enemies 2nd ed. Cambridge Univ. Press Cambridge, 440 pp.

Elzen, G.W., Williams, J. H. and Vinson, S.B. 1984. Isolation and identification of cotton synomones mediating searching behaviour by parasitoid *Campoletes sonorensis*. *J. Chem. Ecol.* **10** : 1251-1264.

Ferguson, J.S. 1992. Biological control : An industrial perspective. *Florida Entomologist* **75** : 421-428.

Guan, X.C., Wu, T.N. and Feng, H. 1978. Studies on rearing of *Trichogramma dendrolimi* Matsumura in Vitro. *Acta Entomologica Sinica,* **21** : 122-126.

Hardin, B. and Hoffman, T.D. 1981. Tiny wasp-key insect in biological control. *Agric. Ecosystem Environ.* **15** : 109-119.

Hoy, M.A., Cave, F.E., Beede, R.H. Grant, J. Krueger, W.H., Olson, W.H., Spollen K.M. Bernett, W.W. and Hendricks, L.C. 1990. Release, despersat and recovery of a laboratory selected strain of walnut aphid-parasite *Trioxys pallidus* (Hymenoptera : Aphidiidae) resistent to ainphos *methyl. J. Econ. Entomol.* **83**:89-96.

Huffaker, C. B. 1977. Augmentation of natural enemies in the People's Republic of China. In Biological Control by augmentation of Natural Enemies. R.L. Ridgway and S.B. Vinson (Eds.) New York, Plenum Press pp 329-339.

Kamalakara Rao, C. 1980. Biological Control of Sugarcane shoot and internode borers in the factory area of the Jeypore Sugar company Ltd. Chagallu, W.G. District Andhra Pradesh. In : Biological Control of Sugarcane Pest in India. S. Sithanantham and A.R. Solayappan (Eds.). The Tamil Nadu Coop. Sugar Federation, Madras pp. 37-38.

Landaluze, P.U. 1950. Application de la genetica al aumento de la eficacia del *Trichogramma minutum* en lucha biologica. *Bolet. Patol. Vegetal Entomol. Agric.,* **18** : 1-12.

Li, L.Y. 1990. Mass production of *Trichogramma* spp. and *Anastatus japonicum* Ashmead with artificial diet in China. *In* FFTC-NAK International seminar on the use of parasitoids and predators to cotrol agricultural pests. Tukuba given Japan Nation Agric. Centre (NARC) spp.

Li, L.Y., Liu, W.H., Chen, C.S., Han, S.T., Shiv, J.C., Du, H.S. and Feng, S.Y. 1988. In vitro rearing of *Trichogramma* and *Anastatus* sp. in artificial eggs and methods of mass production. *Colloque* del INRA No. **43** : 339-352.

Liu, Z.C., Sun, Y.R., Yung, W.H., Liu, J.F., Wang, Z.Y. and Wang, C.X. 1988. Some new improvements in the mass rearing of egg parasitoids on artificial host eggs. *Chinese J. Biol. Control* **4** : 145-148.

Manjunath, T.M. 1992. Biological control by augmentation of natural enemies in India. Retrospect and prospects. In Emerging Trends in Biological Control of Phytophagous Insects. T.N. Ananthakrishanan (Ed.), Oxford and IBH Pub. Co. Pvt. Ltd., New Delhi, pp. 213-222.

Mishra, M.P. and Pawar, A.D. 1981. Release and establishment of *Epipyrops melanoleuca* Fletcher (Lepidoptera : Epipyropidae) on sugarcane in West Bengal. *Indian J. Pl prot.* **11** : 128-131.

Mishra, M.P. and Pawar, A.D. 1984. Use of *Epipyrops melanoleuca* Fletcher (Lepidoptera : Epipyropidae) for the biological control of sugarcane pyrilla, *Pyrilla* perpusilla (Hemiptera : Fulgoridae). *Indian J. Agric. Sci.* **54** : 742-750.

Mishra, M.P., Pawar, A.D. and Srivastava, U. L. 1986. Biocontrol of sugarcane moth borers by releasing *Trichogramma* parasites at Harinagar, West Champaran, Bihar. *Indian J. Pl. Prot.* **14** : 89-91.

Morrison, R. K. 1985. Mass production of *Trichogramma pretiosum* Rilley. *Southwestern Entomologist. Suppl. No.* **8** : 21-27.

Mukthilingam, B., Gangaram, M. N. and Appa Rao, 1983. Mass multiplication of predators of sugarcane scale insects (*Melanaspis glomerata* Green) in Andhra Pradesh. Fourth All India Workshop on Biocontrol. *Univ. Agric. Sci., Bangalore,* Feb., 22-24, 6 pp.

Patel, R.C., Yadav, D.N. and Saramma, P.V. 1979. Impact of mass-releases of *Chelonus* heleopae Gupta and *Telenomus remus* Nixon against *Spodoptera litura* (Fabricius). *J. Ent. Res.,* **3** : 53-56.

Pawar, A.D. 1987. Biological control of sugarcane pest in *India.Pl. Prot. Bull.* **39**: 1-3.

Pawar, A.D., Jagdish Prasad, Ravindra Singh, Yadav, K.P. and Ram Asre, 1980. Introduction and estabilishment of an ectoprasaite, *Epipyrops melanoleuca* Fletcher (Epipyropidae : Lepidoptera) in Bundi (Rajasthan) for biocontrol of sugarcane pyrilla, *Pyrilla perpusilla* Walker. *Indian Sug.,* **30** : 745-749.

Pieulou, D.P. and Glasser, R.F., 1952. Selection for DDT resistance in a beneficial insect parasite. *Science,* **115** : 117.

Pimentel, D., Krummel, J., Gallahan, D., Hough, J. Merrill, A., Schreiner, I., Vittam, P., Koziol, F., Back, E; Yen, D. and Fiance, S. 1981. Cost benefit analysis of pesticide use in U.S. food production, pp. 27-54. In D. Pimentel (Ed.) CRC handbook of pest management in agriculture Vol. 2 CRC Press Boca Ratan Fl.

Ridgway, R.L. and vinson, S.B. (Eds.) 1977. Biological control by Augmental of Natural Enemies. New York, Plenum Press, 480 pp.

Roland, J. and Denford, K. E. 1991. Identification of volatile plant compound for enhancement of biological control by *Cyzenis albicans* (Tachinidae) *Radia* **74** : 135-141.

Rosen, D. and DeBach, P. 1992. Foreign exploration : The key to classical biological control. *Florida Entomologist* **75** : 409-413.

Sailer, R. I. 1969. A taxonomists view of environmental research and habitat manipulation. *Proc. Tall Timber Conf. Ecol. Anim. Control. Habitat Manag.* **1** : 37-45.

Sankaran, t. and Manjunath, T.M. 1988. Biological control. In Forty years of Protection in India, 1946-1986. D. Bap Reddy (Ed.). *Plant Prot. Assoc.* India., Hyderabad, pp. 75-83.

Simmonds, F.J. 1947. Improvement of şex ratio of a parasite y selection. *Can. Entomol.* **79** : 41-44.

Sithanantham, S. 1980. Inundative release of *Trichogramma* and integration with insecticidal methods. In : Biological control of sugarcane pests in India. S.

Sithanantham and A.E. Solayappan (Eds.) The Tamil Nadu Coop. Federation, Madras. pp. 43-60.

Sithanantham, S., Muthuswamy, S. and Durai, R. 1973. Experiments on the inundative release of *Trichogram australicum* Gr. in biological control of sugarcane borer, *Chilo indicus* (Kapur) *Madras Agric. J.* **60** : 547-561.

Smith, S.M., Carrow, J.R. and Laing, J.E. 1990. Inundative release of the egg parasitoid, *Trichogramma minutum* (Hymenoptera : Trichogrammatidae), against forest insect pests such as the spruce budworm, *Choristoneura fumiferana* (Lepidoptera : Tortricidae). *Mem. Ent. Soc.* Canada No. 153.

Solayappan, A.R. 1980. Mass production of *Trichogramma* for released in factory area. In Biological control of sugarcane pest in India. S. Sithanatham and A.R. Solayappan (Eds.). The Tamil Nadu Coop. Sugar Federation. Madras, p. 84.

Szmidt, A. 1972. Studies on the efficiency of various strains of the parasite *Dalbominus fuscipennis* (zett.) (Hymenoptera, Chalcidoidea) under natural conditions. *Ekol. Pol.* **20** : 299-313.

Unquijo, P. 1946. Selection des estirpes de *Trichogramma minutum* de maxima effectividad parasitaria. *Bon. Patol. Veg. Ent. Agric.* **14** : 199-216.

Varadharajan, G. 1976. The scrope and prospects in utilisation of *Trichogramma australicum* Gir. for the control of sugarcane intermode borer, *Chilo indicus* K. in Tamil Nadu. *Madras Agric. J.* **64** : 561-571.

Verma, G.C., Bindra, O.S., Sukhdev Singh and Baljinder Singh, 1979. comparative efficacy of biological mechanical and integrated methods for controlling sugarcane borers in Punjab. *J. Res. Punjab Agric. Univ.* **16** : 45-48.

Vinson, S.B., Nettles, W.C. Jr., Xie, Z.N. and Morrison, R.K. 1988. Potential opportunities and problems associated with mass culture of *Trichogramma pretiosum Colloque de e'* INRA No. **43** : 389-398.

Wickremasinghe, M.G.V. and Van, H.F. 1992. Reaction of adult female parasitoids, particularly *Aphidius rhopalosiphi* to volatile chemical cues from host plants of their aphid prey. *Physiological Entomol.* **17** : 297-304.

White, E. B., DeBach, P. and Garber, M. J. 1970. Artificial selection for genetic adaptation to temperature extremes in *aphytis lingnanensis* Compere (Hymenoptera : Aphylinidae). *Hilgardia,* **40** : 161-192.

Wilkes, A. 1942. The influence of selection on the preferendum of a chalcid (Macroplectron fuscipennis Zett.) and its significance in the biological control of an insect pest. *Proc. R. Soc.* London, Ser. B., 130 : 400-415.

Wilkes, A. 1947. Effect of selective breeding on laboratory propagation of insect parasites. *Proc. R. Soc.* London, Ser. B., **134** : 227-245.

Xie, Z.N., Netteles, W.C. Jr., Morrison, R.K., Irie K. and vinson S.B. 1986. Three methods for the in vitro culture of *Trichogramma Pretiosum* Riley. *Jour. Entomol. Sci.* **21** : 133-138.

Xie, Z.N., Netteles, W.C., Jr., Morrison, R.K. and Vinson, S.B. 1988. In Vitro mass culture of *Trichogramma pretiosum* Riley from eggs collected using salt solution as ovipositional stimulant. Colloques de - e INRA No. 43 : 403-406.

Xing, J.Q. and Li, L.Y. 1990. Rearing of an egg parasite Anastatus *Japonicus* Ashmead in Vitro. *Vcta Entomologica Sinica* **33** : 166-173.

Recent Advances in Indian Entomology
Editor: O.P. Lal
APC Publications Pvt. Ltd., New Delhi, 1996, pp. 245-249

Impact of Semiochemicals in Relation to Host, Host Plant and Natural Enemy Interactions

A.V. NAVARAJAN PAUL

Semiochemicals are important chemical cues influencing habitat location and host selection by insect natural onemies. The effectiveness of natural enemies does not depend on random encounter but on various stimulii present in the ecosystem including semiochemicals.

In orchard ecosystem where annual disturbances are minimum there has been spectacular success reported due to release of natural enemies both in India and abroad. However, in annual agro-ecosystems where annual disturbances are more due to various agricultural practices and pesticidal contamination from the adjacent cropping area, need for repeated augmentation and manipulation of natural enemies have become imminent. Semiochemicals can play a major role in improving the efficiency of natural enemies in biological control by modifying their foraging behaviour and influencing their host selection process.

Habitat location

Habitat location is a very important first step in the selection of hosts or prey. Host specificity found in majority of natural enemies is largely due to the effect of host habitat location. Many types of stimulii such as temperature, light intensity, wind and humidity play a role in habitat location but semiochemicals called synomones released from the food of the host or prey play a significant role (Nordlund *et al.*, 1981).

Semiochemicals useful in host habitat location may originate from host plant of the host insect or damages on the host plant or organisms associated with the host/ pest damage and or host itself.

Read et al. 1970 reported that adults of *Diatriella rapae* are attracted to fresh collard leaves and to very low concentrations of mustard oil (allyl isothiocyanate) which occurs in collard leaves.

Females of *Boisteres longicaudatus* Ashmead, a parasitoid of tephritid fruitfly larvae were reported to be attracted to rotting fruit, a probable habitat of the host. These parasitoids were found attracted to acetaldehyde, ethanol and aceatic acid which are the fermentation products produced by the fungus *Monilinia frueticola* (Hint.) (Creanv et al. 1977).

Sex pheromones of some moths appear to have kairomonal activity for some species of *Trichogramma* (Lewis et al., 1982, Noldus and Van Lenteren, 1985b). These chemicals appear to direct the adult Trichogrammatids to *indicate* areas of moth activity where encounter with eggs is likely. Females of *Campoletis sonorensis* (Cameron) are attracted to flowers of some plants of plant parts even in the absence of hosts. Flowers that elicit such responses release synomones that appear to be used in habitat location by the parasitoid (Elzen et al., 1983).

Ovipositing adults of *Chrysoperla carnea* (Stephens) are attracted by caryophyllene released by actively growing plant tissues (Flint et al., 1979) and indole acetaldehyde (Van Emden and Hagen, 1976) an oxidation product of tryptophan present in aphid honey dew for habitat location.

Host location

In the host location, as and when the parasitoid gets specific cues of the host by various allelochemics, it is arrested towards the host, which is characterised by frequent changes in its direction of flight at large angles resulting finally in its landing very near the host. The chemicals released by the host or prey which assist the entomophage in this process are known as kairomones. Kairomones are known to play many important roles in the host or prey location behaviour of many natural enemies.

Kairomone may attract the entomophage which is not the only or the most important way in which kairomone function. In many cases natural enemies are stimulated into an intensive search behaviour by contact with kairomones. Kairomones of this type have been found in host traces near eggs such as moth scales, host faeces, host mandibular and labial gland secretions and other host products. In close proximity to host eggs, moth scales trigger intensive searching behaviour of *Trichogramma pretiosum*. (Lewis *et al.*, 1972, Noldus and Van Lenteren, 1985). Hexane extract of the cuticle of the larvae of *Corcyra cephalonica* St. containing a kairomone was found to elicit host seeking response by the females of Bracon *brevicornis* (Wesm.) Ram *et al.* (1982).

In *Microplitis crociepes*, walking individuals are influenced by contact kairomones present in frass, salivary secretions and silk of *Heliothis* spp. (Keller and Lewis, 1985).

Chemicals found on the surface of the eggs of *Spodoptera frugiperda* and *Heliothis zea* are reported to elicit response in *Telenomus remus* Nixon and *Trichogramma pretiosum* respectively (Nordlund et al., 1987).

The exuviae, silk, webbing, frass and vomit of Ostrinia nubilalis are found to contain kairmone that attract? adults of *Macrocentrus grandi* (Ding *et al.*, 1989)

In addition to the external kairomones, the role played by internal kairomones in egg deposition is clearly demonstrated very recently.

Kanoh *et al.* (1989) identified several components of an internal kairomone which elicits egg deposition by the braconid parasitoid, *Ascogaster reticulatus* inside eggs of the tortricid, *Adoxophyes* sp. They concluded that the most important component of the kairomone in a host egg is the mixture of several amino acids.

Utilization of semiochemicals in biological control

Semiochemicals can be employed to screen the suitability of the species and strains for a particular agroecosystem soon after potentially useful natural enemies are identified by foreign exploration. This is particularly important when the target agroecosystem differs from the original ecosystems of the exotic natural enemies since the semiochemicals involved in host habitat and host location are likely to differ. Such screening would indicate when introduction of a beneficial insect would not be appropriate, for an agro-ecosystem thereby saving further investment. For example *Microplitis croccipes* (Cresson) attacks *Heliothis zea* larvae feeding on a variety of plants including cotton, soybean but not on maize and maize volatiles are observed to interfere with host location cues (Keller and Lewis, 1985).

Semiochemicals could play a beneficial role in the mass rearing programmes. Nordlund *et al*. (1976) reported increased longivity and fecundity in *Trichogramma pretiosum* when offered the host eggs treated with extract of moth scales containing kairomones.

Frass from larvae reared on artificial diets were observed to have low kairomonal activity as compared to larvae reared on plant materials (Elgen *et al.* 1984). In the mass rearing of natural enemies where artificial diets are used for host rearing, addition of semiochemicals could promote more intensive host searching activity than would occur in their absence. The natural enemies could be stimulated to accept non hosts as in *Telenomus heliothidis*. By incorporating host recognition kairomone, the parasitoids were induced to oviposit into the eggs of Spodoptera *frugiperda* (J.E. Smith) and *Phthorimaea operculella* (Zeller) which are usually rejected by foraging females but are suitable for parasitoid development (Stand and Vinson, 1982).

Hu and Chen (1987) identified trehalose and 8 amino-acids when applied as a mixture acted as kairomone in increasing the host searching behaviour of *Apanteles cypris* and increased its parasitisation by 15-25% in the larvae of *Cnaphalocrosis medinalis*.

Pre-release exposure of natural enemies to host selection cues can promote intensified searching near the site of contact and exposure to kairomones can reduce post release dispersal by stimulating host searching behaviour. Parasitisation was increased in *M. croceipes, T. pretiosum*, when stimulated individuals were used in the release programmes (Gross et al., 1975).

Conclusion

There are several ways in which semiochemicals could be profitably used in bio-control programmes. Importance of synomones from the host plant in enhancing the activity of the natural enemy may be stressed to the plant breeders so that suitable plant varieties favouring intensive natural enemy activity may be developed use of kairomones seem to have promising scope in the inundative release programmes in order to stimulate host searching by the released natural enemies. Research aimed to understand the influence of semiochemicals on the behaviour of natural enemies could lead to development of practical techniques that would increase the efficiency of the natural enemies by means of behavioural manipulation.

References

Ding, D., Sweden Borg, P.D., and Jones, R.L. 1989. Chemical stimulii in host seeking behaviour of *Macrocentrus qrandi*. *Ann. Ent. Soc. Amer.* **82** (2): 232-36.

Elzen, G.W., Williams, H.J., and Vinson, S.B. 1983. Isolation and identification of cotton synomones mediating searching behaviour by parasitoid, *Campoletis sonorgensis*. *J. Chem. Ecol.* **10** :1251-1264.

Elzen, G.W., Williams, H.J., and Vinson, S.B. 1984. Role of diet in host selection of *Heliothis virescens* by parasitoid, *Campoletis sonorensis* (Hymenoptera:Ichneumonidae). *J. Chem. Ecol.* **10** :1585-1541.

Flint, H.M., Salter, S.S., and walters, S. 1979. Caryophyllene. An attractant for the green lacewing. *Environ. Entomol.*, **8** :1123-1125.

Greany, P.D., Tumlinson, J.H., Chambers, D.L., and Bousch, G.M., 1977. Chemically mediated host finding *Boisteres* (Opius) *longicaudatus*, a parasitoid of tephritid fruitfly larvae. *J. Chem. Ecol.* **3** :189.

Gross, H.R. (Jr.) Lewis, W.J., Jones, R.L. and Nordlund, R.A., 1975. Stimulation of *Trichogramma achaeae* with host seeking stimulii at the time of release to improve their efficiency. *J. Chem. Ecol.* **1**(4):431-38.

Hu, J., and Chen, C. 1987. A study of the host searching kairomone of *Apanteles*. *Acta. Entomol. Sin.*, **30** (1):31-90.

Kainoh, Y., Tatsuki, Sugia, H., Tamaki, Y. 1989. Host egg kairomones essential for egg-larval parasitoid, *Ascogaster reticulatus watanable* (Hymenoptera:Braconidae) II. Identification of internal kairomone. *Journal of Chemical Ecology*? **15** (4):1219-1229 (En. 18 ref.) D.V. Entomology National Inst. Agro. Environmental Sci. Kannondai Tsukuba, Ibaraki 305, Japan.

Keller, M.A., and Lewis, W.J. 1985. Behaviour-modifying chemicals to increase the efficiency of predators and parasitoids of *Heliothis* spp. *Proc. Workshop on Biological control of Heliothis : increasing the effectiveness of natural enemies*. 11-15 Nov. 1985, pp. 449-467.

Lewis, W.J., Jones, R.L., and Sparks, A.N. 1972. A host-seeking-stimulant for the egg parasite, *Trichogramma* evanescens. Its source and demonstration of its laboratory and field activity. *Ann. Entomol. Soc. Am.* **65** :1087.

Lewis, W.J., Nordlund, D.A., Gueldner, R.C. Teal, P.E.A., and Tomlinson, J.H. 1982. Kairomones and their use for management of entomophagous insects. XIII. Kairomonal activity for *Trichogramma* spp. of abdominal tips, excretion and a synthetic sex pheromone blend of *Heliothis zea* (Boddie) moths. *J. Chem. Ecol.*, **8** :1323-1331.

Noldus, L.P. J.J., and Ven Lenteren, J.C. 1985a. Kairomones for the egg parasite *Trichogramma evanescens* Westwood I. Effect of volatile substances released by two of its hosts *Pieris brassicae* L. and *Mamestra brassicae*. *L. J. Chem. Ecol.* **11** :781-791.

Noldus, L.P.J.J., and Van Lenteren, J.C. 1985b. Kairomones for the egg parasite. *Trichogramma evanescens* Westwood. II. Effect of contact chemicals produced by two of its hosts, *Pieris brassicae* L. and *Pieris rapae*. *J.Chem. Ecol.* **11** :793-800.

Nordlund, D.A., Lewis, W.J., Jones, R.L., and Gross, H.R., Jr. 1976. Kairomones and their use for management of entomophagous insects. IV. Effect of kairomones on productivity and longevity of *Trichogramma pretiosum* Riley. (Hymenoptera : Trichogrammatidae). *J. Chem. Ecol.* **2** :67-72.

Nordlund, D.A., Lewis, W.J., and Gross, H.R., Jr. 1981. Elucidation and employment of semiochemicals in the manipulation of entomophagous insects. pp. 463-475 in Management of insect pests with semiochemicals E.R. Mitchell, ed) Plenum Press, New York NY, USA.

Nordlund, D.A., Strand, M.R., Lewis, W.J. and Vinson, S.B. 1987. Role of kairomones from host recognition by *Telenomus remus* and *Trichogramma pretiosum*. *Entomol. exp. appl.* **44** :37-44.

Ram, A., Tiwari, L.D., Dass, R. and Mehrotra, K.N. 1982. Evidence for the presence of a kairomone in *Corcyra cephalonica* larvae for *Bracon brevicornis Z. ang. Ent.* **93** :388-41.

Read, D.P., Feeny, P.P., and Root, R.B. 1970. Habitat selection by the aphid parasite Diaretiella rapae (Hymenoptera:Braconidae) and hyperparasite *Charops brassicae* (Hymenoptera : Cynipidae). *Can. Entomol.* **102** :1567-1578.

Strand, M.R., and Vinson, S.B. 1982. Stimulation of oviposition and successful rearing of *Telenomus heliothidis* (Hym. : Seclionidae) on non-hosts by use of a host-recognition kairomone. *Entomophaga*, **27** :365-370.

Van Emden, H.F., and Hagen, K.S. 1976. Olfactory reactions of the green lacewing, *Chrysopa carnea*, to tryptophan and certain breakdown products. *Environ. Entomol.* **5** :469-473.

Recent Advances in Indian Entomology
Editor: O.P. Lal
APC Publications Pvt. Ltd., New Delhi, 1996, pp. 250-259

Insect Predators in India

R. D. GAUTAM

More than 200 years ago role of predators in biocontrol was realised by the Indian Entomologists when Indian Mynah, *Acridotheres tristis* (L.) was shipped to Mauritius during 1762 for the control of red locust, *Nomadacris septemfasciata* Serv. in sugarcane crop (Montia and Mamet, 1946). Thereafter, ladybrid *Cryptognatha flavescens* (De Bach, 1974) and *Chilocorus nigritus* (De Bach, 1996) were sent to USA and Greece' respectively for the control of citrus white fly and citrus scale. Similarly, classical successes in suppressing the Cottony cushion scale, *Icerya purchasi* Maskell, infesting a variety of fruit trees by an exotic coccinellid, *Rodolia cardinalis* (Muls.) during 1929-30 followed by the introduction of another ladybird, *Cryptolaemus montrouzieri* (Muls.) which feeds voraciously on pseudococcids and shield scale have certainly now made dents in the biological pest suppression in our country. These predators including a bug, *Platymeris laevicollis* Distant from Zanzibar which prey on *Oryctes rhinoceros*, a serious pest of coconut have great potential for commercial use.

The impact of predacious arthropods in agricultural crops is receiving more attention in recent years and the importance of predators in the suppression of insect pest is coming into closer focus, based on modern ecological investigations and experiments (Hagen *et al.*, 1976; Whitecomb, 1987). In fact, predators are considered to be more important than parasitoids in determining the population trend of sugarcane pests *viz., Chilo sacchariphagus indicus* (Kapur) in India (Easwaramoorthy and Nandagopal, 1986); *Diatraea saccharialis* F. in USA (Negm and Hensley, 1967; Fuller and Reagan, 1988) and in Brazil (Degaspari *et al.*, 1983, 1988), The success story of National IPM Programme in Indonesia which has brought down 60-70% applications of chemical pesticides in paddy is mainly due to conservation of predatory spiders and laybird beetles.

Advances made on predatory insects and their laboratory multiplication/field utilization are described below.

Predatroy insects

The predatory habits among insects are common in several families. Some of these have potential biocontrol agents while undoubtly many families may contain valuable members that will be revealed by future workers. Out of 29 insect orders at least 11 orders containing around 75 families are known to have predatory insects (Sweetman, 1963; Richards and Davies, 1977) and are placed in Table 1.

While, a few like collembola (entomobryidae) and Thysanura (lepismatidae) are omnivorous. Easwaramoorthy *et al.* (1991) reported around 90 insect predators belonging to nine orders on the sugarcane pests in India.

Table 1. Important families of insect predators.

Order	Family	Remarks
Ephemeroptera (mayflies)		Nymphs of Prosopistoma, Pseudiron and Anopeorus feed on small arthropods.
Odonata (dragonflies)		Adults are generalized predators, usually feeding on suitably sized insect of any kind.
Dictyoptera (mantids)		Both adult and nymphs are predatory on various insects specially caterpillars and flying soft bodies insects.
Plecoptera (stoneflies)	Perilidae, Perlodiade Chloroperlidae, Eusteniidae	Nymphs are exclusively predatory on aquiatic fauna
Hemiptera (plant bugs)	Reduviidae	Both adults and nymphs are predatory on eggs & larvae of several arthropods.
	Nabidae	-do-
	Anthocoridae	-do-
	Hydrocoridae	-do-
	Geocoridae	Both adults and nymphs are predatory on eggs & larvae of insects.
	Lygaeidae	Both nymphs and adults suck eggs and larvae of many insects.
	Pentatomidae	Both adults and nymphs and few species like *Cantheconidia* are predatory on caterpillars.
	Maridae	Predatory on mites and other arthropods.
	Pyrrhocoridae	Antilocus *coqueberti* is predacious on red cotton bugs.
	Ochteridae Naucoridae	Both nymphs and adults are predacious on aquatic life.
	Belostomatidae	Both nymphs and adults are predacious on aquatic life

Contd............

Order	Family	Remarks
		Both nymphs and adults feed on almost all insects exceeding 11 cm in length, small fish, tadpoles, young frogs etc.
Thysanura (thrips)	Aeolothripidae Phlaeothripidae	A few species are predacious on aphid. mites and other smaller insects.
Diptera (flies)	Empidiae	Adults are predatory while larvae rarely carnivorous.
	Mydidae	Adults predatory on several insects while larvae on coleopteran grubs.
	Scinopinidae	Larvae are predacious on insects.
	Therevidae	Adults are predacious.
	Stratromyidae	Larvae are either carnivorous on scavengers.
	Tabanidae	Larvae feed on insects, crustaceans, small earthworms etc. while female flies on the blood of vertebrates.
	Rhagionidae	Adults are predatory on other insects.
	Blephariceridae	Females prey on small dipterans.
	Simuliidae	Adults are predatory on other insects.
	Blephariceridae	Females prey on small dipterans.
	Simuliidae	Females prey on small dipterans.
	Simuliidae	A few species suck blood of other insects.
	Ceratopogonidae	Adults suck blood of moths. caterpillars. dragonflies.
	Asilidae	Adults are predacious while larvae are either predacious or scavenger
	Dolichopodidae	Adults prey on soft bodied insects while larvae are either carnivorous or scavengers.
	Syrphidae	Only larvae are predacious on aphids. caterpillars. sawfly larvae and occasionally phytophagous or scavengers.

Contd............

Order	Family	Remarks
	Scatophagidae	Adults are probably predacious on small insects while larvae are predacious or saprophagous.
	Muscidae	A few species (maggots) are predatory on dipterans larvae while adults suck blood of mammals
	Milichidae	Adults are saprophagous, also seen preying on reduviid and spiders.
	Chloropidae	Larvae prey on root aphids occasionlay.
Lepidoptera (moths)	Noctuidae	Caterpillars of a few species like *Eublema* are predatory on lac insects.
Hymenoptera (wasps)	Sphecidae	Adults prey on arachnids, moths and also hymenopterous occasionally.
	Eumeridae	Adults are tunnel makers and construct tubular nests where caterpillers are offered to their youngones.
	Vespidae	Adults live inside the paper nests. grubs feed voraciously on caterpillars brought by their parents and stored inside the nests.
Neuroptera (lacewings)	Ascalaphidae	Adults are predatory on arthropods.
	Hemerobiidae	Both adults and larvae are predatory on aphids.
	Chrysopidae	Larvae are predatory on spyllids. coccids. aphids. insect eggs etc.
	Nemopteridae	Larvae prey on psocids and other small insects.
	Mantispidae	Larvae prey on aphias.
	Berothidae	Larvae prefer to feed on beetle eggs.
	Neurorthidae	Larvae prey on aquatic forms.
	Osmylidae	-do-
	Ithonidae	Larvae feed on scarabaeoid grubs.
	Coniopterygidae	Larvae prey on coccids. aphids and acarina.

Contd............

Order	Family	Remarks
	Dilaridae	Larvae hide beneath the bark and prey on insects or their eggs.
	Psychopsidae	Larvae hide beneath the bark and prey on insect visting tree for gum.
	Myrmeliontidae	Ant lion burry itself under the debris protruding its large jaws to catch the wondering ants or other insects.
	Nymphidae	No clear information available.
	Stilbopterygidae	-do-
Coleoptera (beetles)	Carabidae	Larvae and adults of predator like *Chlaenius pictus* feed on *Chilo partellus* larvae.
	Amphizoidae	Feed on drowned insects in the aquatic systems.
	Dytisicidae	Adults are carnivorous while larvae prey on molluses, worms, tadpoles, insects, small fishes etc.
	Gyrinidae	In general members of this family are carnivorous.
	Hydrophilidae	Adults and larvae are phytophagous while larvae of *Hydropilus* and its allies are predacious.
	Histeridae	Both adults and larvae prey on several insects specially dipteran larvae.
	Silphidae	Larvae of a few species are predatory on snails and caterpillars.
	Pselaphidae	Larvae are carnivorous while adults occur on moss.
	Staphylinidae	Larvae of few species are predatory while adults live around decaying organic matter.
	Drilidae	Larvae prey on snails.
	Phengodiae	Probably feed on millipedes.

Contd............

Order	Family	Remarks
	Lampyridae	Larvae prey on slugs or snails while adults take little or no food.
	Cantharidae	Both adults and larvae are predatory.
	Lycidae	Larvae of a few species feed on spiders while adults lie under the bark or on flowers.
	Derodontidae	Few species are predatory while many feed on fungus.
	Cleridae	Larvae are predatory mainly on bark eating caterpillars while adutls live on tree trunks or carcasses and skin.
	Cucujidae	Often predacious as well as grain eating.
	Coccinelidae	Majority of the species are predacious on insects. mites and other arthropods during their both larval and adult stage.
	Tenebrionidae	Few species like *Alphitobious diaperinus* are predatory both at larval and adult stage on many stored product insects besides phytophagous.
	Meloidae	Larvae prey on eggs of Orthoptera and Hymenoptera.
	Anthoibidae	Few are predacious on scale insects.

Laboratory multiplication

A critical examination on mass-multiplication reveals that there has always been lack of adequate production and sustained releases of good quality natural enemies to properly judge the impact of releases against crop pests. Although. the list of predators is long hardly a few species belonging to the families *viz.*, cocinellidae. chrysopidae, and reduviidae have been explored for mass-production that too on limited laboratory hosts *viz.*, pseudococcids (*Ferrisia virgata*, *Planococcus pacificus*, *P. citri*, *Maconellicoccus hirsutus*), rice meal moth. *Corcyra cephalonica*, cowpea aphid. *Aphis craccivora* or on scale, *Aspidioutus destructor* (Lakkundi. 1989; Singh and Jalali, 1991; Manjunath. 1991; Gautam. 1994). Predators which can be successfully reared on one of these hosts continuously for several generations are given in Table-2. Gautam (1994) has described detailed technology of mass-multiplication of 17 predators (aphidophagous and coccophagous coccinellids, chrysopid, anthocorid and 14 laboratory hosts for the parasitiod and predators.

Table 2. Laboratory multiplication of common insect predators.

Predator	Family	Laboratory prey	Food for prey
Chrysoperla carnea (Stephens)	Chrysopidae	Rice meal moth	Crushed jowar or maize
Mallada boninensis Okamato	-do-	-do-	-do-
Platymeris laevicollis Distant	Reduviidae	-do-	-do-
Cryptolaemus montrouzieri mulsant	Coccinellidae	Pseudococcids	Potato tubers
Brumoides suturalis (Fabr.)	-do-	-do-	-do-
Nephus regularis (Sicard)	-do-	-do-	-do-
Scymnus brunnescens Motsch.	-do-	-do-	-do-
S. coccivora Ram	-do-		-do- -do-
S. nubilus (Mulsant)	-do-	Cowpea aphid	Cowpea sprouts
S. pyrocheilus (Mulsant)	-do-	-do-	-do-
Adonia variegata (Goeze)	-do-	-do-	-do-
Menochilus sexmaculatus (Fabr.)	-do-	-do-	-do-
Micraspis discolor (Fabr.)	-do-	-do-	-do-
Propylea dissecta (Mulsant)	-do-	-do-	-do-
Pharoscymnus horni Weise	-do-	Scale insect	Pumpkin
Chilocours nigritus (Fabr.)	-do-	-do-	-do-

Fields utilization

Information available on the field utlization of predators is still meagre as compared to parasitoids. Basic components of predator's utilization in pest management are (1) dosage and frequency of releases (ii) release techniques.

Dosage and frequency of releases

The quantity of predators to be released often depends on the intensity of host and feeding potential as well as their reproductive potential. Several workers have advocated a minimum of 1500 beetles per hectare, every season, at the first appearance of pest as these predators are expected to multiply and cope up with the pest population. However, additional releases during early morning or evening may be made over the entire infested areas as warranted by the crop pest species. The details of dosages, target pests and crop habitat of some important predators, are given in Table-3. In general, 2-3 need based releases of predators are essential for effective pest suppression till it gets established in nature.

Table 3. Recommended dosage for insect predators.

Predator	Dosage	Target Pest	Crop habitat
Chrysoperla carnea	6 larvae/plant	aphid	tobacco
	2,500 eggs or larvae/ha.	aphid, jassids	cotton
Cryptolaemus montrouzieri	3-10 adults/tree	mealy bugs, shield scale	aonla, citrus, grapes
	600 grubs or adults/ha	-do-	coffee,
Pharoscymnus horni	1500 adults/ha	sugarcane scale	sugarcane
Chilocorus nigriuts	-do-	-do-	-do-
**Cyrtorhinus lividipennis*	50-75 eggs/m^2 or 100 bugs/m^2	browns plant hopper	rice
**Coccinella septempunctata*	30 grubs/m^2	mustard aphid	mustard

*Mass-multiplication techniques yet to be standardized.

Release techniques

(i) *Adults:* Adult predators may be released by holding the containers partially open and moving all over the field after giving them artificial food like sugar, honey in the captivity eg. *C. montrouzieri, Scymnus* spp., *P. horni, C. nigritus, B. suturalist* etc. Adult releases are helpful where infestation has to be located by the predators themselves. In general, adult releases should be avoided as these require a lot of prey in the laboratory during transiction period.

(ii) *Larvae:* Certain predators like *C. cernea. C. montrouzieri* and *Scymnus* spp. may be gently inoculated on the plants where thick colonies of the pest are seen.

(iii) *Eggs:* Predators *viz., C. carnea* and *M. boninensis* have stalked eggs and can easily be shaved off from the oviposition substratum by gently cutting the base. Similarly, eggs of *C. montrouzieri, S. nubilus, S. coccivora, S. brunnescens S pyrocheilus, N. regularis, B. suturalis* are nicely laid inside the cotton wool used for plugging the beetles. These eggs (about to hatch) alongwith cotton wool may be distributed in the field on infested shoots so that the larvae, on hatching can readily feed on the pests. These eggs being older are not parasitised by the egg parasitoids which are detrimental for the predators. Eggs obtained from older females should be avoided for release as there is possibility of being unfertile.

References

De Bach, P. (1964). *Biological Control of Insect Pests and Weeds*. Reinhold Publ. Corp. New York pp 844.

De Bach, P. (1974). *Biological Control by natural enemies*. Cambridge University Press, Cambridge, U.K. pp. 323.

Degaspari, N. Macedo, N. Botebho, P.S. B., Arauro, J.R., De Almeida, L.C. (1983). Predators and parasites of eggs of *Diatraea saccharalis* (Fabr.). *Ent. Newsl. Int. Soc. Sug. Cane Technol.* **15**:12.

Degaspari, N., Botebho, P.S.M., Arauro, J.R. and De Almeida, L.C. (1988). Predators and egg perasitoids of *Diatraea saccharalis* in sugarcane. *Preguisa Agnopecuaria brasileira,* **22**: 705-792.

Easwaramoorthy, S.; Kurup, N.K. and Santhalakshmi, G. (1991). Arthropod predators in sugarcane pest Management. In *Biocontrol Technology For Sugarcane Pest Management* (eds. David, H. and Easwaramoorthy, 1991 Revised edition) Sugarcane Bredding Institute, Coimbatore, p. 201-248.

Easwaramoorthy, S. and Nandagopal, V. (1986). Life tables of internode borer, *Chilo sacchariphagus indicus* (K.) on a resistant and susceptible variety of sugarcane. *Trop. Pest. Mgmt.* **32**:222-228.

Fuller, B.N. and Reagan, T.E. (1988). Comparative predation of sugarcane borer (Lepidoptera: Pyralidae) on sweet sorghum and sugarcane. *J. econ. Ent.* **81**:713-717.

Gautam, R.D. (1994). *Biological Pest Suppression*. Westvill Publishing House, Delhi : pp. 221.

Hagen, K.S., Bombosch, S. and Mc Murthy, J.A. (1976). The biology and impact of predators. In *Theory and Practice of Biological Control* (eds. Huffakar, C.B. and Messenger, P.S.) Academic Press London p. 93-142.

Lakkundi, N.H. (1989). Assessment of reduviids for their predation and possibilities of their utilization in biological control. Ph. D. Thesis submitted to the Post Graduate School, IARI, New Delhi, pp. 90.

Montia, A.L. and Mamet, R. (1946). A review of 25 years of economic entomology in the Island Mauritius. *Bull. Ento. Res.* **36** : 439-472.

Negm, A.A. and Hensley, S.D. (1967). The relationship of arthropod predators, to crop damage inflicted by arthropod predators, to crop damage inflicted by the sugarcane borer. *J. econ. Ent.*, **60** : 1503-1506.

Richard, O.W. and Davies, R.G. (1977), *Imm's General Text Book of Entomology* Vol. 2 (Tenth edition) Chapman and Hall, London, pp. 412-1354.

Sweetman, H.L. (1963). The *Principles of Biological Control*. W.M.C., Prown Company, Dubuque, IOWA pp. 560.

Singh, S.P. and Jalali, S.K. (1991) Chrysopid predaors, their production and use. *Extension Bull.* No. 2, National Centre for Integrated Pest management, I.I.H.R., Bangalore, pp. 15.

Whitecomb, W.H. (1987). Predator Nature's first line to defence predators like ants and spiders provide all purpose pest control. In "*Research 87*" Gainesville, F.L. 32611 USA (1987) p. 810.

Recent Advances in Indian Entomology
Editor: O.P. Lal
APC Publications Pvt. Ltd., New Delhi, 1996, pp. 260-268

Recent Advances in Insect Physiology

B. G. SRIVASTAVA and MOHINI ANAND

The significance of insect physiology in economic entomology can be best described in the words of Prof. V. B. Wigglesword, " The physiology of insects is to some the handmaid of economic entomology. For although it is not the purpose of physiology to furnish directly the means of controlling insect pests, yet the rational application of measures of control-whether these insecticides of one sort or another, or artificial interferences with the insect's environment is often dependent upon a knowledge of the physiology of the insect in question. Physiology may thus serve to rationalize existing procedures or to discover the weak spots in the ecological armour of a species".

Physiology is a Science of life and concerned with the function of living system and therefore takes into account all those mechanism which imparts the property of life. Physiology is linked with biochemistry and biophysics. Insect physiology is essentially a comparative physiology because of tremendous amount of diversity of function in insects even at a narrow taxonomic range and within species. Due to the differences in physiology and morphology the insects have undergone to dominate the earth by its vast number of species and its capacity to envade environment, inaccessible or hostile to other form of life. Insects, so the major competitors with human being for food, they can be effectively managed, only if we have thorough understanding of their physiology. The work of insect physiology in crop production through insect pest management can be achieved by following means :

By correct identification of the species through physiological approach

Molecular data and numerical methods of analysis for systematics have proved useful in many groups. Bush and Kitto (1978) reviewed on application of genetics to insect systematics and analysis of species differences, so it is now clear that through physiological investigations a slight DNA change can effect the protein synthesis and different protein having different amino acids formed, which work in a different way towards the external expression of organism, while it is phenotypic expression of organism that has been so far taken into accout in taxonomy. Basic factor in biological speciation being a DNA transformation. Taxonomist are now taking this aspect into consideration. Thus the concept so far a vague which are purely that of a phenotypic evolution within a multicellular organism that is basically for the organic evolution. DNA transformation thus form the crux of problem in evolutionary taxonomy.

Insect pest management through nutritional principles

One of methods of insect pest management which is receiving great attention

List of chemicals involved in resistance to insects in crop varieties.

Host Plant	*Chemicals*	*Insect pest*	*Bio-effect*
Rice	1. Oryzanone	*Chilo suppressalis* (Stripped stem borer)	Attractant
	2. Low Asperagine	*Nilaparvata lugene* (B.P.H)	Feeding deterrant
	3. Trans-aconiticacid	-do-	-do-
Cotton	1. Gossypol	*Heliothis* sp.	Growth inhibitor
	2. Quercetine	-do-	-do-
	3. Rutin	-do-	-do-
	4. High gossypol	*Earias* sp. Bollworm	Oviposition deterrant
	5. Total pehnol gummy exudate	*Pemphenelus affinis* (Stem weevil)	-do-
Maize	6MBOA	*Ostria* i.e. *nubilalis* (Europeon com borer)	Feeding deterrant
	DIMBOA	-do-	Growth inhibitor
Cucurbits	Cucurbitacins	*Diabratica undecim-punctata* (Spotted cucumber beetle)	Feeding deterrant
Crucifers	Allylisothrocyanate	Pieris rapae cabbage worm	Attractant
Tomato	Tomatine	*Leptinotarsa decemlineata (Colarodo beetle)*	Repellent
Soyabean	Soponin	*Callosobruchus* sp. (Bruchid beetle)	Oviposition stimulant
	Spogenin	-do-	-do-
Potato	Demissin	*Leptinotarsa decemlineata*	Feeding deterrant
Sorghum	Dhurin	*Locusta migratoria*	Feeding deterrant

by agricultural entomologist is the use of plant varieties resistant to insect pest. A variety may be able to avoid economic damage to greater or lesser extent depending the chemical characteristics of the plants on one hand and feeding and food requirement of insects on the other hand. A knowledge of nutritional behaviour. therefore, holds the key of solving problems related to better utilization of principles of insect nutrition.

Nutritional principles for managing the pest of processed food has great scope in protecting them i. e. by improper ratio of different food constituents that results in impairment of growth and development leading to death of pest. Smallman (1945) reported that sodium bicarbonate in biscuits destroyed sufficient thiamin during baking which has marked effect on normal growth and reproduction of *Tribolium confusum*. Inclusion of certain inorganic salts and sugars qualitatively and quantitatively in food rendered it unsuitable for normal growth and development of insects and other organisms like fungus etc.

Some nutritional factors can act detrimental on the nutrition of insect pest, with respect to plant resistance as shown by Painter (1958), Beck (1965) & others. There are many examples in which agricultural chemicals, fertilizers, insecticides, etc. changed the proportion of nutrient in plant tissues and consequently their nutritional value for pests (Singh, 1970). When different rates of nitrogen and potassium, were supplied to Brussels sprout plants, it was noticed that soluble nitrogen in leaves increased, which resulted in increased fecundity in *Myzus persicae* than in *Brevicoryne brassicae* (Van Emden 1966). Differences in relative concentrations of amino acids and glucose found between varieties of pea plants correlated with resistance of plants to pea aphid *Acyrthosiphon pisum* (Auclair *et al*. 1957). It was clearly shown with synthetic diets, that this aphid choose a diet with certain level of amino acids and sucrose (Cartier 1968). Friend *et al*. (1958) recognizing the importance of amino acids relationship, suggested varieties of plants might be developed that would be nutritionally less suitable than present varieties as food for their pests. Some food components normally possess substances that act as growth inhibitor, that may be toxic also e. g. soyabean or corngerm (Lipke *et. al*. 1954 and Lipke and Fraenkel, 1954). Some fatty acids like linoleic acid may be of dietry value (Patil, 1989 and Hassan, 1995) but certain fatty acids such as capric. caprylic acid are quite toxic (Srivastava and Pant, 1979). Lipke and Fraenkel (1956) pointed out a given carbohydrate may be nutritionally inert or satisfactory as a carbon source but unacceptable in a gustatory sense or may be toxic. Certain vegetable gums, agar are used as food additives in food manufacturing as thickner etc. (Furia, 1968). Among these some include substances or their components such as pentosans which can be deterimental to insects probably owing to their indigestiblility, interferenace in the nutrient absorption or perhaps toxicity.

Physiology led to understand the resistance problem, building of resistance, resistance to insecticide etc. It is very important to study the extent to which insects can metabolize and there by degrade toxic or otherwise deterimental chemicals is of considerable importance to their survival in a chemically unfriendly environment, while all insect probably possess detoxicative capacity. It differs from species to species in the developmental stage and with habitat of the insect. Recent studies of detoxication in insects have revealed that further versatility in the adoptation of insects to their environment is provided by the phenomenon of induction. This is the process in which a chemical stimulus enhances the activity of detoxication system by the production of additional enzyme. The first report of induction in insects was by Agosin and Dinamarca (1963), when DDT was applied to *Triatoma infestans* resulted in increased production of the enzyme NAD Kinase The three most important system of detoxication in insects are the (1) Microsomal oxidase: these enzyme located in the endoplasmic reticulum of most cells insert an atom of oxygen into a variety of functional group of lipophilic molecuies, therefore preparing them for more rapid excretion. This system function through the heme-protein cytochrome P-450 which forms a ternary complex with a molecule of the xenobiotic to be oxidised (2) Glutothione S-transferase is of importance in the metabolism of organophosphate insecticide and (3) Carboxyesterases which degrade carbamate, organophosphate and pyrethroids insecticides as well as juvenile hormone and its analogous.

Studies on the neurophysiology led to understand the mode of action of insecticides. As most of the newer insecticides are neurotoxic, so an understanding of biochemical phenomenon of nervous system is highly rewarding. The biochemical mode of conduction of impulse at the synaptic region between the two nerve endings led to know the action of organophosphate and carbamates in the insect living system. Similarly several endocrine evidence points out at the nervous system as the primary *Locus* of pyrethroid actions. Insects pyrethroids rapidly effect the physiology of nervous system i.e Loss of coordination, period of convulsive activity and ultimately paralysis etc.

Semiochemicals are chemicals which act as specific carrier of message between species and within the species. Allelochemicals are interspecific and are released by individuals of one species and influence the behaviour of individeuals of another species. Allelochemical, play a vital role in insect behaviour and are non-nutritional compounds which interfere with insect feeding and reproduction, they serve as attractants, feeding deterrants, repellents, stimulants and are implicated in host plant selection. They are divided into two groups Allomone chemicals released by the species which influence the behaviour of the releaser. Many plants produce biological active materials that effect the growth and development of other organism. These compounds are considered defensive secretion which are useful to plant and they play an important role in physiology and ecology of the pest. Plants have provided thousands of compounds, among them terpenoids, phenolics, proteinase inhibitors, nitrogen compounds, alkaloids and growth regulator including hormonomimetic compounds are known. Terpenoids acting as attractant for pollinators as well as feeding deterrants and toxicants. Pyrethrum from dried chrysanthemum is a toxic monoterpene, while gossypol of cotton, a sesquiterpene is a major defence chemical against cotton insects. Alkaloids are best known among the toxin against insect attack, solanaceous host produce several toxins like solanine, nicotine, tomatine etc. known to inhibit sugar receptor and hence block sugar responses. Tannins are polymeric phenolic compounds with strong protein absorbing properties. Phenolic compounds which provide pre and post infestation factors involving polymeric phenols and phytoalexins which are allelopathic and allelochemic agents in insect plant interaction.

Kairomones chemicals are released by individuals of one species and excite responses in individual of another species. It is advantagous to the receiver. In general, compounds synthesized by plants become incorporated into body odours of herbivores, and this body odour is used as a searching clue by natural enemies. Corn plants contain the chemical tricosane, and the corn earworm *Heliothis zea* incorporate tricosane unchanged into its eggs and this chemical then is used by the egg parasitoid *Trichogramma evanescens* as a kairomone to find it's host (Lewis et. al. 1972). Herbivores on plants growing in association with other plants species may be attacked more heavily by natural enemies because the second plant species provides attractants for instance, the aphid *Myzus persicae* was much more heavily attacked by the parasitoid, *Diaeretialla rapae;* when the aphid occurred on a crucifer with mustard oil attractants than when it occured on beet with the mustard oil (Boethel and Eikenbary 1986). Kairomones as attractant odors is found in certain predator prey interrelationship. The spider mite *Tetranychus urticae* and *Panonychus ulmi* deposit a volatile compound on their host plant that can be detected by predatory mite such as *Phytosciulus persimilis*. Knowledge of effect of these semiochemicals in nature paves the way for development of new strategies for recent insect pest management. These compounds affect the physiology and act as ovipositon deterrant, feeding deterrants, growth inhibitor and some posses the abilities of repelling the pests. Some of these are listed below in table No. 1

Pheromones are released by individuals of a species and alter the behaviour of another individual of the same species. Pheromones are of different types depending upon the effect they produce like sex pheromones, alarm pheromone, aggregating pheromone and irritation pheromone etc. Use of pheromones is one of several modern technique that shows promise in managing the insect pest. Stored products present a unique pheromone trapping stituation in comparison to forest and field crop that with higher insect tolerance. In storage condition the confined space and uniform habitat led themself favourably to insect management by pheromones. Nearly all the phermones of the important species have been identified. Various new traps to be used with pheromones have been developed successfully monitoring of these insects has been demonstrated for severals species. Physiological studies related to pheromones production, release and factors influencing responses to pheromones have aided in designing monitoring systems. Data obtained from pheromones traps can be used both to timely application of insecticides and to monitor their effectiveness.

Insect growth regulators

With basic research on Insect Physiology we came to know about certain chemicals i.e. insect growth regulators which are specific target oriented, non-toxic to human and its domestic cattles, biogradable, there is less chance of resistance. Insect growth regulators interferes metamorphosis, growth and reproduction of insects. Juvenile hormone and its antijuvenile hormone agents on the basis on studies on various physiological effects of JH in insect led to start work on this hormone and its analogues, which could be used for managing the pest. More than

500 analogues with JH property have been discovered. The biological and insecticidal properties of these compounds have been well studied. Due to limited scope of juvenoids in insect pest management, there is demand to antijuvenoids. As deficiency of JH would also affect the JH dependent processes : such as morphogenesis and reproduction. Most well known among anti JH agents are Precocenes. Precocenes are plant derivated chromomees which cause JH deficiency in treated insects by selectively destroying some cells of corpora allata. Precocenes induce premature metamorphosis and delay in development in a number of hemimetabolous insects.

Ecdysone analogues, the steroid moulting hormone of insects is very important for moulting. Insects must acquire steroids from their food in the form of cholesterol and similar substances for proper growth and development. Non-steroidal compound code number named R H 5849 possess ecdysone mimitic properties. Treatments of insects with minute dose interferes with the feeding activity and induce a premature, lethal moult and some time neurotoxic effect. Azadirachtin, antiecdysteroid agent a Tetranotriter penoid extracted from the seeds of neem (*Azadirachta indica*) is highly effective as an insect feeding deterrant, disrupts growth, moulting and ovipositon (Singh & Srivastava, 1983).

Neurohormone i.e. peptide hormones released from neuron play many diverse functional role in insects as chemical messanger controlling growth, developement and reproduction etc. Research in the area of neuropeptide biology has suggested that neuroendocrine system might be manipulated for better insect pest management. Transgenic plants may be produced having neurohormonal or antihormonal genes. This will cause physiological and biologenetic imbalance in the insects which feed on these altered plants. Neurotransmitters octopamine has been shown to stimulate Adipokinetic hormone secretion, compound like reserpine which can inactivate octopamine and inhibit the release of AKH in locust. Fomanidine have shown the property of octapamine agonist.

Exoskelton, is one of the chracteristics feature of class insecta. It affords excellent mechanism to protect the insect against desication etc. It is unique to them and same as vulnerable target for the action of selective insecticides. chitin synthesis and enzymes associated with the process especially in sclerotization represent potential sites of action for the chitin inhibitors. Acyl ureas are important class of insecticides that inhibit chitin synthesis in insects. The potassium salts of fatty acids have been found to be effective insecticides for the control of aphids & other soft bodied hemipterans on the basis of action on insect cuticle. A combination of inhibitors of biophysical and biochemical sites of action in integument would perhaps potentiate the effectiveness and broaden the activity spectrum of these insecticides.

Role of physiology in reproduction

High fecundity as well as quick rate of development give insect prolific power of reproduction. In most of the insects although nutritional elements acquired during immature or adult stage attain importance and significance in the reproduction. Besides this other factors like physiological as well as environmental also play a crucial and variable role in breeding programmes (Engelmann, 1970; Adiyodi and Adiyodi, 1974). More significantly mating of sexes involved in copulation, temperature, humidity, light cycles, group interaction, quality and availability of the substratum are some factors that effects directly or indirectly the reproductive potential.

1. Mating : It is an decisive factor influencing the egg deposition, however, these do not make clear whether mating merely stimulated laying of mature eggs, whether the production of mature eggs in the ovaries was accelerated or the number of eggs increased or whether all these process were influenced.

2. Age : Variation in the age of male and female undergoing pronouncedly affects the different steps in reproductive programming of the insect e. g. *Dysdercus koenigii, Earias fabia* and *Corcyra cephalonica*.

3. Temperature : Optimum temperature for egg output differ greatly in species to species. Reflecting the temperature the species normally encounter during reporductive period. The temperature range in which given species may oviposit can be rather broad or relatively narrow e. g. *Earias fabia* females were found capable of laying fertile eggs within a temperature range 15°C to 28°C whereas *Pediculus* will not lay eggs below 25°C.

4. Humidity : Effect of humidity is related with temperature since any change in temperature is accompanied by a change in relative humidity. For major number of species 80-90% RH seems to be optimum for egg laying and rate of oviposition decrease below this value.

Beside above in the ovipositional behaviour, physical characteristic like moisture, hair surface and presence of some chemical constituents such as certain essential oils, steroids, terpenoids etc. were found to be the main attributes in stimulating egg laying.

Effect of population density on oviposition interestingly vary in different insects. In crowded situation total egg out put by females of *Schistocerca* was low than that recorded for isolated counterparts. When mixed population of few insects were grouped they laid relatively less eggs then has been attributed to interfere with one & another among individuals in feeding as essential factor in reproduction.

Role of physiology in developing transgenic plants

With the advancement of physiology, molecular biologist have combined improved characteristics of the plants and resistance factors of the plants into a new approach for insect pest managment. They are now trying to insert gene into crop plants. These genetic engineered crop produce insecticidal or antifeedant proteins continuously in the field. From environmental and human health perspective, the use of insect resistant crop engineered promises other attractive benefits. Because the active material produced directly in the crop tissue such concerns as spray drift and ground water contamination etc. are aviated. Two resistance factors already developed. The BT endotoxin and the trypsin inhibitor are both natural protein of high molecular mass. The first successful use of the technology was reported by a Belgian biotechnology company in July, 1987 using a gene from *Bacillus thuringiensis*. This group developed Tobacco plants that produced enough of endotoxin to kill first instar *Manduca sexta* larvae. In United Kingdom Scientist get success with a gene from the cowpea *Vigna unguiculata*. This gene encodes a serine protease inhibitor which imparts resistance to beetle. When introduced into tobacco plants, the trypsin gene imparted resistance to *Heliothis virescens* a pest of tobacco.

Reference

Adiyadi, G. and Adiyadi, R. G. (1974). Comparative Physiology of reporduction in arthropods. In *Adv. Comp. Physio. and Biochem.* **5**: 37-107.

Agrosin, M., Dinamarca, M. L. (1963). The effect of DDT on the level of dietriphosphopyridine nucleotides in *Triatoma infestans. Exp. Parasitol.* **13** : 199-203.

Auclair, J. L., Maltais, J. B. and Cartier, J. J. (1957). Factors in resistance of peas to the pea Aphid, *Acyrthosiphon pisum* (Harr.) (Homoptera : Aphispae). II. *Amino Acids. Cond. Ent.* **89** : 457-464.

Beck, S. D. (1965). Resistance of plants to insects. *Ann. Rev. Ent.* **10:** 207-232.

Boethel, D. J. and Eikendery, R. D. (1986). Interaction of plant resistance and parasitoids and predators of insects. John. Wiley and Sons, New York.

Bush, G. L. and Kitto, G. B. (1978). Application of genetics to insect systematics and analysis of species differences. In Beltsville Symposia in agricultural Research (2) Biosystematics in agriculture. ed. J. A. Romberger. pp. 89-118, New York : Wiley.

Cartier, J. J. (1968). Factors of host plant specificity and artificial diets. *Bull. Ent. Soc. Am.* **14** : 18-21.

Engleman, F. (1970). The Physiology of insect reproduction. Pergamon Press Oxford Y. K. pp. 307.

Friend, W. G., Becks, R. H. and Cass L. M. (1957). Studies on the amino acid requirements of larvae of the onion maggot Hylemya antique (MG), under aseptic condition. *Cand. J. Zool.* **35** : 535-543.

Furia, T. E. (1968). Hand book of food additives. Cleveland, Ohio, Chemical Rubber Co. 771 pp.

Hassan, A.R. (1995). Development of pollen substitute for Italian honey bee, *Apis mellifera ligustica* with special reference to nucleicocidsand pulses. Ph.D. Thesis submitted to P.G. school I.A.R.I., N. Delhi-110 012.

Lewis, W. J., Jones, R. L. and Sparks, A. N. (1972). A host seeking stimulant for the egg parasite *Trichogramma evanescens*. Its source and a demonstrations of its laboratory and field activity. *Ann. Ent. Soc. Am.* 65 : 1087-1089.

Lipke, H. and Fraenkel, G. (1954). The toxicity of corn germs to the mealworm. *Tenebrio molitor. J. Nutn.* **55** : 165-178.

Lipke, H. and Fraenkel, G. (1956). Insect Nutrition. *Ann. Rev. Ent.* **1** : 17-44.

Lipke, H. and Fraenkel, G. and Linner, I. E. (1954). Effect of soyabean inhibitors on growth of *Tribolium confusum J. Agri. Food Sci.* **99** : 410-414.

Painter, R. H. (1958). Resistance of plants to insects. *Ann. Rev. Ent.* **31** : 267-290.

Patil, P. D. (1989). Studies on the nutritional requirement of brinjal shoot and fruit borer, *Leucinodes orbonalis*. Thesis submitted. I. A. R. I., New Delhi.

Singh, P. (1970). Host plant nutrition and composition. Effects on agricultural pests. Information Bull. No. 6, Belleville, Canada, Deptt. of Agri. 19 pp.

Singh, R. P. and Srivastava, B. G. (1983). Alcohalic extract of neem (Azadirachta indica A. Juss.) seed oil as important deterant for *Dacus cucurbitae* (Coq.). *Indian. J. Ent.* **45** (4) : 497-498.

Smallman, B. N. (1945). Relation of insect damage to thiamine content of biscuits. *J. Eco. Ent.* **38** : 106-110.

Srivastava, B. G. and Pant, J. C. (1979). Fatty acid and sterol requirement of *Dacus cucurbitae* (Coq.) maggots under aseptic condition. *Bull. Ent.* **20** : 124-127.

Van Emden, H, F. (1966). Studies on the relation of insects and host plant II. A comparison of the reproduction of *Brevicoryne brassicae* and *Myzus persicae* (Hemiptera : Aphididae) on brussels sprout plants supplied with different rates of Nitrogen and potassium. *Ent. Exp. Applicate.* **9** : 444-460.

Recent Advances in Indian Entomology
Editor: O.P. Lal
APC Publications Pvt. Ltd., New Delhi, 1996, pp. 269-276

*Entomogenous and Entomopathogenic Nematodes : Their Role and Application in Insect Pest Management**

GOPAL SWARUP

Broadly, nematodes can be classified under four groups based on their feeding habits. The four categories are : animal and human parasites, insect parasites, plant parasites and the bacterial and fungal feeders. Animal and human parasites were the first ones to be recognised and thier study has been mainly taken by the veterinarian and medical scientists while the insect parasites have been the domain of zoologists and lately by the trained nematologists. Within the last two decades these organisms have attracted serious attention because of their potential for managing insect pests. This area of research has assumed a significant status because of the awareness about environmental pollution and hazards involved in the increasing use of chemicals to protect crops against insect pests and diseases.

Entomophilic (entos = insect; philos = to like or having an affinity for loving nematodes or entomogenous nematodes) or nematodes parasitising on insects, occur in about 27 families of 9 major group of nematodes : Rhabditiod, Tylenchoid, Aphelenchoid, Strongyloid, Oxyuroid, Ascaridoid, Spiruroid, Filarioid (Secerentea) and Mermithoid (Adenophorea). Besides the insect groups (Lepidopterans, Dipterans, Orthopteran, Hymnopterans) which are directly or indirectly related to crop damage, entomophilic nematodes are also involved in parasitisation of bark beetles, mosquitos, houseflies etc.

Historically, Aldrovandus (1602) recorded presence of 'worms' emerging from bodies of dead grasshoppers. Subsequently, Gould (1747) described emergence of worms, probably mermithid, from ants. Between 1800-1867, Diesing published Systema Helminthum (1851) in which 175 insect nematodes were recorded (including 118 species of *Gordius* and 17 of *Mermis).* The father of present day Nematology, N. A. Cobb in 1927 highlighted the potential of nematodes in controlling insect pests. Filipjev and Schuurmans-Stekohoven in their book, A Manual of Agricultural Helminthology (1941) brought together at one place all the scattered information on insect nematodes. The impetus to researches on insect nematodes was further hightened with works of Dutky and Hough (1955) and Weiser (1955) who recorded a nematode in codling moth. The former group designated it by its accession

*Presentation based on review paper "Entomogenous and entomopathogenic nematodes in Biological Control" by H. K. Kaya, In: Plant Parasitic Nematodes in Temperate Agriculture (Eds. K. Evans, D. L. Trudgill and J. M. Webster (1993); pp. 565-591.

number as DD 136. The scientific name. *Neoaplectana carpocapsae*,was given by Weiser (1955).

Several nematode species occupy the same habitats as insects and therefore, it is not an unusual phenomenon of their association with insects. These nematodes may be found on various parts of the insect exoskeleten or within or on various internal tissues. However, in many cases such associations are not of parasitic nature. It is important that nature of relationship or association is properly understood. In other words parasitism should not be confused with symbiotic relationship or commensalism (one species is benefitted and the other one is neither harmed nor benefitted - a very common feature in insect-nematode relationship). In contrast to commensalism, parasitism by entomogenous nematodes has variable deleterious effects on their hosts. The pathological effects include :

(1) Sterility

(2) Reduced fecundity

(3) Reduced longevity

(4) Reduced flight activity

(5) Delayed development

(6) Other behavioral, physiological and morphological aberrations

(7) In some cases rapid mortality (Entopathogenic : nematodes that kill their hosts quickly - within 48 hrs).

There are about 30 nematode families associated with insects but only 9 of them exhibit biological control potential for insects. These are : Allantonematidae, Parasitylenchidae, Iotonchidae, Tetradonemidae, Sphaerulariidae, Mermithidae, Phaenopsitylenchidae, Steinernematidae and Heterorhabditidae.

Most nematode species belonging in the families of Allantonematidae, Parasitylenchidae and Iotonchiidae have not been exploited for biological control because of their complex life cycles; are difficult to maintain in culture and do not induce significant damage to host. An example of this is the limited use of *Paraiotonchium* (Syn. *Heterotylenchus autumnale parasitisation of Musca autumnale* (face fly) - a major pest of range cattle in USA and Europe. The nematode is an obligate parasite and except for a brief free living gamogenetic stage in fresh cowdung, the life cycle of the nematode occurs entirely in the host. The mated free living female penetrates the haemocoael of the larval face fly but does not produce eggs till the adult fly emerges. Alter adult fly emergence, the female lays several eggs that become parthenogenetic female in a few days Such females produce large quantity of eggs that develop into immature gamogenetic nematodes. These nematodes then invade ovaries of the fly, sterilising the female. Even under laboratory conditions percentage of infection does not go beyond 30 per cent. The nematodes, obligate parasitism and complex life cycle make it difficult to obtain *in vitro* cultures. However, despite drawbacks these nematodes can be introduced in fly infested areas and the fly population can be reduced to some extent.

Tetradonematids have been obtained from Diptera (sand flies), Coleoptera (beetles) and Hymenoptera (fire ants). *Tetradonema plecans* is the most studies species. It is an endoparasite of greenhouse flies. Infection probably occurs after ingestion of eggs or of infective juveniles by the fly larva. The infective juvenile penetrates through the midgut into the larval haemocoel and within a week matures into adults. After mating in the host haemocoel, the female emerges from the larval host through a hole in the body cavity and deposits eggs in soil or the eggs are laid in the host and later liberated in soil after decay of larval cadaver. Older larval hosts may sometimes survive to the adult stage but are usually sterile. These adults aid in the dispersal of the nematode. This nematode has a high pathogenecity and reproductive capacity. Further, it has a good shelf life of 6 months for the egg stage and shows specificity to sciarid flies. The life cycle is synchronised with the host. It is easy to produce *in vivo*. It is effective at high concentrations as a biological insecticide and at low concentration as a long term biological control agent.

Nematodes of sphaerulariidae family rarely kill their hosts but are responsible for sterility, reduced fecundity, delayed development and/or altered behaviour.

Tripius sciariae often causes fatal infections in larval sciarids, pests of mushroom houses and greenhouses. Infection usually results in larval death in 12 days. Infection of older larvae allows pupation and some of these may become adults. These adults flies are sterile and serve as important agents of nematode dispersal. Although Poinar (1965) recorded 100 per cent parasitism and significant decline in fly population within 4 weeks, still according to Gaugler (1987) high effectiveness, excellent dispersal capabilities and relative ease of culture are countered by short survival period of the infective female which hinders its development as a biological control agent.

Mermithids are invariably lethal to their hosts and a few of them have been exploited as biological control agents. These are slender nematodes ranging from 1-30 cm and parasitise a number of insects.

The newly hatched second-stage juvenile is the pre-parasitic stage which penetrates the host cuticle with the help of its stylet, entering the haemocoel, absorbing the nutrients and then stored in its body. Generally, parasitic juvenile remains free in the haemocoel but in a few species it may enter the host brain or nerve ganglion or may cause formation of an internal gall so as to avoid host's defence mechanism. After completion of the parasitic stage, the nematode emerges from the host and begins its free-living stage. Most death occurs rapidly because of the exit hole made by the nematode. The post-parasitic stage enters soil (terrestrial mermithids) matures, mates and oviposits to complete the life cycle.

Two mermithids, *Romanomermis culicivorax* - a parasite of mosquito larvae and *Filipjevimermis leipsandra*- a parasite of larval, banded cucumber beetle, *Diabrotic balteata* have been studied extensively. *R. culicivorax* kills its host within 8 days and has a generation time of 1-2 months. During 1970's, this nematode was put in commercial use but because of poor storage, shipping and economics of mass production it did not gain much favour, specially after success of *Bacillus thuringiensis* subsp. *israelensis* as a potent bacterial mosquito larvicide.

It has been calculated that for 90 per cent host reduction, an unrealistic mermithid : mosquito release ratio of 13 :1 is required.

Filipjevimermithid leipsandra is parthenogenetic species, with each female producing several thousand eggs in soil. The preparasitic stage, released from eggs, penetrates the insect larval cuticle entering the haemocoel. It then develops in the nervous system of the insect. It grows for 12-22 days and ruptures the nervous system of the host. While exiting from the insect body, it kills the insect. This nematode has been mass produced *in vivo,* producing 5 x 10^6 eggs per week. In small plot tests, 78 per cent larval parasitism of *Diabrolica baltesata* has been recorded. However,. *in vitro* production of this nematode has not proved to be very successful.

The positive attributes of mermithids include :

(i) Host specificity

(ii) Lethality of their hosts

(iii) Easy manipulation in laboratory

(iv) Mass production capability

(v) Dissemination through standard spray equipment

(vi) Potential for long-term control

(vii) Environmental safety.

However, these attributes hold true for only a few mermithid species.

The negative aspects are :

(i) Variable efficacy

(ii) Long life cycles of many mermithids

(iii) Obligate parasitism

(iv) Cost of production

(v) Poor storage

(vi) Poor natural dispersal.

An overall lack of knowledge about biology of many merimithid species is an additional constraint for employing them in biological control programmes.

In the Phaenopsitylenchidae family, the nematode, *Beddingia siricidicola* has been used to control wood wasp, *Sirex noctilio* (in Australia). The wood wasp, is normally a benign or at best secondary pest of pine trees in New Zealand and Australia. This insect lays eggs alongwith a toxic mucous and arthrospores of the fungus *Amylostereum aerolatum,* which is mutualistically associated with the insect. The combination of toxic mucous, fungs and infestation by wasp kills the tree. The nematode sterileis the female wood wasp. In Australia, a high rate of parasitism (>90%) was observed in wood wasp population and because the wood wasps are effective flies, this nematode was dispersed to other forests upto 13 kms away. In 1987, outbreaks of *Sirex* occurred in pine plantations in South Australia. About

150.000 trees were inoculated with this nematode and as a result there was significant reduction in tree mortality as well as insect population.

Nematodes belonging to the families of Steinernematidae and Heterorhabditidae are obligate pathogens and are characterised by their association with bacteria in the genus *Xenorhabdus*. Of all the nematodes studied for biological control of insects. these two nematode groups have received the most attention because they possess attributes of effective biological control agents. These are also widely distributed in soil environment.

The thrid stage non-feeding larvae is the infective stage and carries the bacterium in intestine (*Steinernema*) or throughout intestinal lumen (*Heterorhabditis*). In species like *S. carpocapsae* and *S. scapterisci*, the infective stage may come in contact with the host as chance encounter while in some cases like *Heterorhabditis bacteriophora* and *S. glaseri*, the infective stage acturally hunts for its host. Probably, in the latter event, these may be host specific.

Once the host is located, the larvae enter enter host through natural opening and penetrates the haemocoel. The bacterial are released into the haemolymph where they propagate and cause septicaemia, killing the host within 48 hrs. The nematodes feed on the bacteria and host tissues, produce 2 or 3 generations, and emerge from the cadaver as infective larvae and search for new hosts.

Insects killed by steinernematids turn yellowish brown or black and the Heterorhabditid killed insects turn red, brick red, purple, orange or green. In darkness, heterorhabditid - infected insects will luminesce because of the presence of the backterium *Xenorhabdus luminescens*.

In nature, the nematode and the bacterium cannot survive independently from each other. Each species of entopathogenic nematode has a specific natural association with only one Xenorhabdus species. However, a Xenorhabdus sp. may be associated with more than one nematode species. In the nematode-bacterium association, the nematode is dependent on the bacterium for :

(i) Killing its host

(ii) Creating a suitable environment for nematode development (by producing antibiotic compounds that suppress growth of secondary microorganisms).

(iii) Breaking down the host tissues into usable nutrients.

(iv) Serving as a food source.

The bacterium needs the nematode for :

(i) Protection from external environment

(ii) Penetration into hosts' haemocoel

(iii) Possibly, inhibition of host's antibacterial proteins.

Outside the arthropods, only mollusc and two amphibian species have been adversely affected by these nematodes under laboratory conditions. *Steinernema glaseri* infects a snail species and a Steinernematid has been reported to kill young tadpoles.

Genetic selection to improve desirable traits has been successfully accomplished with Steinernematids. Through a selective breeding programme, host finding by *S. carpocapsae to* wax moth, *Galleria mellonella* has been increased from 4 per cent to 80 per cent after 13 passages.

Commercial considerations

The nematode-bacterium complex does not reproduce in nature in the absenmce of host but it has been successfully cultured *in vitro* in laboratory and thus can be mass produced. The symbiotic bacterium grows on a number of proteins (plant/animal) and the nematode feeds on the bacterium alongwith other nutrients in the media. Production in a commercial scale has been accomplished by using inexpensive protein and sterol sources. The nematodes also have been produced monoaxenically in large liquid fermenters using a combination of plant and animal proteins.

A major constraint on commercialisation of these nematodes is their poor storage capability. However, nematodes formulated in various carriers such as alginate, clay, vermiculite and gel forming polyacrylamides store for atleast 6 months under referigeration and 3 months at room temperature.

Application

These nematodes can be applied with ground and aerial spray equipment. Precaustions that need to be taken include removing screens often used with the spray nozzles, using a clean sprayer and ensuring that the spray tank does not exceed 32 C. Nematodes can be also applied through drip or sprinkler irrigation systems. The most appropriate time for nematode application is during early morning, early evening or night or on cloudy days to avoid solar radiation and high temperatures. The nematode-bacterial complex is also compatible with certain chemical pesticides. Incompatibility also exists between some insecticides/ acaricides/nematcides.

Prospects

Most entomogenous nematodes can not be developed as commercial propositions because of (i) difficulty is mass production capability, (ii) resistant stage, (iii) lack of adequate knowledge about their pathogenicity, biology and ecology.

Some may be candidates for classical biological control

Paraiotonchium autumnale has been used to control face fly in USA. It does not regulate the face fly population but the behaviour of the infected flies is changed leading to a non-pest status. The phaenopsitylenchid, *Beddingia siricidicola* was introduced in Australia and controlled the forest pest, *Sirex noctilio* demonstrating that nematodes that sterilise a host can effectively regulate a pest population.

The Steinernematids and Heterorhabditids offer the most promise for commercial development when applied as biological insecticides, under appropriate conditions, against certain insect pests. These nematodes have been most effective in the field. Much more information needs to be gathered on the abiotic and biotic factors associated with the effectiveness of different nematode species and strains against specific target pests so as to obtain reliable and consistent results. Howevers, these nematodes should not be viewed as the sole component of a pest management strategy. Rather, they should represent one tactic in a series of biological, chemical and cultural options available for pest management.

Media for Xenorhabdus species

(1) Nutrient Agar :

Beef extract : 3 g

Peptone : 5 g

Agar : 15 g

Distilled water : 1 L

(2) Nutrient broth :

Same ingredients as above except that agar is not used

(3) Yeast Salts or Yeast broth :

K_2HPO_4 : 0.5 g

$NH_4H_2PO_4$: 0.5 g

$MgSO_47H_2O$: 0.2 g

NaCl : 5.0 g

Yeast extract : 5.0 g

Distilled water : 1.0 L

Galleria media

(1) Distilled water : 100 ml

Honey (warmed to thin) : 100 ml

Glycerol : 100 ml

Deeca Visol or
equivalent vitamins : 5 ml

Pablum or Gerber's
mixed cereal : 1200 ml

(2) Calcium propionate : 1.5 g

Brewer's yeast : 1 tsp

Water : 100 ml

Honey (warmed to thin) : 200 ml

Glycerin	: 200 ml
Polyvisol or other infant mulivitamin preparation	: 12 ml
Bran	: 2400 ml

Mix together the ingredients of A and B separately then combine. finally, pour all liquid ingredients into bran and mix thoroughly. Autoclave 20 mts at 121°C.

References

Dutky, S. R. and Hough, W. S. (1955). Note on a parasitic nematode from codling moth larvae. *Proc. Entomol. Soc. Wash.* **57** : 24.

Filipjev, I. N. and Schuurmans-Stekohoven, J. H. (1941). A Manual of Agricultural Helminthology. E. J. Brill, Leiden.

Gaugler, R. (1987). Entomogenous nematodes and their prospects for genetic improvement. In: Maramorosch, K. (ed.) Biotechnology in Invertebrate Pathology and Cell Culture. Acad. Press, N. Y., pp. 457-484.

Poinar, G. O. (1965). The bionomics and parasite development of Tripius sciarae, a nematode parasite of sciarid flies. *Parasitology* **55** : 559-569.

Weiser, J. (1955). Neopalectana carpocapsae n. sp. Novy Cizopasnik housenek obatece jablecneho Carpocapsa pomonella. *Vestnik Cesk. Zool. Spolecnosti* **19** : 44-52.

Recent Reviews

Gaugler, R. (1987). Entomogenous nematodes and their prospects for genetic improvement. In: Maramorosch, K. (Ed.) - Biotechnology in Invertebrate Pathology and Cell Culture. Acad. Press., pp. 457-484.

Gaugler, R. & Kaya, H. K. (1990). Entopathogenic nematodes in biological control. CRC Press, Boca Raton, Fla.

Kaya, H. K. (1985). Entomogenous nematodes for insect control in IPM systems. In: Hoy, M. A. & Herzog, D. C. (Eds.) Biological control in agricultural IPM systems. Acad. Press, N. Y., pp. 283-302.

Kaya, H. K. (1987). Diseases caused by nematodes. In: Fuxa, J. R. & Tanaka, Y. (Eds.) Epizootiology of Insect Diseases. John Wiley & Sons, N. Y., pp. 453-470.

Kaya, H. K. & Gaugler, R. (1993). Entopathological nematodes. *Ann. Rev. Entom.* **38** : 181-206.

Popiel, I. & Hominick, W. M. (1992). Nematodes as biological control agents. Part II. *Advances in Parasitology* **31** : 381-433.

Recent Advances in Indian Entomology
Editor: O.P. Lal
APC Publications Pvt. Ltd., New Delhi, 1996, pp. 277-282

NPV as Environment Friendly Insecticide

S. CHAUDHARI

In recent years there is a demand for biological suppression of insect pests because of insect have developed resistance to a number of commonly used insecticide and over and above adds to the pollution problem in the environment. There is a need to develop efficient and ecologically safe method of pest control at an acceptable cost.

Quite naturally attention was now turned to insect viruses, nuclear polyhedrosis viruses (NPV) in particular as safe and effective approach for controlling the insects-pests, which account for more than 40 per cent of the described arthopod viruses. Nuclear polyhedrosis viruses under trial have given encouraging results in controlling the insect in laboratories and field trials which causes a disease after feeding the virus sprayed foliage leading to certain death of the pest.

The disease affected larvae turns pinkish red and ruptures releasing the virus, called polyhedral inclusion bodies (PIB) in the field which helps in infecting other healthy caterpillar feeding on contaminated foliage. A single larva will had as many as 200 crores disease causing PIB. The virus can be produced from the diseased larvae or by inducing the disease in the healthy caterpillars by feeding them virus coated leaves. The highly specific NPVs may be regarded as a ideal bio-pesticide either for use alone or in integrated pest management programme. Although NPV will not replace the chemical insecticide, this will definitely reduce the use of chemical insecticides against which the pest are becoming resistants and more and more insecticides is required to kill the them adding to pollution problem, it is responsible for elimination of natural enemies (parasite and predators) also allowing rapid rebound of insect population. Hence the need for such biological pathogen will increase and there is very reason to hope that these NPVs will find a place as environment friendly insect controlling agent during the coming years.

The NPVs of *Heliothis* spp. (Elcar), *Lymantria dispar* (Gyp-check), *Orygia pseudosugata* (Bio-control-13, *Neodiprion sertifer* (Neo-check) and *N. leuconti* are produced in vivo commercially and are registered for insect control in U.S.A. and Canada.

Nuclear polyhedrosis virus belongs to family Baculoviridae. The main feature of this group is rod shaped virus particles.

The NPV which has many virions (enveloped nucleocapsids) occluded within intra nuclear protein, crystals called polyhedra (PIB) or inclusion bodies (IB). These PIB are usually 1-10 μm in size and composed of a single protein, polyhedrin, which accounts for about 95% of the mass of the polyhedra.

The virion are present either as single (SNPV) or multiple (MNPV) nucleocapsid within the envelops.

Polyhedra are mere vectors of virions when generation of larve is temporarilly spaced from the next generation. The actual virus particles is the enveloped nucleocapsid which comprises DNA-protein complex bond by protein shell called capsids. The PIB stabilize virion allowing them to remain viable for long.

Nuclear polyhedrosis biology and replication

The nuclear polyhedrosis life cycle involves two structurally distinct phenotypes with each having a specific role (Blissard and Rehrmann, 1990). The budded virus phenotype is produced during early stages of infection and spreads the infection from cell to cell within the insect and is highly infectious for cultured cells. The polyhedra-derived virions (PDV) pheontype (also known as occluded virus (OV) is produced during terminal stages of viral infection becomes concentrated in the nucleus and is occluded within polyhedra. After ingestion by a susceptible insect, polyhedra are dissolved releasing PDV which initiate infection within the midgut cells, PDV are relatively non-infection for cultured cells. Although the nucleocapsids of the two phenotypes appear to be identical, their envelops are derived from different sources. During both the early and late phases of viral gene expression of glycoprotein, gp 64 is synthesized and transported to the margin of the cell where it is incorporated into the host cell plasma membrane. During the budding process, the nucleocapsids become enveloped by the modified membranes. In contrast, PDV, become surrounded by an envelope that assembles within the nucleus subsequently they become occluded and are released upon the death of the cell.

The non-cellular midgut membrane of insect larvae serves as a mechanical barrier to the virus. It is overcome by producing on viral coded protein that degrades the peritrophic membrane, thereby enhancing infection. Some proteins mediating the adsorption and fusion of virion envelops with membrane of gut cells have also been reported.

The molecular basis of nuclear polyhedrosis pathogenesis is poorly understood, NPV infection is polyorganotrophic and soon after infection in midgut cells general dispersion of virus occurs.

Viral progeny will be first observed 8 hr after inoculation and the developmental cycle of the virus was essentially completed 24 h after inoculation.

Normally the larval stages causing damage are most affected and infection develops after PIB contaminated food is ingested. Usually larvae take 4-14 days to die depending on age temperature and dose inoculum. During this period the insect may continue to feed causing damage to the crop. To minimize the damage, timing of application is critically important, so that the insect is killed as early as lifestage as possible before peak.

Transmission of virus from generation to generation can acheived in several ways. Transovarian transmission seems to be common, eggs from infected females often became contaminated externally. There may be significant carry over of virus in the soil or on the food plant, larva to larva transmission usually occur through the eating of contaminated foliage with virus and this contamination can come from the soil or from exudates from diseased larvae through faecal matter and parasite may transmit virus via the oviposition.

There are three ways of using insect viruses in biological control system.

1. Insecticidal approach. The virus is applied like a chemical insecticide by spraying the PIB suspension on dusting the foliage with wettable powder virus formulation.
2. With introduction of NPV by the release of infected larvae.
3. The virus is most suited for integration into IPM. Since nuclear polyhedrosis viruses are compatible with chemical insecticides and other entomopathogens either as mixture or used sequentially (Benz, 1971; Chaudhari and Ramakrishnan, 1980; and Chaudhari, 1987).

The only drawback with the viral pesticide is the loss of original activity under direct sun exposure. This problem of viral inactivation due to sun-exposure can be minimised by addition of UV-protectants before spraying the crop, like activated charcoal, black ink and amino acids etc. (Chaudhari and Ramakrishnan, 1980; and Ramakrishnan et al., 1981). The virus should be sprayed in the evening time only with putting virus sprays to underside of the leaf surface. It will protect the virus from direct sun-exposure. The following is the brief account of some NPV as direct control agents experimented in our country. The NPV of S. litura @ 93 x 10^6 PIB/ml and half the dose of carbaryl 0.02 per cent with the activated charcoal per cent as UV protectant significantly reduced the leaf damage in tobacco nurseries and the number of larvae recovered from treated plots is significantly reduced from control plots (Ramakrishnan et al., 1981) (Table 1). The percentage of seedling damage after virus treatment (3.2 x 10^8 PIB/ml) was significantly reduced compared to that of the control. It was also similar to results obtained with endosulfan (0.05 per cent) and carbaryl (0.1 per cent). The reduction of damage in terms of larval equivalent

Table 1. Efficacy of the nuclear polyhedrosis virus in foliar protection against *Spodoptera litura* on tobacco crop.

	Nursery	
Treatment	% leaf damaged	Mean No of larvae recovered (100 Ist and 2nd stage larvae were released)
NPV 93 x 10^6 PIB/ml	28.45	14.3
NPV 46.5 x 10^6 PIB/ml + 0.02% carbaryl	17.60	7.3
NPV 46.5 x 10^6 PIB/ml + 1% activated charcoal	28.75	11.0
Control	42.93	37.3
CD at 5%	7.231	6.32

is two healthy larvae equal to ten diseased larvae in cauliflower under artificial infestation conditions (Table 2) (Chaudhari and Ramakrishnan, 1980). The successful use of NPV of *Amsacta albristriga* a pest of groundnut in Tamil Nadu has been reported by Jacob (1972). The efficacy of NPV in combination with 1/10th the recommended dose of insecticide caused 70 per cent mortality while the insecticide at this concentration caused only 10.3-21.3 per cent mortality indicating higher effectiveness of the disease under physiological stress induced by the chemicals. To be more preicise quantitative estimate of foliar protection with application of NPV against Bihar hairy caterpillar *Spilosoma obliqua* the consumption studies indicated 40 per cent reduction in foliar damage when treated with NPV (Table 3). Santharam and Balasubramaniam (1983) found a single spray of the combination of reduced doses of carbaryl (0.05 per cent) and NPV at 125 LE/ha was on par with three sprays of NPV at 375 LE/ha and one spray of carbaryl 0.1 per cent. It was found that tanic acid (0.25 per cent) and boric acid (0.5 per cent) are compatible with NPV and increased the virosis significantly when mixed with virus before feeding.

Table 2. Foliar protection by NPV on cauliflower under artificial infestation condition.

Treatment/larvae per plant	Average area eaten in sq. cm.	Average no. of larvae recovered	Remarks
2 Larvae	19.89	2.0	Healthy
4 Larvae	35.20	4.0	-do-
6 Larvae	38.37	5.5	-do-
8 Larvae	47.64	7.25	-do-
10 Larvae	64.04	9.00	-do-
10 Larvae 26.2 × 10^6 PIB/ml	27.92	7.25	died with diseased symptom
10 Larvae 26.2 × 10^6 PIB/ml + charcoal	22.75	7.5	-do-
CD at 5%	8.91	-	-

Table 3. Effect of NPV disease on food intake and excretion.

Treatment		Days after infection (Values expressed in cm) 1	2	3	4	5	Total
Control	Intake	0.0267	0.0156	0.0565	0.0546	0.0627	0.2142
	Excretion	0.0042	0.0041	0.0255	0.0395	0.0475	0.1208
Virus 2	Intake	0.0313	0.0246	0.0456	0.0394	0.0285	0.1694
	Excretion	0.0046	0.0048	0.0216	0.0301	0.0221	0.0833
Virus 1	Intake	0.0291	0.0235	0.0452	0.0227	0.0173	0.1378
	Excretion	0.0037	0.0045	0.0171	0.0134	0.0096	0.0483
CD at 1% feeding		N.S.	N.S.	N.S.	0.0141	0.0274	

NPV infection in the later stages of the larva increased the susceptibility of the final instar to fenvalerate, cymermethrin, endosulfan and monocrotophos. The enhance susceptibility was maximum in cypermethrin followed by endosulfan (Rabindra and Jayaraj, 1990).

NPV and 0.05 per cent endosulfan were compared for the control of H. *armigera* on *Cicer arietinum in Maharastra*. The virus applied at 100, 250, 350 LE/ha alone or with endosulfan. The Lowest pod damage (3.84 per cent) and highest yield 1379 kg/ha were observed in plots treated with 2 sprays of endosulfan alone. Similar yields were obtained with highest concentration of virus applied alone or incombination with 1 spray of endosulfan (Pawar et al., 1990).

Rabindra and Jayaraj (1988) evaluated the efficacy of certain adjuvants for NPV of *H. armigera* on chickpea. They found that jaggery (0.5 per cent) cotton seed kernel flour 1.0 per cent and chick pea flour 1 per cent) singificantly increased mortality due to virosis. They adocated the use of these adjuvants which serve as phagostimulants for increasing the efficacy of the virus. Dhandapani et al. (1987) found that four round of virus spray at the rate 450 LE/ha with endosulfan 350 g ai/ha with adjuvant like larval extracts of *H. armingera* (4%) or whole milk (10%) crude sugar (15%) or cotton seed kernel (2.5%) were as effective as endosulfan at the rate of 700 g ai/ha is reducing larval population. Spraying of HaNPV 500 LE/ ha thrice at an interval starting from bud stage on sunflower along with release of *Chrysopa scelestes* @ one first instar larva/head thrice at 10 days interval starting reduce the larval population significantly regarding yield HaNPV 500 LE/ha recorded 1200 kg/ha followed by 1200 kg/ha in Chrysopa released plots as against 1013 kg/ha in the treated check.

Safety evaluation

The virus was administered to test animals like albino rats, guinea-pigs and rabbits in single dose as well as multiple massive dose exposure. The doses administered varied from 4 x 10^9 to 90 x 10^9 PIB/animal clinical investigation on haematology, urology, carbohydrate metabolism recorded at various intervals of time showed no differences in the various parameters between the virus treated and untreated animals, food consumption and weights gain of the treated animals were relatively the same as that of untreated animals. No gross pathological condition of any of the body organs could be deduced in any of the animals. The maximum total score for the skin reaction in the acute dermal toxicity tests was found to be zero as per Draize scoring. The score in the primary irritation to eyes and skin was found to be zero in both the experiments. The direct injection to the body of albino rats did not result in any abnormality or death of the animals. Hence it could be concluded that SINPV is not pathogenic to the three species of vertebrate tested.

Summary

Although the production of polyhedral inclusion bodies of NPV is expensive since it is produced on living hosts, added cost often can be rationalised on the basis of safety and specificity relatively to chemicals. From the performance of the nuclear polyhedrisis viruses it was concluded that this in future would be on ideal biocontrol agent for main insect group of lepidopterous and Hymenopterous pests. The inherent advantages of the NPV are as follows.

1. NPV can offer alone and in combination with other control measures such as chemical insecticides, neem kernel extracts etc.
2. NPV are transovarially transmitted and hence will be cariied over to subsequent generation of insects.
3. NPV do not interfere adversely with parasite and predators on the contrary these parasites and predators feeding on diseased insects spread the disease to healthy insects also.
4. Farmers can extract, handle, store and apply the virus easily.
5. As per safety tests conducted with SINPV it is not pathogenic to vertebrates.

References

Benz, G. 1971. Synergism of micro-organisms an chemical insecticides in "Microbial Control of Insect and Mites" eds. H.D. Burges and N.W. Hussey, pp. 327-353.

Blissard, G. W. and Rohrmann, R.F. 1990. Baculovirus diversity and molecular biology. *Ann. Rev. Entomol.* **35** : 127-155.

Chaudhari S. 1987. Combination effect of Baculovirus of *Diacrisia obligua* Walker with low doses of insecticides and fertilizers on Larval virosis. *Indian J. Ent.* **49** (4) : 515-519.

Chaudhari, S. and Ramakrishnan, N. 1980. Field efficacy of Baculo-viruses and its combination with sublethal doses of DDT and endosulfan on cauliflower against tobacco caterpillar, *Spodoptera litura* (F.). *Indian J. Ent.* **42** (4) : 592-596.

Dhandpani, N. and Jayaraj, S. and Rabindra, R.J. 1987. Efficacy of ULV application of NPV with certain adjuvants for the control of *Heliothis armigera* on cotton. *J. Biol. Control* **1** (2) : 111-117.

Jacob, A. 1972. Studies on nuclear polyhedrosis of three species of lepidoptera, Ph.D. Thesis submitted to Tamil Nadu Agricultural University, Coimbatore.

Pawar, V.M., Chudurwar, R.D., Kadam, B.S., Thombres, U.T. Dhawandar, S.D. and Seeras, N.R. 1990. Field efficacy of NPV against *Heliothis* (Lepidoptera : Noctuidae) on gram *Cicer arietinum* in Maharastra, *Indian J. Agric. Sci.* **60** (4) : 287-289.

Rabindra, R.J. and Jayaraj, S. 1988. Evaluation of certain adjuvant for NPV of *Heliothis armigera* on Chickpea. *Indian J. Expt. Biol.* **26** : 60-62.

Rabindra, R.J. and Jayaraj, S. 1990. Effect of nuclear polyhedrosis virus infection on the insecticide susceptibility of *Heliothis armigera* and *Spodoptera litura* larvae. *J. Biol Control.* **4** (1) : 31-34.

Ramakrishnan, N., Chaudhari, S. Kumar, S. Rao, R.S.N. and Satyanarayan S.V.V. 1981. Field efficacy of nuclear Polyhedrosis virus against the tobacco caterpillar, *Spodoptera litura* (F.). *Tob. Res.* **7** (2) : 129-134.

Santharam, G. and Balasubramaniam, M. 1983. Effect of nuclear polyhedrosis virus used alone and in combination with insecticide in controlling *Heliothis armigera* on Bengal gram, *Cicer arietinum*. *J. Ent. Res.*, **6** (2) : 179-181.

Recent Advances in Indian Entomology
Editor: O.P. Lal
APC Publications Pvt. Ltd., New Delhi, 1996, pp. 283-288

Insect Immunity and Role of Insect Growth Inhibitors as Immunosuppressants

G.T. GUJAR

Insects, about a million strong group of animals, have evolved over a long period of about 500 million years. During this period, they have come across a wide range of other animals and plants that determined and influenced the evolutionary life processes. Of the living beings that insects have interacted with; microorganisms constitute an important group. These microbes showed a wide variety of association; some beneficial; some pathogenic and many indifferent. Insects have evolved an ability to live with these microbes; either by integrating their physiology to that of microbes for mutual benefit as in insect-symbiont relationship or by developing immunity (and resistance) to the pathogenic and nonpathogenic ones. In view of simplicity of insect haematologic system; they appear to have evolved a nonspecific immunity to the invading microbes, unlike in higher animals, so that enemies are taken care of with the least biological stress. Insect immunity is thus the reaction of cellular and humoral factors present in the haemolymph of insects.

Historically, studies on insect immunity were initiated in late 19th century. Metalnikov in early part of this century gave impetus by emphasizing importance of cellular defense reactions to the invading microbes. At the same time, Pailot realised importance of humoral factors which are induced as a result of infection and provide a back-up for haemocytes in detoxication and clearing up of foreign antigens. Thereafter, Boman of Stock holm University, Sweden carried out a pioneering work in the 70s on the characterization of immune proteins in *Hyalophora cecropia* and isolated a number of antibacterial proteins like attacins, cecropins. Our investigations in the 80s on protein metabolism showed the evidence of immune system in the desert locust that recognised and cleared the presence of human serum albumin at a very fast rate (Gujar and Mehrotra, 1983).

Cellular defense mechanisms

Haemocytes that constitute about 5-10 % of haemolymph are basically of 6 different types viz., prohaemocytes, plasmatocytes, granulocytes (coagulocytes), spherulocytes, oenocytoids and adipohaemocytes. Of these, granulocytes are a plesiomorphic cells that are found in all arthropods, and are reported to have evolved into other types in high arthropods, and are reported to have evolved into other types in higher arthropods. Granulocytes are structurally similar to that of coagulocytes and also perform the function of coagulation. Granulocytes and plasmatocytes are grouped and designated a immunocytes in view of their ability to recognize, digest and clear the foreign antigen. These haemocytes are chiefly involved in the following defense reactions (Pathak. 1993; Gupta. 1994).

Phagocytosis

One of the main fuctions of haemocytes in insects is their ability to phagocytosize the foreing object like protozoa, bacteria. yeast cells, erythrocytes. fungi and even autolysing tissues. Once these foreign objects are recognised by the immunocytes. either randomly or with the presence of chemotactic factors present on them or by the release of host factors by immunocytes; all other immunocytes converge on the foreign target. The foreign object is then ingested through various processes like pinocytosis, encirclement with pseudopods or enveloping with spreading cytoplasm and is later broken down with the help of enzymes to clear off the products of foreign antigen. Phagocytotic ability of insect depends upon the antigen. frequency of attact and phagocytic index. Gnerally this ability is more in mature larvae than in young ones. Although a wide range of foreing objects are phagocytosized. some like *Bacillus thuringiensis, B. cereus,* microsporidia. *Nosema* spp. and certain fungi. *Sorosporella livella* and *Coelomomyces* spp. are not phagocytosized and hence proves pathogenic to the insects.

Encapsulation

Encapsulation is a defense mechanism for inactivating foreign biotic and abiotic substances that are too large to be phagocytosized. These include bacteria, fungi, nematodes, eggs, latex, sephadex beads, araldite, cellophane thread, glass needles, etc. The encapsulation of foreign antigen is carried out by haemocytes, esp. granulocytes by putting on several layers of cells, 20-75 distinguishable in 2-3 regions. This also leads to melanization, more often at the inner core of a capsule, so that products like quinones show antimicrobial activity to the antigen.

Besides cellular encapsulation; humoral encapsulation has been reported in mosquitoes, chironomid larvae and certain other dipterans against antigens like nematodes, fungi, bacteria, protozoa, injured homologous and heterologous tissues, some organic polymers. The process is first initiated with the binding of activated phenoloxidases and other sticky proteins to the foreign antigen and later solidifiction and tanning of the capsule and further its sclerotization. The humoral encapsulation is fast, usually completed in minutes and is efficient defense mechanism in certain dipterans (Vey, 1993).

Nodule formation

Nodule formation is similar to the cellular encapsulation and may be involved simultaneously in an insect. Typically, a capsule is formed around a melanized core of coagulum produced by granulocytes that surrounds the antigen. The nodule formation occurs in response to the both biotic and abiotic objects that can not be phagocytosized and is fast taking 1-5 minutes as in *Galleria melonella* and *Pieis brassicae* challenged with *B. cereus* and *Escherichia coli.*

Coagulation

Coagulation is in response to the wound or injury to the insect and involves formation of coagulin (clot) at the site from coagulogen (lipoproteins) with the help of clotting enzymes (serine proteases).

Secrettion of humoral factors

A number of humoral factors are present or secreted in the haemolymph of insects to provide the back-up to the cellular defense in response to the invading antigen. These are mainly Hacmagglutinins (lectins).

Haemagglutinins (lectins)

These are the high molecular weight proteins having multiple binding sites for binding with microbes and vertebrate erythrocytes, leading to agglutination. These are used for recognition, opsonization of the foreign object, and also mediating in the cell-to-cell communication during would healing. These help in proliferation of immunocytes.

Prophenoloxidase activating system

This system is supposed to recognise the presence of the foreign antigen and also produces quinones that are toxic to them.

Antimicrobial factors

Many low molecular weight poteins like attacines, cecropins, lysozymes have been isolated from the lepidopterans like *H. cecropia,* and many other insects and have been found to have antimicrobial activity. In our investigations on antibacterial factors, we observed atleast 10 major bacteria induced proteins including 2 lysozyme like proteins with molecular weights of 14-77 kDa in the haemolymph of *Helicoverpa armigera*; some with high antibacterial activity (unpublished work).

Immunosuppression as a mechanism for developing pest control strategy

Although insects have an ability to fight back the attact of invading microbes, they could be made more susceptible to the infection by suppressing various immune mechanisms. Such methods could then be combined with other methods of pest control to devise an appropriate pest control strategy. Bettini *et al.* 1951) and Stairs (1964) proposed that insects become susceptible to the infection if the activity of phagocytosing cells is blocked in *vivo.* Gupta (1994) similarly found that the insect growth regulators could be used to suppress the haemocytes in insects that could help in increasing insects' susceptibility to microbes.

Insect growth regulators as immunosuppressants

Juvenoids

Juvenoids are the most effective chemicals that interfere with the morphogenesis in insects and thereby, with growth and developmental processes. Thesse are found as useful insecticides against dipterans, hemipterans and recently against orthopterans like cockroaches. Hazarika and Gupta (1987) studied the effect of juvenile hormone I on the total and differential haemocyte counts in the German cockroach and found that THC were affected with plasmatocytes showing significant reduction.

Ecdysteroids

Similarly, investigations on the effect of ecdysteroid treatment on the haemocytes are carried out. Ahmad and Khan (1988) reported extensive damage to the granulocytes, plasmatocytes, and even prohaemocytes in the grasshopper, *Hieroglyphus nigrorepletus* on treatment with makisterone A.

Azadirachtins

Azadirachtins are the important tetranortriterpenoids isolated from the kernels of the neem tree, *Azadirachta indica* A. Juss (Butterworth and Morgan, 1968). these have a wide range of biological activity and are the important constituents in the neem products used for the control of insect pests (Gujar, 1992, 1994). The effect of azadirachtin on haemocytes was first studied by Qadri and Narsiah (1978) who found 20% reduction in the total haemocyte counts of the American cockroach, *periplaneta americana* treated with azadirachtin. Later, Ayyanger (1987) reported drastic reduction in total haemocyte counts of larvae of *Spodoptera litura* by 37.4%, 83.75%, and 68.5%, 24, 48 and 72 hr after treatment with Azadirachtin at 1 µg/g dose. Azambuja *et al.* (1991) investigated the immune depression in *rhodnius prolixus* induced by Azadirachtin. The last instar larvae treated with Azadirachtin @ 1.0 µg/m1 through blood meal showed immune depression in terms of 1) reduction in number of haemocytes and ability to form nodules following challenge with *Enterobacter cloacae* B12, 2) reduced ability to produce antibacterial and lysozyme activity of Azadirachtin treated insects to destroy the primary infection caused by the inoculation of *E cloacae* cells. The mechanism of action of Azadirachtin on the immune system may be through its effect on the endocrines as the primary mode of action seems to be on the inhibition of brain hormone release (Rembold, 1989). Ecdysteroids are known to stimulate production and differentiation of haemocytes and Azadirachtin is thus anatagonistic to the action of ecdysteroids. Secondly, Azadirachtin may also be inhibiting general insect metabolism thereby leading to the depression of lysozyme and other antibacterial proteins. In view of limited understanding of interaction between endocrine and immune system, further studies are necessary to know the mode of action of Azadirachtin as immunodepressant.

Plumbagin

Plumbagin is chemically 2-methyl 5-hydroxy 1,4-naphthoquinone isolated from the roots of the medicinal plant, *Plumbago* spp. It has pharmaceutical and pesticidal activity (gujar, 1990). Plumbagin has been extensively studied for its morphogenetic and toxicity effects on *Dysdercus* and has been reported to inhibit the synthesis of ecdysteroid, makisterone A in *Dysdercus cingulatus* (Joshi and Sehnal, 1989). Recently, Sexena and Tikku (1990) reported effect of plumbagin on haemocytes of D. koenigii and found that total and differential haemocytes are drastically reduced with 24 hr. The adverse effect of plumbagin is seen on the filopodial elongations of plasmatocytes and granular haemocytes (immunocytes). Further investigations of ultrastructure of haemocytes (Tikku *et al.*, 1992) showed the acute vacuolization of cells and a gradual destruction of the organelles, followed by the dissolution of plasma membrane and passing out of the internal organelles like mitochondria and endoplasmic reticulum. This is responsible for lowering of the resistance and mortality of the insects. Similar studies carried out by us also showed that total haemocyte count of last instar nymphs of *D. koenigii* were drastically reduced by 63% within 24 hr of treatment of plumbagin at 15 μg/insect (5 day old nymph) unpublished results).

Miscellaneous chemicals

The larvae of *Aedes aegypti* (L.) treated with phynyl thiourea were found to have significantly reduced melanization as compared to control exposed to *Neoplectana carpocapsae* at 1000-4000 juveniles/m1 for 1 hr. Pheny1 thiourea may also inactivate toxin produced by the nematode or affect its associated bacterium, *Achromobacter nematophilus* (Beresky and Hall, 1977).

Jarosz (1985) used several metabolic inhibitors (actinomycind D, chloramphenicol, cycloheximide, mitomycin C, fluorouracil, cyclophosphmide and hydrocortisone hemisuccinate) to inhibit inducible dense system, lysozyme, in the larvae of *G. mellonella.* Actinomycin D and cycloheximide depressed significantly synthesis of lysozyme as induced in larvae in response to *Pseudomonas aeruginosa;* showing induction required mRNA and ribosomal protein synthesis.

Cyanides (10^{-3} m), azides (10^{-2} M), EDTA (5 x 10^{-3} M), Thiourea (10^{-10} M), Cycloheximide, ascorbic acid are the inhibitors of coagulation. Some of these like ascorbic acid, glutathione, thiourea, ouabain are found to have ovicidal and sterility effects on *D. koenigii* (Datta and Banerjee, 1988). N-acetyl galacosamine, N-acetyl neuramic acid, hexuronic acid, sialic acid are the inhibitors of agglutinating property of haemolymph.

These immunosuppressants could be further explored for their toxicity *per se* and in combination with microbial agents to achieve an ultimate aim of insect pest management.

References

Ahmad, A. and Khan, M. A. (1988). Effect of triol and makisterone A on the haemocytes of *Hieroglyphus nigrorepletus* Boliver (Orthoptera : Acrididae). *Proc. Indian Acad. Sci. (Anim. Sci.),* **97** : 203-210.

Ayyangar, S. G. S. (1987). Physiological effects of neem (*Azadirachta indica* A. Juss) fractions on tobacco caterpillar, *Spodoptera litura* (Fab.). Ph.D. thesis, I. A. R. I., New Delhi.

Azambuja, P., Garcia, E. S., Ratclifee, N. A. and Warthen, J. D. Jr. (1991). Immune depression in *Rhodnius prolixus* induced by the growth inhibitor, Azadirachtin. *J. Insect Physiol.* **37** : 771-777.

Beresky, M. A. and Hall, D. W. (1977). The influence of phenyl thiourea on encapsulation, melanization, and survival in larvae of the mosquito, *Aedes aegypti* parasitized by the nematode, *Neoplectana carpocapsae. J. Invert. Pathol.* **29** : 74-80.

Bettini, S., Sarkar, D. S. and Patton, R. L. (1951). Observations on the fate of vertebrate erythrocytes and hemoglobin injected into the blood of the American cockroach (*Periplaneta americana*). *Science* **113** : 9-10.

Butterworth, J. H. and Morgan, E. D. (1968). Isolation of a substance that suppresses feeding in locusts. *J. Chem. Soc., Chem. Commn.* 23-24.

Datta, S. and Banerjee, P. (1988). Ovicidal and insect sterilization effect of ascorbic acid, glutathione, ouabain, and thiourea. *Annals of Entomology* **6** : 49-56.

Gujar, G. T. (1990). Plumbagin, a naturally occurring naphthoqui-none. Its pharmacological and pesticidal activity. *Fitoterapia* **61** : 387-394.

Gujar, G. T. (1992). Neem (*Azadirachta indica* A. Juss) : A natural insecticide, status and Need for bioassay standards. *Pestic. Res. J.* **4** : 69-79.

Gujar, G. T. (1994). Insect Growth Regulators. Action, Metabolism and pest control status. In Recent Advances in Insect Physiology and Toxicology, Agricole Publishing Academy, New. Delhi.

Gujar, G. T. and Mehrotra, K. N. (1983). Turnover of proteins in haemolymph of the desert locust, *Schistocerca gregaria* Forsk. *Proc. Indian natn. Sci. Acad. B* **48** : 476-480

Gupta, A. P. (1994). Insect Hemocytes : Classification and immunologic functions. In Recent Advances in Insect Physiology and Toxicology (ed. Gujar, G. T.), Agricole Publishing Academy, New Delhi. pp. 106-206.

Hazarika, L. K. and Gupta, A. P. (1987). Variation in hemocyte population in various developmental stages of *Blattela germanica* (L.) (Orthoptera : Blattellidae). *Zool. Sci.* **4** : 307-313.

Jarosz, J. (1985). Attempts to depress the inducible defence system of Galleria mellonella larvae using diverse metabolic inhibitors. *Biologisches Zentralblatt* **104** : 193-203.

Joshi, N. K. and Sehnal, F. (1989). Inhibition of ecdysteroid production by plumbagin in *Dysdercus cingulatus. J. Insect Physiol* **35** : 737-741.

Pathak, J. P. N. (1993) Insect Immunity, Oxford and IBh Publ. Co. Pvt. Ltd., New Delhi, pp. 192.

Qadri, S. S. H. and Narsiah, J. (1978). Effect of azadirachtin on the moulting processes of last instar mymphs of *Periplaneta americana* (Linn.). *Indian J. exp. Biol.* **16** : 1141-1143.

Rao, C. G. P., Ray, A. and Ramamurthy, P. S. (1984). Effect of ligation and ecdysone on total haemocyte count in the tobacco caterpillar, *Spodoptera litura* (Noctuidae : Lepidoptera). *Can. J. zool.* **62** : 1461-1463.

Rembold, H. (1989). Azadirachtins : their structure and mode of action. In insecticides of Plant Origin (Eds., Arnason, J. T., Philogene, B. J. R. and Morand, P.), ACS Symp. Sr.. 387, Washington, D. C. pp. 150-163.

Saxena, B. P. and Tikku, K. (1990). Effect of plumbagin on haemocytes of *Dysdercus koenigii* F. *Proc. Indian Acad. Sci. (Anim. Sci.).* **99** : 119-124.

Stairs, G. R. (1964). changes in the susceptibility of *Galleria mellonella* (Linnaeus) larvae to the nuclear polyhedrosis virus following the blockage of phagocytes with India ink. *J. Insect Pathol.* **6** : 373-376.

Tikku, K., Saxena, B. P., Satti, N. K. and Suri, K. A. (1992). Plumbagin-induced ultrastructural haemocytic response of *Dysdercus koeingii* F. *Insect Sci. Applic.* **13** : 787-791.

Vey, A. (1993). Humoral encapsulation. In Insect Immunity (ed., Pathak, J. P. N.), Oxford and IBH Publ. Co. Pvt. Ltd., New Delhi, pp. 59-68.

Recent Advances in Indian Entomology
Editor: O.P. Lal
APC Publications Pvt. Ltd., New Delhi, 1996, pp. 289-290

Role of Insect Growth Regulators in the Pest Management Strategies

B. SUBRAHMANYAM

The widespread use of synthetic organic pesticides has obviously led to problems like pest resurgence, insecticide resistance, residues in food and fooder and hazards human health. Realizing this, man has always been exploring various safer methods of pest control and infact the concept is now changed to management. Williams (1956, 1967) first suggested that endocrine imbalance might be an effective tool for control of pest populations through disruption of developmental and reproductive processes. Interfering with the functioning of endocrine system can therefore introduce transient toxicological effects leading to infertility or death. As a part of the group discussion, various developments in this approach, basic as well as applied, would be brought to the attention of the trainees and emphasis would be laid on the role of this strategy in IPM.

Following the initial discovery of the chemical nature of the insect endocrinal secretions involved in morphogensis, namely juvenile hormone and ecdysone, several other neurohormones, prostaglandins, biogenic amines were identified. Two decades of basic research in this area has brought out sufficient evidence to universally agree that artificial manipulation of the neuroendocrine centres of insects can open up a new avenue in pest management.

Juvenile hormone analogues and mimics

Of the several neurohormones, juvenile hormone has received utmost attention in insect control and its analogues and mimics are now popular. These analogues have short residual life, low cost of production and short sensitive stage on the insect. Though juvabione (paper factor) was the first compound discovered from nature; synthetic studies have yielded many products more potent than the endogenous JH of insects. Examples of such juvenoids are: Hydropene (Altozar); methoprene (Altosid); kinoprene (Ensar); epofenonane; phenoxycarb etc. They are at present used, though on a restricted scale, for control of mosquitoes, greenhouse pests, stored grain pests have shown potential for the control of phytophagous species.

The chief requirement for the successful use are: high biological activity, stability under field conditions, low toxicity to biocontrol agents. These compounds need special formulation technology to suit to various application need. Molecules with JH-activity would cause disturbance in metamorphosis leading to formation of intermediate forms that fail to reproduce. Since JH also stimulates overian development in adult stages, sufficient caution should be exercised in their use on maturing insects. The presence of chromene compounds in plants like Ageratum sp., that are selectively toxic to corpora allata is yet another promising area. Though, their potential is yet to be evaluated, the concept and the availability of such molecules is yet another promising area deserving detailed investigation.

Sinre JH controls a variety of physiological processes, exogenous application can cause disruption of embryogenesis, morphogenesis of larval forms, induce sterility, disrupt diapause, reduce pheromone production. Availability of stable analogues certainly would go a long way in the use of these mimics in agricultural pest control.

Ecdysteroids

Ecdysones promote the moulting process and are steroidal in nature. They are also found in plant kingdom. They are more polar and hence penetrate very less into insects via cuticle. Besides, they are pharmacologically active on mammals. These two factors thus hamper the use of moulting hormone analogues less promising in pest control.

Azadirachtin and neem

Pure azadirachtin derived from neem seed kernels is both a strong antifeedant and growth regulator. It is one natural product that attained the status of "Classical molecule par excellence". More than 200 species of insects are sensitive to this molecule with profound IGR activity at 1 ppm and antifeedant activity in the range of 10-20 ppm. Neem products are gaining popularity in pest management, the overall effect noticed being antifeedant activity. Margosano, a formulation of neem extract with known azadirachtin content is the first product in U.S.A. India, being the natural home of neem, has now come forward registering formulations viz., Margocide, Godrej Achook, Nimbicidin, Repellin. Increasing emphasis is to be laid on the quality of formulation, the actual content of the active principle azadirachtin and neem oil before reproducible results are expected on field scale use.

Research needs

In addition to the above, knowledge of endocrinology of insects can be exploited in several ways for pest management.

1. Development of antagonistic or agonistic analogues that block or mimic the action of specific hormone receptor sites.
2. Using recombinant DNA technology to develop insect-specific microorganisms having insects of antihormonal peptides. The recombinant genes would cause hormonal deficiencies following their entry into the insect.
3. Inhibition of neurophrmone-activating enzymes or degradative proteases would result in abnormal hormone titres (keeley & Hayes, 1987).

Recent Advances in Indian Entomology
Editor: O.P. Lal
APC Publications Pvt. Ltd., New Delhi, 1996, pp. 291-303

Recent Trends in Insect Taxonomy

R. K. ANAND

Nature is full of different kinds of diversified forms which man has always been curious to know. He is a biological species and his activities have been affecting the conditions of earth more and more. If we want to understand the effect of his present and future activities on our local and global ecological system, we must first know what species make up our biological communities, and what processes permit both the stability and directed change. This information combined with the knowledge of how evolution operates, helps us to understand and anticipate the effects of todays human activities on the future of world. Here, the taxonomists plays the major role. He orders and classifies organisms. In this way, the world is so usually classified that we often have to make special efforts to recognise. Moreover, correct identification is a pre-requisite for international references.

The main purpose of this article is to give a brief review of our present knowledge of systematics of Indian insects and to summarize in brief the various lines of investigations on the recent trends in the taxonomy of insects.

It is considered necessary at the outset to specify what are exactly meant by the term 'Systematics' and 'Nomenclature', which are entirely different from one another. Systematics deals with classification. The subject of systematics, today more properly termed "Taxonomy' (Taxis = arrangement, Nomos = law) deals with classification of living organisms under various categories,or taxa, in accordance with evolutionary hierachry. Nomenclature (Nomen = name, calare = to call) on the other hand deals with naming of the various taxonomic categories. Systematics, a science concerned with establishing durable classification has itself undergone tremendous change in the last 2-3 decades. Although traditional approaches still dominates practice of most systematists, molecular data and numerical methods of analysis have proven useful in many groups.

Taxonomic investigations may start with populations that show apparently similar characters between individuals on morphological grounds which may tentatively be taken as constituting a species. The characteristic differences between various such species should then be assessed as accurately as possible. This is the analytical stage or 'Alpha' taxonomy. The 'Beta' taxonomy is a synthesis of these characters and takes into consideration the common characters presented by the different species for grouping them under various super-specific categories. The final stage, the 'Gamma' taxonomy goes into details of infra specific categorisation.

Taxonomic measurements must naturally start with a unit and the accepted unit for the purpose is the species (defined by Alfred Emerson as an evolved or evolving, genetically distinctive, reproductively isolated natural population).

From the ideas of taxonomy, it is obvious that many aspect of investigations in biological sciences must be complimentary to taxonomic investigations. A student of Zoology, investigating any other aspect of animal life such as genetics, ecology, ethology, biology, bionomics, physiology, toxicology, economic zoology, zoogeography etc. will be profited very much through a knowledge of taxonomic position of the species.

It is practically impossible to give a fairly accurate estimate of the number of insect species that may be encountered within the country. Beeson (1941) estimated 40,000 and presently it may be assumed to be 2-2.5 times. Of the recorded species, the description are widely scattered in literature, though a good attempt is made in Fauna of British India, Genera Insectorum and Catalogue of Indian insects. But, still the taxonomic identities of many of the species are not firmly established. There is a general feeling among biologists that taxonomists are whimsical in changing names. Such persons would readily agree that complications are likely to be involved, if a species is known by different names from various parts of the world. It is to serve this purpose that names are changed. Anand (1977) discussed all these aspects. The recent publications of valid and rejected names by the International Commission of Zoological Nomenclature is a tremendous advance towards nomenclatural stabilization. Nomenclatural confusion in specific indentities have been mainly because in several cases species have been established on unique type material or specific identities of subsequent material determined on rather superficial comparison with type or printed literature. It has to be borne in mind that whereas a holotype has to be necessarily fixed while describing a species as new and establishment of identities of others with it, a holotype is only just one of a population that represents that species and that naturally cannot cover all variations, that are to be met with among other members of that population that form a species before the specificity can be accurately established.

Researches in taxonomy during the past and establishing the identity of the species concerned were so far mainly based on morphological characters. It is only during the later part of this century, that newer trends emerged which will be dealt in brief, in the following pages.

1. Studies on external genitalia and its taxonomic significance

During the second half of this century, stress has been on the study of male and female genitalia, besides in some groups internal anatomy and reproductive system has been used in segregating various taxa. The normal procedure adopted for study of genitalia in pterygotes, is to use both fresh and dried specimens, last 3-4 abdominal segments are cut and boiled in 10% KOH, for 1--20 minutes depending on the chitinization of the cutical. washed in water, passed through 70% alcohol, cleaned in xylol. In species with poorly scleritized parts, the genitalia are occasionally stained in acid fuchsin and studied in 70% alcohol.

In various groups of insects, different components of the genitalic parts are reported to be of taxonomic value at generic or specific level. Within the insects, six general types of male genitalia can be recognised.

(1) Single median penis present, situated posterior to, but between the bases of stylus, bearing limb bases of segment IX; the latter limb bases may bear a gonapophyses in addition to stylus, and a similar structure may be present on segement VIII (Thysanura).

(2) Paired penis usually present (sometimes united to form a single organ in which the ejaculatery duct remains separate), situated cauded of the stylus bear plates (clasper) and sternum IX. (Ephemeroptera). In Dermaptera, the primitive forms have paired penis, but in the higher earwigs, the two penis are united basally but apically may be somewhat divided.

(3) Grylloblattoidea have paired phallic lobes, between which at the base is situatited the gonopore, they may represent paired penis from which the ducts have been withdrawn. In Grylloblattoidea, a distinct sternum IX is present, together united to form a composite sclerite with styli in Blattaria or the structure may lack styli as in Acridiidae. The phallic lobes in most Orthopteroids are divided to form several lobes which either remain separate (Blattaria, Mantodea, Grylloblattoidea) or become united to form a single primary phallic organ (Gryllidae, Tettigoiidae, Acrididae).

(4) In Hemiptera, there is some sign of appendage plates bearing styli. Paired claspers (parameres) are usually present together with an aedeagous which is somewhat variable in structure. In Coleoptera and Lepidoptera, the plan is much the same.

(5) In Mecoptera and Megaloptera, the plan is similar to Hemiptera but the parameres are 2-segmented.

(6) In Diptera, Trichoptera and Hymenoptera, there is a mediam aedeagus, 1-2 jointed parameres and in addition usually a pair of structures (paraphyses) lateral to the aedeagus. These paraphyses are termed clasperes in Diptera, volsella in Hymenoptera and titilators in Trichoptera, and are attached usually to a basal plate.

In most of the groups, male genitalia have more taxonomic significance than the female genitalia. In females, ovipositor is an organ for oviposition and is basically of two types : (i) an extension of abdomen itself or (ii) a ventral process from segements VIII and IX. The oviposition of first type is present in Thysanoptera, Mecoptera, Lepidoptera, Coleoptera Diptera and Trichoptera. The distal segments of the abdomen are tapered and are capable of being retracted telescope like. Such an ovipositor shows little variation from comparative morphology point of view. The second type of ovipositor is more widely distributed. Special structural modifications associated with peculiar functional requirements are present in certain groups (Blattaria, Acrididae, Aculeate Hymenoptera).

In Coleoptera, the site of sperm storage in female insects, the spermatheca or receptaculum seminis, is primarily an invagination of integuement at the posterior end of the venter of 8th abdominal segment. Typically, there is a duct, which may be expanded to form a copulatory bursa, leading from the vagina to the sparmathecal chamber. A tubular gland associated with the latter structure may be present. A study of the present type serves as an addition to our knowledge of internal structure of Coleoptera and in a number of cases is of taxonomic and phylogenetic significance. Further more, in population studies, such data aid the assessment of the proposition of females fertilized and therefore of those contributing to population growth, a point of considerable importance when density is excessively low or tending to decline because of control measures.

2. Chaetotaxy

Another aspect which has been used in segregating different taxa is that of chaetotaxy, particular in certain groups of Coleoptera, Lepidoptera, Neuroptera and Diptera. This is more useful in case of larval recognition. It is seen that setae have a definite arrangement on the head, thorax and abdomen and have been assigned names or numbers. Setal arrangement is used extensively in determining the families of caterpillars. Setal maps have also been used for various larval recognition. A setal map shows arragement of setae on one half (left) of any thoracic or abdominal segment. In Coleoptera, the last or next to the last abdominal segement may bear structures useful in identification. Subapical spines, spurs or asperites may exist on the lateral and ventral aspect of caudal segment among some larvae of Coleoptera. In Dermestidae, Coccinellidae, Anobiidae, the arrangement of spines and setae have taxonomic value. In certain groups of Coleoptera, particularly Carabidae, Curculionidae, there is superficial clothing on the elytra referred to as 'elytral vestiture' which may be in form of scales, setae, hairs or fine pubescence. These vary greatly in size, shape and structure and their is a regular pattern of their arrangement and have taxonomic value. Ramamurthy & Ghai (1988) studied the elytral vestiture in *Myllocerus* spp. and demonstrated its importance as a character in establishing taxonomic identity of various species. In Diptera, chaetotaxy on the thorax is of great taxonomic value particularly in Tipulidae, Calliphoridae and on the head in the muscoid files.

3. Scanning

The third aspect of recent trends is the use of scanning microscope. It has been used in various groups of Coleoptera, Thysanoptera, Hymenoptera, Diptera, to scan various parts of the insect body and to analyse the structures present. In one such group of Chrysomelidae, the scanning studies were conducted to find out to what extent differences in number and arrangement of the digitiform pegs of maxillary palpi correspond to systematic and phylogenetic divisions and to what extent they are correlated with differences in adult habitats. A comparison was also made with Bruchidae, Cerambycidae and Curculionidae. These studies led to the conclusion that ancestral Chrysomelidae had actively flying floriclous adults.

4. Numerical taxonomy

With the continuing discovery of the new taxa, the systematist slides into a pattern of description and revision, subsequent redescription and revision, each time adding a little more information to a system whose information content is so low that it reflects little more than the most recent revisions, speculation concerning the group. Questions concerning the theoretical basis of classification and the committant issue of the utility of taxonomists product have been raised a number of times in the past, coupled with criticism of the inherent inefficiency of traditional working methods in taxonomy. Out of these theoretical tactonics, there has emerged a body of thought and accompanying techniques which are now being synthesized into an approach to systematics known as numerical taxonomy.

Numerical taxonomy is the "Numerical evaluation of the affinity or similarity between taxonomic units and the ordering of these units into taxa on the basis of their affinities". It is an attempt to reduce the subjective element in taxonomy. The principles of numerical taxonomy are :

(i) The only value of a classification is in its information content. The general classifications are those based on overall resemblance or maximum correlation of attributes.

(ii) A classification is a hierachy of forms arranged according to differences and similarities in their attributes.

(iii) Comparison of differences & similarities of organisms must be based on large number of attributes - a phenetic sample of the genotype.

(iv) All attributes must be regarded as equal in value.

Techniques of numerical taxonomy

(i) Studies by traditional methods, the assembling of material and the selection of attributes for study. But, here one attempts to utilize all available attributes. As a working rule, the number of attributes studied should be more than sixty preferably about hundred. The choice of attributes should be random.

(2) The next step is analysis of computation of some measures of similarity or affinity for each possible pair.

The majority of work in numerical taxonomy to date has concerned itself with classification, in the group recognition sense although note worthy progress in computerized method of identification has been made. A major benefit of numerical taxonomy has been the introduction of more questioning critical attitude into the general area of systematic theory and methodology.

5. Morphometry

It is the study of the measurements of the relevant parts in a population of species and study their significance. Morphometric studies have been carried out in a number of groups of insects. In *Schistocerca gregaria* Forskal, morphological variations is phase polymorphism has been worked out by several workers. The most sensitive characters which indicate significant differences in two phases in width of head at genal level and 20 different body parts and their indices. (Bhanotar & Mahto, 1977; Mahto *et a1*., 1977).

Another important group studied is aphids where identification rests on morphological characters of apterous and alate vivipaous parthenogenetic forms. Some aphids show variation according to geographical regions. In addition to morphological characters, ratios of lengths of some characters such as antennal segments IV/III, VI/III base/III, VI flagellum/III, cornicile/III prove useful in taxonomy. Statistical analysis of external dimorphism in apterous and alate parthenogentic viviparous forms may afford clue in separation of one species from the other. These studies may also indicate the evolutionary trend of particular character and perhaps throw light on the interrelationship of different species, genera, tribes and families.

6. Use of computers in taxonomy

To adopt numerical taxonomy in total would involve changing of the basic concept of biology, which would be entirely unacceptable. But, computers are creeping into every sphere of life. Biologists anticipate computer studies rapidly, became a feature of degree classes and some are running schemes based on taxonomy and ecology. The types of exercises include indentification, descriptions, diagnosis, distribution, keys scanning, mapping, storing information, new species determination, ecology, biology and classification, which are summed up below.

(i) *Identification* : A contribution which a computer can make is in the sphere of insect identification. A number of satisfactory programmes have been developed which can identify a taxon from more than 100 taxonomic units. The operator carrying out identification is provided with a list of diagnostic characters from which to select, say initially any three or four. From these, the computer provides the identification or indicates which taxa remain to be separate by further selection of characters. It will also advise on the best characters for doing this and will give other chacacters for checking against the specimen, once the identification has been given. This method of identification has the advantage that characters may be selected in any order and thus of short circuiting of most of information necessarily when working through a dichotomous key. It is quicker to use than the written key and the operator may not well be so much of an expert. In a computer identification routine, if some characters are not avaiable, the computer can be instructed to avoid using them - pherhaps only male specimens collected, then female characters can be ignored. When a number of identification programmes exist, the correct matrix catalogue must be selected and these may use habitat or ecology as well as taxonomic distinctions.

(ii) *Descriptions, diagnosis and distribution* : These can all be produced by computer in a standard form and logical sequence throughout and are ideally suitable for publication, in a taxonomic work. The information provided to enable this to be done may be in the same form as that needed to produce phenetic classification. As in a principal component analysis, ratio of charcters or association of characters may be important.

(iii) *Keys* : Dichotomous key can be produced by computer either with single or multiple character couplet. They can be based on equalization of the taxa or by grouping according to the most easily recognizable characters. If more than one character is to be used in each couplet, the computer should be programmed to print out in the order of reliability or usefulness.

(iv) *Scanning* : The development of visual input and output display units has made biologists ask how they can take advantage of the facility. The visual output display units has many uses, the almost instanteneous production of groups and illustrations. It enables the computer to recognize broad differences between specimens and two dimensions. we envisage its development for the recognition of leaf mines in insect groups where the species can be diagnosed by configuration of the mine.

(v) *Mapping* : The computer has been extensively used for producing species distribution maps by CAB. There has also been progress in using mapping methods for driving chromosome maps and understanding of genetic make up. A third use for mapping is in location and orientation of a new point (new taxon) in relation to others.

(vi) Storing : One of the greatest computer assets has in its ability to store up-to-date minute information about everything pertinent to the subject. Not only information leading to keys, descriptions, indexing etc., but also synonymies, location of specimens in museums, biology, methods of control, references etc. A worker could quickly obtain any required information in a much shorter time than he would spend in library. One such system is CD-Rom.

Another valueable use of storing capacity relates to compiling of bibliographies with a summary. An efficient computer aided information center whose job would be to acquire, store for retrieval and distribute information relating to particular biological discipline is being provided by Commonwealth Institute of Entomology and Washington University.

(vii) *New species determination* : When the characters of a new taxon are provided for identification, the computer will recognize the taxon as new only when characters are supplied which do not correspond with, or are in different combination from those already in the matrix. It is always adviseable to check a number of characters even though identification may be possible on the basis of few. Having established that the taxon is new to matrix, it is possible to find its phenetic affinities, by supplying the characters given in the similarity matrix. The affinities can be expressed in different ways of ten as percentage similarities to the remaining taxa or by giving the nearest neighbours or by production of two or three dimension photographs.

(viii) Ecology and Biology : Ecological studies can be facilitated by electronic sorting of the data in order to find the relevant information. Simulation techniques have been used to infer behavioural pattern of insects.

7. Cytotaxonomy

Chromosome cytology has contributed to insect systematics in a big way. Chromosomes can be viewed simply as especially long DNA molecules (plus proteins). It has been shown that several quite distinct types of genetic systems characterize certain major groups eg. Hymenoptera are genetically haploid, while those of higher Diptera are diploid. The cytological method of sex determination vary greatly within the insects. In aphids, special cytogenetic mechanisms control the complex alteration of sexual and perthenogentic generations. Such differences in genetic mechanisms are obviously of great significance in population genetics and evolution. They are not merely morphological characters like those used in defining systematic categories since they concern whole of genetic apparatus on which speciation, evolution and adaptive radiation depend.

The amount of genetic recommendation varies greatly from species to species, depending on the number of chromosomes. If we know how the major differences in genetic mechanisms have arisen, we can use the knowledge in establishing the relationships of insect orders. Cytotologists are also concerned with cytotaxonomic differences which exist between related species. These consist of differences in chromosome numbers, size and shape. Such differences can be used for segregating sibling or cryptic species.

In most groups, each chromosome has a single short region (centromere or kinetochore) which serves to attach it to spindle at mitosis. In some groups, chromosomes do not have centralized centromere. In few species of parathenogenetic insects which lack males, evolution of chromosome sets by polyploidy has taken place. In aphids, special cytogenetic mechanisms control the complex alternation of sexual and parthenogenetic generations.

The extent to which cytotaxonomic difference between species can in practice be detected varies with the material The analytical study of cytological differences between species has only proved possible in groups where chromosomes are not too numerous. Cytotaxonomic studies have been carried out to a great extent in Orthoptera and Diptera. The gall midges have two distinct types of chromosomes, both of which are present in the nucleaus. Such differences in genetic mechanisms are of great significance in population genetics and evolution and in many instances also physiological genetics.

For understanding the relationships of intraspecific categories of races, microgeographic races and local populatition, Cytology is likely to become increasingly important but cytogenetics is not to be taken as an automatic solution to the systematic problems.

8. Physiological and molecular systematics

Insects appeared on this earth some 250-500 million years ago. This early establishment has helped these organisms in diversifying, and modifying to a great extent. Such modifications, greater adaptability, immense degreee of evolutionary matureity and change, in the physiological processes are so acute and pronounced that it has been able to outnumber all other forms of animals put together and invade diversified environmental niche on ground, air and water. Recent advancements in scientific studies have indicated that analysis of these physiological processes may give some clues for dealing with diverse and complex problems in establishing insect identities.

There are two ways in which organic evolution or taxonomy may be approached through biochemical methods. One of this is primarily descriptive. This is the field of comparative biochemistry. The literature on biochemical attributes of different taxa is voluminous

Taxonomic problems particularly those involving the clustering of species and the estimation of phytogenetic relationship have intrigued chemically oriented biologists. With biochemical characters, as with ordinary characters, one has the immediate problems of assuring that truly comparable attributes are compared. This is the problem of establishing homology of parts.

The term molecular systematics and biochemical systematics are used in conjuction with smaller, secondary systematics, secondary chemicals such as alkaloids or terpenes, but these chemicals are of restricted distribution in insects.

Incompatible characters have caused untold numbers of taxonomic arrangements. The example of four species of tephritid flies and their morphological incompatiblity is well nown. Tradionally, character incompatibility are dealt with by ignoring some character or by demonstrating that some characters have been misinterpreted Molecular data can also produce incompatibilities e. g. in tephrotid fly, the two enzymes aldolase and superoxide dismutase, are mutally incompatible characters since both groups together. The amino acid sequence of aldose in species may be different despite the fact that electrophoretic mobility of this enzyme is identical. Electrophoretic data can be reinterpreted by use of more sensitive electrophoresis, by protein sequencing or by DNA sequencing. Molecular incompatibilities result from long recognised evolutionary phenomena of reversal, parallelism & convergence.

The most fundamental and complete molecular data available are DNA and RNA, sequences. Unfortunately, lack of interspecific comparison makes much of available insect nucleotide data irrelevant to systematics, several approaches permit DNA sequence difference to be estimated. Restriction endonuclease mapping relics on the existance of DNA hydrolyses that attack only particular short DNA sequence. Another indirect method of obtaining estimates of DNA sequence divergence is DNA hybridization.

Protein synthesis : The amino acid sequences of protein in insects are very few. Partial lyzozyme sequences for three species of Lepidoptera have been determined by Jolles *et a1.* (1979). Protein sequence difference can be estimated by producing an antibody against one protein and measuring the antibody response to a homogenous protein from another species. Such studies have been done in *Drosophila* and other dipterans.

Gel Electrophoresis : The gel electrophoresis of proteins is by far the most widely used molecular technique in insect systematics. The technique relies on the fact that identical proteins migrate the same distance under the electrical force applied to an electrophoretic gel while non-identical protein usually migrate different distances. The proteins used in systematics are usually enzymes. the techniques is very simple because only crude homogenates are run on the gel, the desired enzyme is than distinguished from thousands of proteins on the gel by its unique response to the stain. The enzyme specific stain also establish homology of the proteins across species. The general term for any enzyme variant revealed by electrophoresis is 'isoenzyme' (isozyme) whereas those alternate molecular forms of an enzyme produced by different alleles of an enzyme gene are referred to "allozyme" (alloenozyme).

Gel electrophoresis has proven useful at all the three of the most important systematic tasks - species discrimination, species identification and hierachical classification. Electrophoresis is usually useful within genera. Several genera have been studied in Odonata (*Gomphus*) Othoptera (*Gryllus, Chlorothippus, Troglophilus*), Hemiptera (*Megicaude, Lygus, Gerris, Enchenopa*), Coleoptera (*Myllocerus Henosepilachna, Tribolium*), Lepidoptera (*Ypomanta, Rhagoletis, Heliconius, Chroristoneura, Enchenopa*), Hymenoptera (*Tridomyrmex, Formica, Bombus. Rhytidophora*) & Diptera (*Drosophila, Aedes, Simulium, Phlebotomus, Dacus*).

Isoelectric focussing is a new technique. This method separates proteins solely on the basis of net charge, unlike conventional electrophoresis. It produces much finer bands.

Another technical development has been the use of two dimensional (2-D) gels, with a single sample separated first by isoelectric focussing and then by molecular weight.

The problems associated with electrophoretic data are quite well kinown. The use of laboratory cultured insects in electrophoretic systematics is risky because of genetic alteration of cultures by inbreeding and inadvertant selection. Another problem is non-genetic environmental modifications of electrophoretic migration distance. Still another problem is existence of enzyme modification genes.

Electrophoresis can be powerful aid in identifying immature and sibling species. A detailed description of use of electrophoretic key has been published for *Rhagoletis Anopheles* and *Dacus*.

Paper chromatography : It is a tool for separation of population categories in taxonomic work. The use of spectrophotometric methods for separation of biochemically different but morphologically similar forms in various groups of insect has helped to a great extent in solving complex problems.

Biochemical methods provide access to a great deal of high quality information, which can be applied to the solution of taxonomic problems at all levels. some special biochemical approaches noteably from protein chemistry, DNA hybridization and immunochemistry, promise to be of extreme importance to the development of our understanding of the relationship between higher taxa.

9. Zoogeography

The study of geographical distribution of insects is another subject related to taxonomy, which has received attention from many entomologists. It is important enough to know the ravages of the species that occur in an area, in searching for biological control agents in evaluating pest potentials, crop selection, and danger from diseases of plants, animals and man; but zoogeography is much more. It is a field of interpretation of ecology, physiology, systematics, phylogeny, palaeontology, geography & serology, It is not static but dynamically relates to the tolerances of individuals and populations to their origin, evolution and dispersal. It nas been possible to know world fauna in certain groups. The application of insect zoogeography to phylogenetics and systematics cannot be denied The extent of geographical ranges of a species can be marked on a map.

Ross (1956) has presented a plan of procedure tor interpreting evolution and distribution of a group. First, one must study the characters of living and fossil species and work out phylogeny and disribution pattern. These patterns should then be correlated with geological time scale. Comparison may then be made with known history of other groups. In determining routes of dispersal, qualitative approach is essential rather than statistical. In integrating phylogeny and distribution, changes in environmental, extinction of lines and variation in both rate of evolution and rate of dispersal must be borne in mind. That the living species of insects have not evolved continously in the same area they now inhabit is generally accepted.

Six zoogeographical regions - Palearctic, Ethoapian, Oriental, Australian, Nearctic & Neotropical are known.

10. Taxonomic revisionary studies

Last, but not the least, is the importance of revisionary studies in various groups of insects cannot be ignored. During the past, the earlier workers described species, designated them by various names as cotypes, synypes etc., besides in some cases designated them as Holotype, Allotype, Paratype, Topotype etc. Moreover, same species was known by different names; which has been synonymized by subsequent workers. This has necessitated to carry out revisionary studies, but one of the

Table. General value of systematics to other fields.

Role	Notes
Stimulate ideas	
Generates major concepts in Biology	Establishes generalization that have led to major discoveries e. g. in evolution and nature of populations.
Generates specific theories	Emgenders testable hypotheses e. g. about life cycle, habitat & behaviour.
Makes informations available	
By association	Provides the means to collect together & retrieve all information of a given taxon.
By prediction	Properties of many species can be estimated from known properties of related taxa.
For comparison and synthesis	Exposes natural diversity and contributes directly to studies on faunistics, zoogeography and other fields.
Allows Identification :	Identification is the key to available information on organisms and hence to the understanding of nature and to practical application.

problems faced in such studies is that the orginal collcection sometime is not available.

Conclusions

In the end, is briefly outlined the role of systematics in support of Entomology. Systematics is the scienctific study of organic diversity including relationship. It includes the theory and practice of describing, naming and classifying organisms. Identification is a pre-requisite for all other enquiries about the natural world. Systematics has an important conceptual role in biology. The importance of systematists has only been realized in a limited way. Some have included philosophical statements, others have emphasized the relevance to applied work. But because of underfunding of systematics and limitations of opportunities, it is lagging far behind the other branches of Entomology.

Systematic work is carried out chiefly by workers of particular insect groups who are associated with major collections that require time consuming curatorial tasks. The conception thus engendered that systematics work depends on narrow specialist, together with erroneous view that identification is a service rather than experitse. But relationships between systematic and entomology cannot be denied, the general role is summerized in the table

Systematics aids all kinds of entomological endeavours. It has key parts in faunistic, zoogeographic & ecological studies and provides essential support to many other fields. The value of systematics has become more visible following the development of concern for the environment. Systematics permits basic identification, makes information available, assembles information from a comparative perspective and thus follows synthesis. Moreover, systematics is a rewarding scientific pursuit. It is a pursuit kept dynamic by recent philosophical and methodological innovations (eg. biochemical and genetical techniques). Systematics cannot be isolated from other studies, nor can any kind of study be isolated from systematics.

References

Anand, R. K. 1977. The stability of scientific names. *Everyday Science,* **22** : 5-7.

Beeson, C. S. C. 1941. The ecology and control of the forest insects of India and neighbouring countries. Publ. Vasant Press, Dehradun.

Bhanotar, R. K. and Mahto, Y. 1977. Variations in the known variation stable (Phase gregaria) population of the desert locust, Schistocerca gregaria Forskal. *Rec. Zool. Surv. India,* **72** : 341-361.

Jolles, J. , Schoentgen, F. , Croizier, G., Croizier, I. and Jolles, P. 1979. Insect lysozymes from three species of Lepidoptera, their structural relatedness to the C. (Chickpea) type lysozyme. *J. Mol. Evol.,* **14** : 267-71.

Kapoor, V. C. 1991. Theory and practice of Animal Taxonomy. Publ. Oxford & IBH, Delhi pp. 234.

Menon. M. G. R. 1964. Systematics of Indian Insects. Entomolgical Soc. of India, supplement pp 70-87.

Mahto, Y. N. , Bhanotar, R. K. and Bhatnagar, R. K. 1978. Concentration of Locusta migratoria L. in Delhi and morphometric studies on certain body characters. *Ent. Mon. Mag.*, **113** : 105-112.

Pradhan, S. 1964. Some thoughts on speciation in insects. Proc. XIII. Int. Congr. Ent., London.

Ramamurthy, V. V. and Ghai, S. 1988. A study of the genus Myllocerus (Col.: Curculionidae). *Oriental Ins.*, **22** : 377-500.

Ross, H. H. 1956. Evolution and classification of the mountain caddisfly. Univ. of Illinois Press 213 pp.

Sokal, R. R. and Sneath, P. H. A. 1964. The Principles of Numerical Taxonomy.

Tuxen, R. L. 1956. Taxonomists Glossary of genitalia in insects. Pub1. Rjnar Munksgaard, Copenhagen.

Recent Advances in Indian Entomology
Editor: O.P. Lal
APC Publications Pvt. Ltd., New Delhi, 1996, pp. 304-306

Present Status and Future Strategies of Insect Identification in India with Particular Reference to Chalcidoidea

S. I. FAROOQI

The significance of taxonomy and identification has already been emphasized in the past by several workers such as DeBach (1964), Mayr (1968) and Rosen and DeBach (1973) to such an extent that now practically every entomologist appears to be fully convinced that accurate identification of pest or parasitoid alone provides a clue to evolving a successful bio-control progrramme. But unfortunately there are only few, who are really serious about the question of how and from where to get these identifications done. It is this question that I wish to place before you for purpose of discussion with the hope that atleast a thinking will be generated among fellow entomologists in order to find a solution.

Although this problem exists with regard to all groups of insects but for the present, it will be discussed with reference to Hymenoptera, mainly the Chalcidoidea. As mentioned by Noyes (1978), there are some 16,541 species of Chalcids throughout the world. In contrast to this, the number of known species from India is approximated to be 1039 by (Mani, 1989). This does not indicate fewer chalcidoids from here but points out towards insufficient explorations within the region. As a matter of fact a more rich fauna is expected, keeping in view the presence of all types of ecological conditions and greater variety of flora.

Incidently Chalcidoidea is not only one of the largest groups within parasitic hymenoptera but also stands at the top with regard to its importance in biological control. As given by Clausen (1978) and later analysed by Noyes (1985), the number of Chalcidoidea used in biocontrol programmes was 408 as against 374 in respect of parasitic hymenoptera. The number of introductions where some control was achieved was 233 in chalcidoidea against 122 in others and similarly the number where full economic control was obtained figured 119 in chalcidoidea and

Group	Species involved in bio-control programme	Introductions where some control was achieved	Introductions where full economic control was achieved.
Chalcidoidea	408	233	119
Rest of parasitica	374	122	52
Aculeata	86	19	6

52 in rest of parasitica. The number of species involved out of *Aculeata* was still less.

Summary of the data given by Clausen (1978) and analysed by Noyes (1985) on the used parasites and predators in bio-control programmes.

Within the country there are just three main centres for identification in general, namely IARI, ZSI and FRI, out of these the last one has become most inactive since there is no taxonomist on any group. ZSI participates in identification programme but there is only one hymenoptera specialist who works on Inchneumonidae. At IARI, a regular identification service is somehow maintained with the help of six group specilists, out of which only one is in hymenoptera who deals with chalcids. However, looking on a countrywide basis, the situation is not very disappointing with regard to Chalcioidea, as there are 18 Chalcidlogists available at different places. Out of these atleast 10 are active workers who could be approached for purpose of identification. Since there does not exist a system for organising and bringing about coordination among them there is less possibility of making use of these talents. However, whenever these experts are approached on personal basis they are always ready to extend their cooperation.

Position of Chalcidoid workers in the country.

S. No.	Name of place	Total no. of workers	Name of active workers	Advisory specialists or worker not willing to identify
1.	Aligarh	4	Dr. M. Hayat Dr. M. Yousuf	* Dr. S. M. Alam * Dr. M. M. Agarwal
2.	Agra	2		** Dr. G. Maheshwari ** Dr. M. K. Mukerjee
3.	Bangalore	2		*** Dr. S. Nagarkatti *** Dr. H. Nagraja
4.	Bareilly	1	Dr. M. Verma	
5.	Calicut	3	Dr. T. C. Narendran Dr. U. C. Abdurahiman	* Dr. K. J. Joseph
6.	New Delhi	3	Dr. S. I. Farooqi Dr. Beche Lal Dr. Debjani De	
7.	Madras	1		* Dr. M. S. Mani
8.	Pant Nagar	1	Dr. Majid Khan	
9.	Srinagar	1	Dr. T. Husain	

* Advisory
** Not willing to identify
*** Not in India.

It is quite encouraging to see an upward trend in the number of insects that are received for purpose of indentification. The data maintained at IARI for the past three decades is evident of this fact. The number of specialists engaged is however, very disappointing.

With respect to hymenoptera, the data from 1981-90 shows that approximtely 500 specimens were received for identification on an annual basis. A still deeper analysis may indicate that 70-80% of hymenoptera were belonging to parasitica alone, clearly showing an upward trend for utilizing biological agents in pest management programmes.

The seriousness of the problem can be visualised by seeing that while 20-25% of all insects received at IARI are Hymenoptera and 70-80% of all hymenoptera are parasitica, the number of specialists available to deal with this is only one. Further there is no procedure available in the present system to have his replacement in the event of his retirement or any other eventuality. The situation puts forward a taugh challenge before the biological control workers. Should they abandon their researches or continue without proper identifications? Should they always depend on I. I. E., London or US Nat. Museum, Washington for identifications or mobilize all their resources to establish a strong certre of identification in the country. The problem requires immediate attention, if at all sincere progress in the field of biological control is intended.

Keeping in view the above problem, let us try to put our brains and evolve ways and means to solve it. It is hoped that the outcome of the present discussion may provide atleast some guideline to formulate our future strategies for becoming self-sufficient in our identifications, not only of parasitic hymenoptera but of whole of Insecta.

References

Clausen, C.P. (Ed.), 1978. Introduced parasites and predators of arthropod pests and weeds : A world review. U.S. Dept. Agr., Agr. Handbook, 480 : 1-551.

DeBach, P. 1964. Biological control of insect pests and weeds, London, Chapman and Hall. pp. 844

Noyes, J.S., 1985. Chalcidoids and Biological control. Chalcid Forum, No. 5 : 5-7.

Noyes, J.S. 1978. On the number of genera and species of Chalcidoidea of the World. *Entomol. Gaz.*, **29** (3) : 163-164.

Mani, M.S., 1989. Fauna of India and the adjacent countries. (Chalcidoidea Hymenoptera). Vols. I and II, pp. 1933.

Mayr, E., 1968. The role of systematics in biology. Science (Wash., D.C.), 159 : 595-599.

Rosen, D. and P. DeBach, 1973. Systematics, morphology and biological control. *Entomophaga,* **18** : 215-222.

Recent Advances in Indian Entomology
Editor: O. P. Lal
APC Publications Pvt. Ltd., New Delhi, 1996, pp. 307-308

Recent Advances in the Taxonomy of Lepidoptera

S. L. GUPTA

Lepidoptera is one of the biggest order of class Insects next to order Coleoptera. The insects belonging to this order are commonly known as butterflies and moths. It is estimated that about 2 lacs species of butterflies and moths are known from the world including 15,000 species of butterflies. In this way it is nearly 10% of the total animal kingdom and are roughly equivalent to known plant species. The size varies from 2-3 mm to 300 mm of wing span.

The adults are characterized by their broad scaled and typically patterned wings and developed or vestigial proboscis which is used for sucking liquid nourishment, particularly nectar, honeydew and the juices of fruits but also used for obtaining dissolved salts from urine, ash, carrion and the lachrymal secretions of some mammals.

Economically they are harmful as well as beneficial. Commonly the larvae may be defoliators, leaf rollers and leaf tiers, leaf miners, stem borers, root feeders, flower feeders, seed and fruit feeders, fruit piercer, eye frequenter and blood suckers and stored product pests. Some very small moths feed on hair, wool, horn and feathers. The activities of the insects of this order are not always detrimental to man but they may be beneficial also as pollinators, silk producers, agents of biological control, predators and aesthetic values.

Basically the butterflies and moths are identified by their external morphological characters such as size and shape of wings (well developed, reduced or divided into plumes), colour pattern of wings, type of antenna (simple, hooked, clavate and bipectinata etc.), type of labial palp (porrect, upturned, recurved and ascending etc.), proboscis (reduced or well developed), fore legs (developed or reduced), hind tibia (with single or paired spurs or without spurs), frenulum (well developed, reduced or absent) and the wing venations. Out of these, the colour pattern of the wings and arrangement of veins are two most important characters for the identification of various species.

As the above characters used by earlier workers are very much variable within a species, now a days the taxonomist are using male and female genital characters for differentiations and confirmation of the identity of the species. This technique is very widely used throughout the world in all the families of order Lepidoptera.

Since the butterflies and moths are mostly harmful in their larval stage, its identity is very important from applied aspect. For the identification of larva (caterpillar), the arrangement of setae on head capsule and body of the larva alongwith the arrangements of the crochets on the lower surface of prolegs are examined. This type of study is known as larval chaetotaxy. This is also used throughout the world in almost all the families but only in few species.

Recently with the advancement of science and scientific appliances the approach for taxonomic studies has changed but in very limited cases. Now the following new techniques are also used alongwith the earlier aspect of morphological studies :-

(1) Use of electron microscope

For the study of very minute morphological characters in details now the taxonomists have started using Electron microscope. Since the above microscope is not easily available to every one, it is used only in specific cases.

(2) Use of type materials

Earlier it was a common practice to describe new texas without comparing them with the type of already known taxas. The result was that the same species was described from different localities under different names. As a result of the study of the type materials, now it has become possible to segregate or differentiate number of species. This type of study is conducted in every part of the world in almost all the major families of order Lepidoptera.

(3) Use of electrophoresis

This is biochemical method for identification of various species is applied in very specific cases in families Noctuidae, Pieridae, Tortricidae, Papilionidae, Pyralidae, Gelechidae and Nepticulidae in Cuba, U.S.A. England, Germany, France and Japan. However, in India no work has been done on this aspect.

(4) Use of haemocytology

In Canada this type of work has been done for recognizing final instar larvae of some closely related species of genus *Euxoa* of family Noctuidae.

(5) Use of chromatography

Thin layer chromatography was used to determine semi-quantitatively the amino acids in overwintering larvae of *Coleophora dahurica* belonging to family Coleophoridae in China. This method is suitable for comparative study of insect classification.

In England gas chromatography and pattern recognition methods were used to develop a potential method for differentiating *Helicoverpa armigera* from Helicoverpa zea. However, in India no such technique have been applied.

Conclusion

Out of the above techniques, for identification of species, external morphology including that of male and female genitalia is the most important tool. The remaining techniques are important but practically they are applied only in very limited manner.

Recent Advances in Indian Entomology
Editor: O. P. Lal
APC Publications Pvt. Ltd., New Delhi, 1996, pp. 309-310

Recent Trends in the Biosystematics of Curculionoidea and Scarabaeoidea

V.V. RAMAMURTHY

Biosystematic research on insects is undergoing rapid changes in the whole world. These changes are dramatic and unsurmountable in the developed world. This is essentially due to their advancement in science and because of the reason that the inventory of the biodiversity of insects had almost been accomplished in full in these countries. In such countries the recent trends in the biosystematics of CURCULIONOIDEA and SCARABAEOIDEA comprise the following:-

1. Utilising morphological, ecological, other biological attributes of immature stages and interpreting these for biosystematics.
2. Exploring the internal anatomy in depth and interpreting these for information on the classification, phylogeny and evolution of insects.
3. Analysis of morphometrics employing numerical taxonomy techniques — especially the parsimony analysis and interpreting these for inferences on phylogeny and evolution.
4. Character evaluation and assessment involving sophisticated computational techniques so as to select taxonomic characters for formulating keys for classification of taxa.
5. Karyological studies leading to the analysis of chromosomes/DNA strands and exploring these for biosystematic studies.
6. Biochemical/molecular biology studies involving
   (a) Proteins/polypeptides/amino acids of haemolymph, ovaries, testes, other tissues with the aid of electrophoretic techniques.
   (b) Enzymes/other metabolic biochemicals, especially those involved in digestion, respiration, reproduction, etc. with the help of HPLC, CMS, electrophoresis and other analytical techniques.
   (c) Hydrocarbons of cuticular structures with the aid of GLC, GCMS and other techniques.
   (d) Defensive secretions and their components, and
   (f) Molecular biology studies involving PCR/RFLP/REN analyses to bring out the DNA configuration and genetic mapping.

If we undertake a comprehensive surveillance of these recent trends in India, it could be observed that such studies are almost lacking. This is purely due to the reason that we are yet to prepare an inventory of the existing biodiversity even at a macrolevel. Nevertheless attempts had been made on need based situations - like *Trichogramma*, (Manickavasagam *et al.*, 1991), aphids, mosquitoes, coccinellids and leaf-hoppers. At IARI such attempts had been made on the species of *Lepropus* (Poorani, 1994) and intraspecific populations of *Myllocerus undecimpustulatus, M. dentifer and Deporaus marginatus* of Curculionidae (Ramamurthy, 1988; Ramamurthy and Ghai, 1993).

References

Manickavasagam, S., Paul, A.V.N. Farooqui, S.I. and Ganguly, A.K. 1991. Isozymes and protein profiles of four species of Trichogramma Westwood: (Trichogrammatidae: Hymenoptera). *Journal of Biological Control* **5**(2): 70-73.

Poorani, J. 1994. Biosystematic studies on the genus Lepropus Schoenherr (Coleptera: Curculionidae) from the Orient. Ph.D Dissertation, P.G. School, IARI., 1104 ix pp.

Rāmamurthy, V.V. 1988. Insect Plant interactions - a role in the speciation of insects. *Phytophaga* **2**: 159-164.

Ramamurthy, V.V. and Ghai, S. 1993. Biodiversity and conservation biology in agroecosystem's —our entomologists, biosystematic point of view. *Hexapoda* **5** (2): 127-131.

Singh, R. P. and Srivastava, B. G. (1983). Alcohalic extract of neem (Azadirachta indica A. Juss.) seed oil as important deterant for *Dacus cucurbitae* (Coq.). *Indian. J. Ent.* **45** (4) : 497-498.

Smallman, B. N. (1945). Relation of insect damage to thiamine content of biscuits. *J. Eco. Ent.* **38** : 106-110.

Srivastava, B. G. and Pant, J. C. (1979). Fatty acid and sterol requirement of *Dacus cucurbitae* (Coq.) maggots under aseptic condition. *Bull. Ent.* **20** : 124-127.

Van Emden, H. F. (1966). Studies on the relation of insects and host plant II. A comparison of the reproduction of *Brevicoryne brassicae* and *Myzus persicae* (Hemiptera : Aphididae) on brussels sprout plants supplied with different rates of Nitrogen and potassium. *Ent. Exp. Applicate.* **9** : 444-460.

Recent Advances in Indian Entomology
Editor: O. P. Lal
APC Publications Pvt. Ltd., New Delhi, 1996, pp. 311-316

Pest Management in Cereal Crops

VIJAI S. SINGH

The national strategy of achieving self-sufficiency in food for our ever-increasing population rests mainly on the production of cereal crops like wheat, rice, sorghum, maize and barley. During the past three decades of green revolution, considerable success has been achieved in increasing production of these crops through cultivation of newer high yielding genotypes and related agrotechniques envisaging higher use of fertilizers and irrigation. The altered crop canopy of newer cultivars with more juicy and thicker foliage coupled with their extended cultivation in non-traditional areas/seasons, although helped in increasing production, has resulted in emergence of newer pest problems and accentuation of some existing ones. Full realisation of the production potential of these high yielding genotypes at farmers level still remains a far cry, mainly because of inadequacies in management of pest problems, which constitute the major constraints in production.

Cereal crops are damaged by a large number of insect and mite pests right from sowing till harvest. In general the *Rabi* crops have fewer pests as compared to *Kharif* ones, obviously due to more conducive weather factors prevailing during *Kharif* season. Rice accounts for the longest list of pests while wheat the shortest, yet it is difficult to enumerate and talk of all these pests in one lecture. Hence for the sake of brevity and in view of limitation of time, only brief mention of key pests and salient management strategies against these are attempted here.

Key pests

Termites continue to be traditionally serious pests of wheat, barley, sorghum and maize especially under rainfed cultivation. Out of above two dozen species involved two viz., *Odonototermes obesus* and *Microtermes obesi* are most important causing damage from sowing till crop maturity. Being polyphagous in nature and their peculiar underground existence termites are difficult to control except through prophylactic chemical control measures. Gujhia weevil *(Tarrymecus indicus)* used to be serious on wheat and barley in seedling stage but since the advent of high yielding varieties, its incidence has declined and it has become a non-pest. Different species of shootfly *(Atherigona spp.)* infest the cereal crops in seedling stage and these have gained in importance recently owing changes in sowing time and cropping sequences, *Atherigona soccata* infests sorghum and maize, A. naqvii attacks wheat, barley and maize while *A. oryzae* infests rice. Gall midge (*Orseola oryzae*) is a serious pest of rice right from nursery stage to panicle emergence, the characteristic damage being 'silver shoot'.

Several insect species slick plant sap during growth phase of these crops, causing considerable damage to plant vitality and yield, besides some of these acting as disease vectors. Foremost among these are brown plant hopper *(Nilaparvata lugens)*, green leaf hopper *(Nephotettix spp.)* and white backed plant hopper *(Sogatella frucifera)* on rice, aphid *(Rhopalosiphum maidis)* on barley, maize and sorghum, *Sitobion avenae* on wheat and brown wheat mite (*Petrobia latens*) on wheat and barley. BPH and GLH have assumed serious threat to rice cultivation. Aphids on barley have been traditionally serious while brown wheat mite has become serious on wheat under unirrigated conditions recently.

Incidence of leaf eating is another problem in cereal crops. These include armyworm *(Mythimna separata)* and cut worms (species of *Cirphis, Euxoa and Agrotis)* which cause sparadic but serious damage to rice, wheat, barley and maize. Rice has several other foliage feeders including case worm *(Nymphula depunctalis)*, *Hispa armigera*, whorl maggot *(Hydrillia spp.)* and leaf folder *(Cnaphalocrocis medinalis)*, which take a heavy toll of the crop occasionally.

Stem borers, however, remain most serious bottlenecks in successful cultivation of the cereals like rice, sorghum and maize causing stem tunnelling, lodging of plants and yield loss.The most important species are *Scirpophaga incertulas* and *Chilo suppressalis* on rice, *Chilo partellus* on sorghum and maize and *Sesamia inferens* on sorghum, maize and wheat. Varietal resistance, conservation of natural enemies and chemical control measures are important tools in their management.

The cereal crops suffer severe damage in their earhead stage also due to several pest species. Thus sorghum suffers heavily by gall midge (*Contarinia sorghicola*) during flowering and grain formation while the ripening sorghum grains are damaged by *Calocaris angustatus*, *Eublemma silicula* and *Cryptoblabes* sp. Rice crop suffers through gundhi but (*Leptocorisa* sp.) while gram pod borer (*Heliothis armigera*) causes earhead damage in maize, rice, wheat and sorghum.

Losses

Thus all the cereal crops have pest problems throughout their growing season suffering considerable losses in yield. Various estimates of losses by different pests in different cereals have been given ranging from 3 to more than 50 per cent, majority of these coming from differences recorded in yield from chemically protected and unprotected crops. During the National Seminar on crop losses due to insect pests (1983) it was stated that on All India basis losses sustained in foodgrains worked out 6.37 per cent, which included 6.63 per cent in cereals and 3.13 per cent in pulses". Taking this figure of 6.63 per cent loss in cereals and the current cereal production, the country incurs considerable loss of cereal grains.

Management strategies

Under the circumstances, efficient protection from ravages of insect and mite pests is of utmost importance for stability of production of these important crops. Thus, emphasis has been on developing integrated management strategies so as to keep the pest populations below economic injury levels through utilization of multipronged approach including host plant resistance, need based insecticidal use, appropriate cultural practices and conservation of natural enemies of the pests. Efforts have therefore been made to identify sources of resistance to key pests of the different crops, identification of genetic and biochemical mechanisms involved and to develop cultivars resistant to key pests. Besides, determination of vulnerable crop stages, economic injury and threshold level of different pests have been done in several cases and efforts are in progress regarding others. Refining of chemical control methods, their timing of applications for judicious integration with other pest control methods, especially for safety to natural enemies have been undertaken. In this context efforts are on for better utilization of botanicals particularly. neem formulations to be suitably incorporated in overall management practices. especially in pests like BPH when resurgence has become a problem. In the field of bio-control besides conservation of the natural enemies, attempts are being made for inoculative and inundative releases of specific parasitoids against pests like stem borers of rice and sorghum, gall midge, leaf hoppers and plant hoppers of rice. Utilization of cultural practices like sowing date variations, seed rate variation, fertilizer management, crop sequence, summer ploughing and weed management, some of which are known since long, have imminent usage for reducing/avoiding pest infestation.

While all these methods individual being about encouraging results, their harmonious integration in overall integrated pest management is essential to achieve desired pest suppression for stability of cereal production. A brief account of pest management practices for cereal crops are given below where these are dealt with in three groups owing to similarity of pest problems viz., wheat and barley, sorghum and maize and rice.

1. *Management of pests of wheat and barley*

1. Destroy termitaria (termite mounds) in the vicinity of fields and treat the spot with lindane or chlorpyriphos sprays. This should be practiced on community basis in villages/farms.
2. Use only well rotten manure otherwise termite incidence is aggravated.
3. For prevention of termite damage, treat the seeds with chlorpyriphos or endosulfan using, respectively, 90 and 240 ml a.i. with 5 litres of water and spraying over one quintal seed spread on floor and periodically turned over to ensure proper treatment. This should be done 1-2 days before sowing. For sowings in fields having previous history of termite incidence, this operation should invariably be undertaken.
4. Avoid late sowings. Normal season crop is not attacked by shootfly (*A. naqvii*). In case of barley, early sowing (1st fortnight of Nov.) usually reduces aphid attack.
5. Use aphid resistant barley varieties viz. DL 117 or DL 200. For brown wheat mite, wheat variety C 306 may be preferred.
6. In case preventive seed treatment for termites has not been undertaken before sowing and termite incidence is noticed in standing crop, mix 0.8 1 chlorpyriphos with 50 kg sand and broadcast in one hectare field. This treatment is not as effective as seed treatment but appreciable reduction in termite damage could be achieved.
7. In case late sowings (Dec. end or January) have been undertaken and shootfly

damage is found approaching 20 tiller infestation, spray cypermethrin twice at fortnightly intervals in seedling stage. The most vulnerable period of crop being upto 60 days of germination.

8. For sucking pests like brown wheat mite, aphids and jassids, spray formothion or oxydemeton methyl or phosphamidon @ 250 ml a.i./ha when pest incidence is noticed and repeat, if necessary, after a fortnight.
9. For caterpillar pests, like armyworm and cut worms have sporadic occurrence, spray fenitrothion @ 500 ml a.i./ha when pest incidence is noticed.
10. At harvest, remove stubbles and burn them to kill borer stages present inside.

2. Management of pests in sorghum and maize

1. Sow crop early, immediately after onset of monsoon, to reduce damage by shootfly and midge in sorghum and to reduce borer damage in maize.
2. Use pest resistant varieties like Gangas, Decean hybrid, Kundan, Chandan in maize and CSH5 or CSH9 in sorghum.
3. For shootfly control in sorghum (i) in areas having regularly high shootfly infestation apply carbofuran 3% granules @ 1.5 g/m row in seed furrows, and (ii) in areas with mild pest incidence the seed rate should be increased to 8-10 kg/ha and infested plants to be uprooted after 10-12 days of germination, through a mechanical campaign. In maize, spring sowings in northern India are usually infested by shootflies, seed furrow application of carbofuran granules as recommended above may be undertaken.
4. For stem borer control in sorghum, apply endosulfam 4% or carbofuram 3% or lindane 1% granules in whorls at 25 and 35 days after germination, or when granules not available, endosulfam 4% dust may be applied in whorls @ 8 to 10 kg/ha at the same interval as given above. In case of maize, spray endosulfan 0.05% 15 days after sowing and if need be apply endosulfan 4% or lindane 1% granules in whorls 15 days after the spray.
5. For midge in sorghum, spray endosulfan @ 0.44 1' ha or carbaryl @ 1 kg a.i.' ha or dust endosulfan 4% @ 12 kg' ha. The application should be done at 50% flowering and repeated after 5 days.
6. For earhead worms in sorghum, organise a campaign of shaking the cobs over kerosenized water. If the infestation is severe use insecticidal treatments recommended for midge control.
7. Collect and burn the stubbles and panicle residues, after harvest to kill hibernating pest stages.
8. The stems kept for fodder purposes should be used up well before the next monsoon season.

3. Management of pests of rice

1. During summer plough the fields and scrap the bunds to expose pest stages

to sun and predators. Practice clean cultivation. Removal of weeds from fields and adjoining areas reduces incidence of borers, grasshoppers, leaf hoppers, leaf folder and gall midge etc.

2. Organise campaigns in nurseries to collect egg masses on leaves, silvershoots, infested leaves and netting of plant bugs to reduce their incidence.

3. Avoid late sowings to reduce gall midge infestation.

4. Use resistant varieties for key pests of the area.

   (i) gall midge - Karna, Lalat, Sarasa, Abhaya, Ruchi, Kshira, Phalguna, Surekha, Shakti, Dhanyalakshmi, Pothan, Mahaveer IR 36.

   (ii) WBPH - HKR 120

   (iii) Stemborers - Ratna, IR 20, IR 26.

   (iv) BPH - Sonasali, Mansarovar, Vajram, Chaitanya, Nagarjuna, Krishnaveni, Chandana, Pratibha, Kakatiya, Aruna, Makom, Remya and HKR 120.

   Some of the varieties having resistance to more than one pest are?

   (a) Suraksha - Gall midge, BPH and Blast

   (b) Shaktiman - Gall midge and Rice tungro virus

   (c) Lalat - Green leafhopper and Rice tungro virus

   (d) Ananga - BPH and gall midge

   (e) Rasmi - Gall midge, BPH and Blast

   (f) Daya

   (g) Kshira - Gall midge, BPH, WBPH, GLH and Rice tungro virus.

   (h) Bhuban - Gall midge, BPH and Bacterial leaf blight.

5. Flooding of infested fields kills caterpillars and pupae of cutworms and armyworms, pupa of *Atherigona* and grasshopper eggs.

6. In the nursery spray 0.04% methyl parathion or 0.05% dimethoate after 10 days of germination and repeat after 15 days to control various pests.

7. At the time of transplanting (i) clip infested leaves, galls, egg masses etc. and (ii) dip the roots of the seedlings in 0.02% chlorpyriphos + 1% urea for 4 hr before planting.

8. After 10-15 days of transplanting, apply mephosfolan or quinalphos granules in standing water @ 1kg a.i./ha.

9. For borers, spray quinalphos/phosphamidon/sumicidine @ 0.5kg a.i/ha at 7 days internal twice after 60-65 days of planting.

10. In areas endemic to brown plant hoppers, spray twice with carbaryl (1 kg a.i./ha) or malathion (0.025%) or endosulfan (0.04%) at 7-10 days interval, whenever hopper population reaches above 5-10 per hill. The spray should be directed towards the base of the plants.

11. If gundhi bug population during November reaches more $1/m^2$ or 10 bugs/ 100 earheads, dust 4% endosulfan @ 10 kg' ha.
12. For gall midge, spray 4-5 times with phosphamidon/carbaryl/lindane during vulnerable phase i.e. 20-45 days after transplanting.
13. For armyworm, spray DDVP (0.5 kg/ha) or carbaryl (1kg/ha), if earhead damage reaches to 5%.
14. After harvest, plough the fields to expose eggs and pupae to sun and predators. Also uproot and burn the stubbles to reduce pest incidence in subsequent crop.

Recent Advances in Indian Entomology
Editor: O.P. Lal
APC Publications Pvt. Ltd., New Delhi, 1996, pp. 317-322

Integrated Pest Management in Maize

K. K. MARWAHA

The concept of integrated pest management implies the rational integration of various methods of insect pest control *viz.* chemical, biological, cultural etc. to suppress pest populations to a level below economic injury levels. The formulation of suitable artificial diet for the mass rearing of maize stalk borers *viz. Chilo partellus* (Swinhoe) and *Sesamia inferens* walker, the two key pests of maize, the former throughout Indian Union during *kharif* season and the later in pennisular India especially during *rabi* season; has greatly helped the maize Entomologists to determine the economic threshold of these *(borers)*. Another tissue borer of maize which is gaining impetus as a serious pest during spring season in the northern regions particularly, Punjab, Rajasthan, Delhi and Terai regions of U. P. are shootfly species belonging to genus *Atherigona*. The release of variable numbers of borer eggs at variable stages of maize crop growth led to determine the economic threshold of borer which depicts a sigmoid relationship between economic injury level and crop-age. This constituted a ready reckoner as to when the control measures may be initiated. An important implication of this study which may be highlighted is that the severity of damage by the maize stalk borer was significantly more in 10-day old crop as compared to the later ages of the crop. The lower tail of sigmoid curve represents the economic threshold i.e. the pest density at which control measures should be determined to prevent the increasing pest population from reaching the economic injury level. The economic threshold according to Stern *et al.* (1959) is lower than the economic injury level to permit sufficient time for the initiation of control measures and for these measures to take effect before the population reaches the economic injury level. The middle steep portion of curve represents the economic injury level as the lowest population density that will cause economic damage i.e. the amount of injury that will justify the cost of artificial control measures. The upper portion of the curve represents the economic damage. The sigmoid curve therefore signifies the importance of economic threshold amongst the three components in providing the guide lines as to when the control measures should be initiated. It is clear that threshold lies between 10 to 15 day old crops as with little increase in age, the plants begin to tolerate infestation. The prophylactic control measure at the critical age of the crop thus becomes an obvious necessity in areas demarcated by regular occurrence of maize stalk borer year after year. Any control measure after 30 days of germination may not prove to be successful as well as economical. This basic information on economic threshold provide the guidelines which have greatly helped in minimizing the overall number of insecticidal applications to control the key pests of maize.

Cultural control paves the way for manifesting genetic yield potential of maize crop

It may be highlighted that the marked genetic yield potential of maize crop not yet exploited by the plant breeders manifests as is evidenced by 29.0 percent increase in yield in unprotected early sown crop (mid-June) as compared to unprotected normal sown crop (Table-1.) The protection of the early sown crop with endosulfan further increased the yield to about 50.0 percent when compared to late sown unprotected crop. This has made pest management a practicability for small farmers under various conditions prevailing in the states of Haryana, Punjab and Western U. P.

Table 1. Yield potential of maize manifests in early sowing.

Maize crop	Yield (q/ha) in different sowings: Mid June (early)	Ist week of July (normal)	Per cent increase in yield over normal sowing
Protected	57.2	48.6	17.7
Unprotected (Control)	49.1	38.1	29.1
Per cent increase in yield over control	16.5	27.5	

Table 2. Effect of integrated control for the pests infesting maize crop (variety-Ganga 5).

Plant population Per hectare	Treatment*	Percent infestation including dead-hearts	Grain yield (q/ha)	Percent Increase in yield over respective controls	Net profit (Rs.)
	1	39.1	34.9	------	------
	2	25.2	35.6	2.0	105.00
52,800	3	13.0	38.9	11.5	460.00
	4	0.7	44.5	27.5	1188.00
	1	37.4	37.4	------	------
	2	33.1	37.4	0.0	------
66,000	3	13.3	43.4	16.0	760.00
	4	0.3	48.7	30.2	1443.00

*T_1 - Untreated control, T_2 - Mechanical uprooting of dead-hearts, T_3 - Mechanical uprooting of dead-hearts + 4.0% granular application of endosulfan in the whorl, T_4 - Mechanical uprooting of dead-hearts + 0.1% endosulfan spray followed by 4.0% granular application in the whorl fortnight later.

The cultural control involving the planting at two population levels provided a clear guideline to the farmers that higher plant population i.e. 66000 plants per hectare resulted in higher yield rather than 52,000 plants per hectare (Table-2) can be profitably adopted by farmers by sowing maize seeds 20 cm apart rather than 25 cm apart. In the region where seeds, of released hybrids are not easily available, the farmers can cultivate local varieties at higher plant population. This required less expenditure in terms of costly inputs and being early maturing can fit in the cropping system of farmers.

The mechanical control involving removal of dead hearts contrary to expectation appears to be unnecessary in the planning of integrated control of the maize stalk borer.

In the exploitation of polyphagy for the borer control, sorghum has been established as a preferred host when compared to maize. The dead heart formation in sorghum in early stage of crop growth leads to tillering which facilitates further trapping of the maize stalk borer as also the latter brood of this pest. This ensures lower borer damage in maize. Also sorghum can be used as fodder, to some extent grain yield may be obtained indicating that there is no total loss of buffer sorghum crop. It is imperative that sorghum as buffer crop is sown in rows only on all sides of maize crop along with few rows in the centre of the plot. The optimum percentage of sorghum should vary between 5 to 15. The trap of sorghum can provide safest, economical and easily adoptable method of pest control by the Indian Farmers.

Evolution of control schedules for the pests of maize

Two control schedules involving the use of lindane and endosulfan have been evolved. The foliar application of lindane emulsion (0.05%) 15 days after sowing followed by second granular appliction of 1% lindane granules @ 15 kg/ha in the plant whorls a fortnight after the first application or foliar application of endosulfan emulsion (0.1%) 15 days after sowing and second application of 4% granules @ 15 kg/ha in the plant whorls a fortnight after the first appliction have been evolved. The lindane application accounted for 10.0% and endosulfan 16.5% increase in yield. The earlier practice of adopting calender schedules involving up to five insecticidal applications for the control of maize pests have been completely dispensed with.

For the benefit of maize seed producers where the aim is to eliminate the pest and increase the grain yield, the suggested schedule involves the side dressing of 3% carbofuran or 5% mephosfolan (cytrolane) granules depending upon the availability, in 10-day old maize crop near the exposed roots of maize plants @ 15kg granules/ ha. The treatment of maize crop with mephosfolan and carbofuran granules separately resulted in 44.0 and 33.0 per cent increase in grain yield (Table-3). Similarly control schedules based on methyl parathion, malathion, phosphamidon and ekalux have been evolved. The residues of aforesaid insecticides were non-detectable at the time of harvest in grains and fodder.

The seed treatment is not only economical but fits in the farming practices by lowering the plant protection cost: The seed treatment can be easily adopted in

Table 3. Impact of chamical control ıneasures on the yield of maize.

Treatment	Dead-hearts (%)	Yield per hectare (kg)	Per cent Increase in yield due to treatment	Increase in yield (kg) per hectare due to treatment	Net profit per hectare (Rs.)
Lindane	2.1	5946.3	9.8	532.5	638.75
Endosulfan	3.8	6301.3	16.4	887.5	1077.00
Mephosfolan	0.6	7810.0	44.3	2396.3	3258.45
Carbofuran	3.0	7188.8	32.8	1775.0	2326.50
Control	11.8	5413.8	-	-	-

maize crop where the returns per unit area are low. Lal and Co-workers 1961 evaluated seed treatment with BHC dust @ 5, 10 and 20 gm per 100 gm seed and finely powdered lindane as 1,5 and 10 gm/100 gm seed using gum arabic as sticker and reported that BHC 5 gm band lindane 10 gm/100 gm seed protected the maize crop for 11 and 15 days respectively from borer attack. However higher dose of BHC affected germination of maize seeds. Marwaha *et al.* 1984 reported the significance of seed treatment with carbofuran 35 ST having 1/3 the active ingradient than the application of conventional carbofuran granular formulations in the seed furrow at sowing for borer control. Marwaha *et al* 1986 also highlighted the effectiveness of furathiocarb 40 SD (1.0 per cent) used as seed treatment for maize stalk borer control under manual release of borer eggs. The stimulating effect of carbofuran on plant growth, maize root development and higher mortality of pest has been reported by various research workers. Ten different species of spiders *viz., Araneus sinhagadensis* Tikader, *Araneus* sp, *Leucage decorta, cyclosa insulana, Pardosa* sp, *Castianeira* sp, *Cheiracanthium* sp, *oxyopes, Salticus ranjitus* and *Myrmarachne platelcidae* (Camb) belonging to nine genera and five families were collected from the whorls of maize plants. The feeding propensity of *oxyopes pandae* and *Araneus sinhagadensis* on various stages of *C. partellus* has been studied in captivity. It has been reported that the aforesaid species of spiders consume 10-12 freshly hatched larvae and 2-3 second and third stage larve of *C. partellus* (Table-4) in 24 hours especially during night time. Also the higher safety margin for the three species was observed when 0.1% phosalone and to some extent 0.1% endosulfan were some what safer to the aforesaid spider species when exposed to the films of insecticides in petridishes. More work on identifying selective insecticides relatively safer to spiders and parasitoids need be done. However rationale behind this approach is that selectivity of insecticides coupled with feeding propensity should allow the maximum interplay of predatory spiders in integrated control. The feasibility of integrated control for the maize stalk borer was thus explored by the rational integration of cultural control involving the use of tolerant variety Ganga-5, early sowing of maize where irrigation facilities are available, higher plant population, mechanical control involving removal of dead hearts and

Table 4. Relative feeding propensity of different species of spiders per day on the larvae of *C. Pertellus.*

Spider species	Stage of host larvae (*C. partellus*)	Average number of larvae consumed
Oxyopes pandae	Freshly hatched	11.40(1.038)
	10-day old	6.60 (0.791)
	18-day old	4.80 (0.667)
Araneus sinhagadensis	Freshly hatched	9.80 (0.975)
	10-day old	6.00 (0.730)
	18-day old	3.20 (0.476)
cheiracanthium sp.,	Freshly hatched	7.60 (0.863)
	10-day old	4.00 (0.571)
	18-day old	2.40 (0.336)
	S. Em. +	0.083
	C.D. at 5 per cent	0.239
	C.D. at 1 per cent	0.323

The figures in parentheses represent the log transformed values.

chemical control using a selective insecticide exerting least disruptive effect on biotic fauna.

The cultural control involving the planting at two plant population level provided a clear guide line to the farmers that higher plant population i.e. 66,000 plants per hectare resulted in higher yield than 52,000 plants per hectare and thus can be profitably adopted. The chemical schedules involving the reduced number of applications of selective insecticides coinciding with the critical stage of crop growth coupled with its higher safety margin for biotic fauna will allow the maximum interplay of the beneficial parasitoids/predators by way of lateral entry. Thus, the pest management involving cultural and chemical control methods for the control of key pest *C. partellus* is quite feasible.

References

Chatterji, S. M., Sarup. P., Bhamburkar, M. W. Marwaha, K. K., Panwar, V. P. S. and Siddiqui, K. H. 1972. Evolution of control schedule for the pests of maize with particular reference to the stem borer, *Chilo zonellus* (Swinhoe) *Indian J. Ent.* **34** (2) : 142-147.

Marwaha K. K. and Sarup, P. 1978. Assessment of carbofuran residues in maize and its implications in biological efficacy against *Chilo partellus* (Swinhoe). *J. ent. Res* **2** (1) : 73-78.

Sarup. P., Panwar, V. P. S. Marwaha, K. K. and Siddiqui, K. H. 1977. Exploitation of polyphagy for the control of *Chilo partellus* (Swinhoe) infesting maize crop. *J. ent. Res.* **1** (2) : 184-192.

Sarup P. Sharma V. K., Panwar V. P. S., Siddiqui K. H., Marwaha K. K. and Agarwal K. N. 1977. Economic threshold of *Chilo partellus* (Swinhoe) infesting maize crop. *J. ent. Res.* **1** (1) : 92-99.

Sharma V. K. and Sarup P. 1979. Predatory role of spiders in the integrated control of maize stalk borer *Chilo partellus* (Swinhoe) *J. ent. Res.*, **3** (2) : 190-202.

Marwaha K. K. Siddiqui K. H. & Sarup P. 1984. Newer carbofuran and carbosulfan formulations to control the maize stalk borer *Chilo partellus* (Swinhoe) *J. ent. Res.* **8** (1) : 93-97.

Marwaha K. K. Siddiqui K. H. & Sarup P. 1986. Impact of one application of some insecticidal formulation for the control of maize stalk borer *Chilo partellus* (Swinhoe) under artificial infestation *J. ent. Res.* **10** (2) : 187-190.

Stern V. M., R. F. Simth, R. Vanden Bosch and K. S. Hagen 1959. The integrated control concept *Hilgardia* **29** (2) : 81-85.

Recent Advances in Indian Entomology
Editor: O. P. Lal
APC Publications Pvt. Ltd., New Delhi, 1996, pp. 323-326

Breeding for Insect Resistance in Maize

V.P.S. PANWAR

The ultimate goal of research work on plant resistance to insects is the development and continued release of resistant and otherwise useful crop varieties. By its very nature, the task is cumulative and progressive, which makes it demanding in terms of funds and cooperative efforts of well-nit teams of scientific workers at least a well-trained entomologist and a competent plant breeder. The breeding for insect resistant plants involves entomological specifications and the foremost requirement is to know your pest, its ovipositional and feeding behaviour in all instar stages etc.

Techniques for evaluating resistance

Adequate techniques for measuring the degree of resistance are probably the key factors to determine the success or failure of any insect resistance programme. For determining the techniques to be applied in insect resistance plants, three fundamental factors must be considered irrespective of crop or insect involved.

(i) What are the best for identifying the gene(s) for resistant to the major insect pests of crop under consideration.

(ii) How can the genes for resistance to an insect be incorporated into commercial cultivars, inbreds, composites with introducing adverse effects.

(iii) How can insect resistance be incorporated so that the cultivar remains resistant for as long period as possible.

Natural infestation

Natural infestation of maize pests seldom occur at a time (stage of the plant development), intensity or uniformity adequate for improvement of genetic resistance. Hence, we have to depend on mass rearing of the pests and insects produced retain their ability to attack and damage plants so that differences in resistance become apparent.

Artificial infestation

Progress in identifying or building levels of genetic resistance to insect pests depends on the ability to distinguish, in each cycle of selection the most resistant genotype. Uniform infestation levels at appropriate stage of plant development are required for selecting resistant genotypes, reducing or eliminating 'escapes' or accumulating genetic resistance. This will be possible only when the pest is mass reared in the laboratory under controlled conditions.

Infestation techniques

Before doing infestation we should keep the points in mind (i) plant age to infest; (ii) number of eggs or larvae per application; (iii) number of applications; (iv) time of day to infest and (v) egg versus larval infestation and source of insect material.

Infestation is done (1) by manual distribution of egg masses either dropping into whorl of plants or affixed (usually pinned) to the plant or (2) by manually placing eggs or larvae on various plant parts usually with a camel-hair brush.

A manual larval dispenser, Bazooka calibrated to deliver uniform amount of larvae/corn cob grits mixture to each plant has been used successfully for four maize pests. However for the maize stalk borer, *Chilo partellus*, the release of larvae with the help of 'Bazooka' is yet to be standardised.

Evaluating plant reaction to insect damage

The reaction of plants exposed to insect attack must be measured either by visual observation or actual measurements at the proper stage of growth. For evaluating large amount of material, visual rating scales are being used. Most rating scales include 5 or more classes that describe the insect damage or plant response. Different rating scales have been recommended for different insect species. The rating scale (1-9) used for evaluating damage by the maize stalk borer, *C. partellus* is as follows:- (1) Apparently healthy plant. (2) Plants showing slightest damage on leaf or a few pin holes on 1-2 leaves. (3) Plants showing more pin holes or short holes on 3-4 leaves. (4) Plants showing injury (pin holes, shot holes, slits) in about one third of total number of leaves and mid-rib tunnelling on 1-2 leaves if any. (5) Plants showing about 50% leaf damage (pin holes, shot holes, slits, streaks) and mid-rib damage, if any. (6) Plants showing varied types of leaf injury in about two-third of total number of leaves. (7) Plants with every type of leaf injury and almost all the leaves damaged. (8) Entire plant showing maximum leaf injury and likely to form dead-heart (such plants usually show stunted growth). (9) Dead-heart (represent extreme type of leaf injury).

Mechanisms and inheritance of resistance

Knowledge of resistance mechanisms makes it easier to select plant characteristics that confer resistance. Inheritance of resistance studies estimate the number of genes controlling resistance and determine type of gene action. This information indicates the best breeding system to increase resistance levels.

Breeding methods

Maize is highly heterogeneous species with 10 pairs of chromosomes. Because maize is naturally cross-pollinated, artificial control of pollination is necessary both for self-fertilization or for crossing of selected plants. Multilocation tests assist in the monitoring of levels of resistance, their buildup or breakdown and shifts in biotypes of races of pests.

The standard breeding procedure for hybrid maize consists of development of inbreds and their subsequent use in hybrid combinations. The chances of obtaining inbred lines resistant to a particular pest depend on the frequencies of genes conditioning resistance. If the resistance is controlled by additive effects, involving as many resistant inbred lines as possible in the hybrid will result in a high level of resistance. When resistance is governed by recessive genes, it is necessary to have them in a homozygous recessive condition in all lines, where resistance is governed by dominant genes all lines need not necessarily be resistant.

Resistant lines can also be used to transfer resistance to susceptible by agronomically desirable lines by crossing a susceptible and a resistant line and extracting from the progeny the most resistant and agronomically best lines.

Resistance can be transferred through a backcrossing programme when only one or two genes are involved. When many genes are involved, the back cross procedure may result in reduced resistance with each successive back cross.

Gene action conditioning resistance to most insect pests in maize appears to be additive. This indicates that population improvement procedures, such as mass selection and various recurrent selection schemes are effective in accumulating desirable genes for resistance to various maize insect pests.

The following points are important in breeding for insect resistance in maize:-

1. Knowledge of gene action conditioning resistance is available for few maize insects and more work is required to provide a clear understanding of the modes of inheritance.
2. To the extent possible, resistance should be improved in adapted genetypes. Introduced genotype that may have resistance can also be used. Transfer of resistance from related species should be the last resort because of the long process required in recovering good agronomic characteristics.
3. The levels of insect resistance are generally low. This requires the evaluation of resistance in an array of germplasm sources.
4. Escapes occur under both matural and artificial infestation. If the proportion of escapes is not great, they can be eliminated in the successive cycles of evaluation and selection.
5. Insect population and infestation level, whether natural or artificial should be regulated depending on the level of resistance in plant materials. If resistance levels are low, expression of existing genes for resistance may be masked by heavy infestation and all genotype may appear equally susceptible. As resistant genes are accumulated, infestation level can be increased accordingly.
6. When natural insect levels are adequate and uniform, effective selection is possible, if appropriate plot techniques are used. Environments known to have relatively high insect levels over seasons may be utilized for this purpose.
7. If rapid and sustained progress is desired, successful artificial infestation is necessary to identify plant reaction.
8. Rating scales to measure differences in resistance should be practical, rapid and reliable.
9. Through the use of appropriate breeding procedures the components of resistance (non-preference, antibiosis and tolerance should be exploited singly or in combination.

10. Breeding progress depends on the cooperation of scientists specializing in different disciplines. Isolated efforts by breeders entomologists and pathologists will prove varieties deficient in one or more aspects. Therefore, an integrated, multidisciplinary approach is of the utmost importance.

References

Maxwell, F.G. Jennings, P.R. 1980. Breeding Plants Resistance to Insects. John Wiley and Sons Inc. New York, 683 pp.

Painter, R.H. 1951. Insect Resistance in Crop Plants. Macmillan New York, 520 pp.

Panda, Niranjan, 1979. Principles of Host-Plant Resistance to Insect pests. Hindustan Publishing Corporation (India) Delhi-386 pp.

Sarup, P., Marwaha, K.K., Panwar, V.P.S. and Siddiqui, K.H. 1978. Evaluation of some exotic and indigenous maize germplasms for resistance to *Chilo partellus* (Swinhoe) under artificial infestation. *J. ent. Res.* **2** (1) : 98-105.

Siddiqui, K.H. and Marwaha, K.K. 1993. The Vistas of Maize Entomology in India. Kalyani Publishers, Ludhiana, 184 pp.

Recent Advances in Indian Entomology
Editor: O.P. Lal
APC Publications Pvt. Ltd., New Delhi, 1996, pp. 327-334

Artificial Diets for Tissue Borers of Maize

K.H. SIDDIQUI

The availability of insects in the laboratory is an important pre-requisite for a sound integrated pest management. Rearing of insects on their natural hosts is not only time consuming due to regular change of plant material involving more man power, but also becomes rather impracticable when the natural food is not available in the field. Further, not only a lot of host plant material is wasted but also frequent handling of the culture results in high mortality of the insect. These drawbacks have led to evolve artificial diets.

The formulation of artificial diets has made feasible the mass rearing of insects in the laboratory. Obviously, the ease with which the diet could be prepared economically with known ingredients restricts the use of plants as such and hasten the mass rearing uninterruptedly in short space and time. The availability of different developmental stages of an insect in large number through rearing on artificial diet has added new dimensions in investigating diverse entomological problems. This has assumed special significance in the case of tissue borers which usually require live plant material to maintain their culture for multiplication. However, the success of mass rearing on artificial diets would depend, to a large extent, on the use of minimum number of indigenously available diet ingredients, easy preparation techniques as also low cost of insect production.

Concerted efforts were made during the past three decades for compounding artificial diets pertaining to three principal tissue borers of maize in India *viz.*, (i) stalk borer, *Chilo partellus* (Swinhoe), (ii) pink borer, *Sesamia inferens* Walker and (iii) maize and sorghum shoot fly, *Atherigona soccata* Rondani.

The maize stalk borer, *Chilo partellus* (Swinhoe) (Crambidae: Lepidoptera)

Pant, Gupta and Nayar (1960) formulated an artificial diet containing both carboxy-methyl cellulose and dry maize leaves in which newly hatched larvae of *C. partellus* established, fed and gave rise to normal moths laying viable eggs (Table 1). However, this diet proved to infructuous when tested by Chatterji *et al.* (1968). The first successful artificial diet to rear *C. partellus* was based on wheat germ (Chatterji *et al.*, 1968 - Table 1). The composition of this diet was the same as formulated earlier by keaster and Harrendorf (1965) for the rearing of South-Western cron borer, *Zeadiatraea grandiosella* (Dyar). Due to high cost and non availability of certain ingredients, the wheat germ based diet could not be adopted for mass rearing of *C. pertellus* at different locations in India. This necessitated the search of alternate diets for *C. partellus*.

Table 1. Artificial diets for the mass rearing of *C. partellus*.

Ingredient		(Pant *et al.*, 1960)	(Chatterji *et al.*, 1968)	(Dang *et al.*, 1970)	(Lakshmi-narayana & Soto, 1971)	(Siddiqui & Chatterji, 1972)	(Siddiqui *et al.*, 1977)	(Sharma & Sarup, (1978)	(Seshu Reddy & Davies, 1978)	(Singh & Sarup, 1987)
Red *rajmah* grain power	(g)	-	-	-	-	74.80	-	-	548.0	-
Kabuli Gram grain powder	(g)	-	-	420.00	100.00	-	-	-	-	-
Green gram grain powder	(g)	-	-	-	-	-	37.50	-	-	35.00
Dew gram grain powder	(g)	-	-	-	-	-	37.50	-	-	35.00
Lentil grain powder	(g)	-	-	-	-	-	-	75.00	-	-
Wheat grain powder	(g)	-	-	-	-	20.00	20.00	20.00	-	15.00
Wheat germ	(g)	-	15.00	-	-	-	-	-	-	-
Casein	(g)	5.50'	17.50@	-	-	-	-	-	-	-
Brewer's yeast	(g)	2.20		40.00	13.30	4.00	5.00	5.00	40.00	5.00
Sugar	(g)	5.50[A]	11.50[B]	-	60.00[C]	-	-	-	-	-
Cellulose	(g)	2.50[D]	2.50	-	-	-	-	-	-	-
Salt mixture (Wesson's No. 2)	(g)	-	5.00	-	-	-	-	-	-	-
Total mineral salts	(g)	0.44	-	-	-	-	-	-	-	-
Vitamin fortification mixture	(g)	-	7.00	-	5.00	-	-	-	-	-
Vitamin E (1 capsule = 100 mg)	(g)	-	-	-	-	0.10	0.20	0.10	5.20 (20 capsules)	0.20
Cholesterol	(g)	0.22	-	-	-	-	-	-	-	-
Ascorbic acid	(g)	-	1.70	13.00	4.30	1.30	1.70	1.50	13.00	1.70
Leaf power (dry)	(g)	1.80[M]	-	-	40.00[S]	-	-	-	200.00[S]	10.00[S]
Methyl *p*-hydroxybenzoate	(g)	0.10	0.75	8.00	2.70	0.80	0.80	0.90	8.00	0.80
Sorbic acid	(g)	-	-	4.00	1.50	0.40	0.40	0.50	5.00	0.40
Aureomycin	(g)	-	0.07	-	-	-	-	-	-	-
Choline chloride	(g)	0.09	-	-	-	-	-	-	-	-
Agar powder	(g)	3.50	12.50	51.00	25.00	5.10	6.00	6.00	51.00	6.00
Formaldehyde (40%)	(ml)	-	1.00	8.00*	2.70*	1.00	1.00	1.00	1.00	1.00
KOH solution (4 molar)	(ml)	-	2.50	-	-	-	-	-	-	-
Water	(ml)	125.00	450.00	3120.00	1000.00	380.00	390.00	390.00	4500.00	390.00

@=Vitamin free, A=Glucose, B=Dextrose, C=Sucrose, D=Carboxy-methly cellulose (CMC).
M=Corn, S=Sorghum, *=10%

Dang et al. (1970) formulated an artificial diet using *kabuli* gram (*Cicer arietinum* L.) as base-ingredient (Table 1). The poor performance of this diet led Lakshminarayana and Soto (1971) to increase the number of ingredients (Table 1). Such diet contained sorghum leaf factor, sucrose and vitamin fortification mixture not only increased the overall cost of insect production but also involved cumbersome preparation technique.

Later, Siddiqui and Chatterji (1972) developed an artificial diet based on kidney bean (red *rejmah)* grain power (*Phaseolus vulgaris* L.) for the mass rearing of *C. partellus* (Table 1). The reduced number of ingredients in the diet without host plant factor and its easy preparation technique made feasible the mass production of egg masses of *C. partellus*. Due to superior performance of such diet over those based on *kabuli* gram (Dang *et al.* 1970; Lakshminarayana and Soto, 1971) its adoption at different locations in the country was suggested. Vitamin E was introduced for the first time in the artificial diet for rearing *C. partellus*. The obvious difficulty to procure kidney bean at different maize entomological centres of the All-India Coordinated Maize Improvement Project necessitated the search for alternative and easily available pulses as principal-base ingredients. This led to exploitation of genus *Vigna* which constitutes a rich source of protein.

Major break-through in mass rearing of *C. partellus*

Siddiqui *et al.* (1977) achieved a major breakthrough by compounding artificial diets involving either mixture of green gram, *Vigna radiata* L. Wilczek (*mung)* and dew gram, *Vigna aconitifolia* (Jacq.) Marichat (*moth)* grain powder or a mixture of dew gram and cowpea, *Vigna unguiculata* (L.) Walp. (*lobia)* grain powder as principal-base-ingredients (Table 1). The intensity of damage caused by the larvae of *C. pertellus* reared up to either one or two generations on these two diets was comparable with those obtained from natural food comprising maize plants. Thus, these two diets based on mixture of two pulses were the most suitable and offered not only promise for mass rearing of maize stalk borer in the laboratory but shall led to success in locating real source (s) of resistance to *C. partellus* infesting maize in the field. This probe in an hitherto unexplored area of differential response of *C. partellus* precluded the inadvertent entry of mis-identified source of resistance in the maize population improvement programme (Siddiqui and Sarup, 1978).

Later studies revealed that lentil, *Lens culinaris* Medic (*masoor)* (Sharma and Sarup, 1978) (Table 1), sorghum leaf factor replacing wheat flour in the kidney bean based artificial diet (Seshu Reddy and Davies, 1978) (Table 1) were used as principal base-ingredients of artifical diets for the rearing of *C. partellus*. Singh and Sarup (1987) compounded an improved artifical diet by adding sorghum leaf factor with green gram + dew gram as principal base-ingredients (Table 1).

Significane and optimal quantity of ingredients in artificial diets

Besides, beans/pulses as principal base-ingredients which provide protein source in the artificial diet(s), other ingredients, though used in relatively small quantities in the diet, significantly affected various developmental parameters of

C. partellus. Thus, the essential and supplemental ingredients (Siddiqui, 1977) were categorized on their relative significane *viz.*, impact of vitamin E in egg production (Siddiqui and Sarup, 1980), significane and optimal quantity of ascorbic acid for normal development (Siddiqui and Sarup, 1981; Singh and Sarup, 1986), and the deficiency of formaldehyde affecting development and its optimal dose in the diet (Siddiqui and Sarup, 1982). But the deleterious effect of streptomycin (Siddiqui *et al.*, 1985) and aureofungin an anti-microbial agent, was evident in artificial diets (Siddiqui *et al.*, 1986). Contrary to expectations, the incorporation of vegetable oils *viz.*, soybean oil or cotton seed oil in the artificial diet did not appear to be suitable natural source of vitamin E for either substituting or supplementing synthetic vitamin E (Siddiqui *et al.*, 1988).

Hence, in formulating an artificial diet for the mass rearing of *C. partellus* not only the essential ingredients are premarily required but also the supplemental ingredients provide the balanced diet.

Pink stem borer, *Sesammia inferens* Walker (Noctuide: Lepidoptera)

As early as 1969, Chatterji and his co-workers reared *S. inferens* on wheat germ based artificial diet (Table 2). This diet did provide the quidelines for further improvement by replacing some costly ingredients.

An artificial medium based on bean seed kernel was developed by Kalode *et al.* (1970). Lingappa (1978) formulated an artificial diet based on french bean flour and finger millet flour (Table 2). Due to very low percentage of *S. inferens* larvae reaching adult stage, he modified the same in 1987 (Table 2). Later, Butcheswara Rao in 1983 reported *kabuli* gram flour based diet for its rearing in the laboratory. This diet was earlier developed at Common Wealth Institute of Biological control, Bangalore (India) (Table 2). Taking the clue from *rajmah* based diet developed for mass rearing of *C. partellus* (Siddiqui and Chatterji, 1972), the artificial diet developed at CIBC, Bangalore for *S. inferens* was slightly modified by replacing *kabuli* gramm flour with *rajmah* grain powder (Butcheswara Rao, 1983) (Table 2). This modified diet contained 18 ingredients and reported to be suitable for rearing *S. inferens* in the labortory though validity was rather questionable due to lack of data on various biological parameters.

An artifical diet having green gram grain powder, dried maize leaf and whorl powder and sucrose along with other nine ingredients was formulated by Siddiqui *et al.* (1983) for successful rearing of *S. inferens* (Table 2). Easwaramoorthy and Shanmugasundaram (1988) reported that the whole life of the pink stem borer was completed on two different artificial diets. They reared *S. inferens* larvae up to 3rd instar on one diet (Diet 1- Table 2) whereas from 4th instar till pupation the larvae were reared on another diet (Diet II - Table 2). Their whole rearing methodology appeared to be very cumbersome, time consuming, requiring more man power with high cost of insect production. The frequent changing of larvae from one diet jar to another might also result in mechanical injury to the developing larvae. Senthilkumar and Siddiqui (1993) successfully reared *S. inferens* on artificial diet using green gram grain power, maize grain powder and dry maize leaf whorl and tender stem powder. According to them the whole rearing methodology was quite simple with low cost of production of healthy insects (Table 2).

Table 2. Artificial diets for the mass rearing of *S. inferens*.

Ingredient		(Chatterji, (et al., 1969)	(Lingappa, 1978)	(Lingappa, 1987)	(Butcheswara Rao 1983)	(Butcheswara Rao 1983)	(Siddiqui et al., 1983)	(Easwaramoorthy & Shanmugasundaram. 1988) I	II	(Senthilkumar & Siddiqui 1983)
Green gram grain powder	(g)	-	-	-	-	-	70..0	-	-	50.0
Kabuli gram grain powder	(g)	-	-	-	50.0	-	-	60.0	100.0	-
Red rajmah grain powder	(g)	-	-	-	-	50.0	-	-	-	-
Maize grain powder	(g)	-	-	-	50.0	50.0	-	-	100.0^{G}	50.0
Wheat grain powder	(g)	-	-	-	-	-	18.0	-	-	-
Wheat germ	(g)	15.0	-	-	-	-	-	-	-	-
Portion of maize plant	(g)	-	-	-	4.0*	15.0^{X}	5.0^{+}	-	-	5.0^{+}
Rice stem powder	(g)	-	-	-	-	-	-	-	-	-
Casein	(g)	17.5	-	-	-	5..0	-	50.0	-	5..0
Sugar	(g)	11.5^{B}	-	-	15.0^{A}	15.0^{A}	2.0^{A}	22.5^{A}	60.0^{A}	15.0^{A}
Milk powder	(g)	-	-	-	10.0	-	-	-	40.0	-
Brewer's yeast	(g)	-	4.0	4.0	10.0	8.0	5..0	22.5^{E}	20.0	8.0
Cellulose	(g)	2.5	-	-	-	-	-	-	-	-
Salt mixture (Wesson's No. 2)	(g)	5.0	-	-	0.25	0.25^{C}	-	-	2.0	0.3^{0}
Vanderzant's fortification mixture	(g)	7.0@	-	-	-	-	-	-	-	
Vitaminette forte (Tablet)		-	-	-	-	One	-	-	-	-
Multivitaplex Odumex capsule		-	-	-	Two	Two	-	Two	-	-
Vitamin E (1 capsule=100mg)	(g)	-	0.2	0.2	0.1	0.1	0.1	0.1	0.2	0.1
Cholesterol	(g)	-	0.1	0.2	1.0	1.0	-	-	2.0	1.0
Ascorbic acid	(g)	1.7	1.3	1.3	2.25	1.5	1.7	10.0	6.5	1.5
Methyl p-hydroxybenzoate (Nipagin)	(g)	0.75	0.8	0.8	2.0	2.0	0.8	2.0	4.0	2.0
Sorbic acid	(g)	-	0.4	0.4	1.0	1.0	0.4	1.0	2.0	1.0
Aureomycin	(g)	0.07	-	-	-	-	-	-	-	-
Agar powder	(g)	12.5	5.1	5.7	12.0	12.0	6.0	10.0^{F}	23.0^{F}	12.0
Linseed oil	(ml)-	-	-	-	1.5	1.0	-	-	3.0	-
Formaldehyde (40%)	(ml)	1.0	1.0	1.0	2.0^{D}	2.0^{D}	1.0	1.0	4.0	1.0
KOHsolution (4 molar)	(ml)	2.5	-	-	-	-	-	-	-	-
Water	(ml)	450.0	350.0	210.0	750.0	650.0	390.0	450.0	1100.0	650.0
Fench bean flour	(g)	-	74.8	74.8	-	-	-	-	-	--
Finger millet flour	(g)	-	20.0	20.0	-	-	-	-	-	--
Hostacycline	(g)	-	-	-	-	-	-	1.0	2.0	-
Sugarcane shoot powder	(g)	-	-	-	-	-	-	75.0	-	-
Saffola oil	(ml)	-	-	-	-	-	-	5.0	-	-

*=Stem powder. A=Sourcose, @=Vitamin fortification mixture, X=Leaf and stem powder, B=Dextrose, += Leaf whorl and stem powder, C=Common salt, D=10%, E=Tablet form, F=Powder or shred, G=Coarse flour.

Maize and sorghum shoot fly, *Atherigona Soccata* Rondani (Muscidae : Diptera)

In India, a begining has been made to rear shoot fly, *A. soccata* on artificial diets in the laboratory. The earliest artificial diet for shoot fly using agar, yeast, propionic acid and water was formulated by Dang *et al.* (1971) (Table 3). Sukhani and Jotwani (1979) formulated two diets based on (i) black gram (*Vigna mungo* Hepper) grain powder, and (ii) banana pulp (*Musa paradisiaca* Linn.) (Table 3). According to them the former diet was most suitable in respect of important biological parameters. However, according to Singh *et al.* (1983) the diets developed by Dang *et al.* (1971) and Sukhani and Jotwani (1979) proved to be unsuitable rearing media for *A. soccata* in the laboratory as about 5 per cent files emerged and that too abnormal. Evidently, artificial diets for shoot fly are still in the formative stage for adoption universally.

Table 3. Artificial diets formulated for the rearing of *A. soccata*.

Ingredient		(Dang *et al.*, 1971)	(Sukhani and Jotwani, 1979)	(Singh *et al.*, 1983)
Black grain grain powder	(g)	-	30.0	-
Casein	(g)	-	3.0	15.0[A]
Brewer's yeast	(g)	50.0	50.0	20.0
Shredded p-hydroxybenzoate	(g)	-	2.0	0.75*
Agar powder	(g)	20.0	20.0	-
Propionic acid	(ml)	3.0	-	-
Water	(ml)	930.0	895.0	200.0

A=Vitamin free; *=Dissolved in 3 ml of 75% ethyl alcohol.

References

Butcheswara Rao, A. 1983. Techniques of scoring for resistance to maize stalk borer (*Sesamia inferens*) pp. 16-26. In Joginder Singh (ed.), *Techniques of scoring for resistance to the major insect pests of maize*. All-India Co-ordinated Maize Improvement Project. Indian Agricultural Research Institute. New Delhi.

Chatterji, S.M., Sharma, G.C., Siddiqui, K.H., Panwar, V.P.S. and Young, W.R. 1969. Laboratory rearing fo the pink stem borer, *Sesamia inferens* Walker, on artificial diet. *Indian J. Ent.*, **31** (1) : 75-77.

Chatterji, S.M., Siddiqui, K.H., Panwar, V.P.S. *Sharma, G.C. and Young*, W.R. 1968, Rearing of the maize stem borer, *Chilo zonellus* Swinhoe on artificial diet. *Indian J. Ent.*, **30** (1):8-12.

Dang, K., Doharey, K.L., srivastava, B.G. and Jotwani, M.G. 1971. Artificial diet for the mass rearing of *Atherigona varia soccata* Rondani. Entomologist's *Newsl.*, **1** *: 65.*

Easwaramoorthy, S. and Shanmugasundaram M. 1988. Mass rearing of *Sesamia inferens Wlk. and Chilo sacchariphagus indicus* (Kapur). pp. 103-110. In H. David and S. Easwaramoorthy, (eds.). *Biocontrol technology for sugarcane pest management*. Sugarcane Breding Institute, Coimbatore.

Kalode. M.B., Misra, B.C., Verma, A. and Das, R.C. 1970. Rearing of rice pests on synthetic and natural diets. Technical Report. Central Rice Research Insitute, Cuttack, 203 pp. (unpublished).

Keaster. A.J. and Harrendorf, K. 1965. Laboratroy rearing of south-western corn borer, *Zeadiatraea grandiosella* (Dyar) on wheat germ medium, *J. econ. Ent.*, **58** (5) 923-924.

Lakshminarayana, K. and Soto, P.E. 1971. A technique for mass rearing of sorghum stem borer, *Chilo zonellus (=pertellus). Sorghum Newsl.*, **14**: 41-42.

Lingappa, S. 1978. Development of artificial diet for mass rearing *Sesamia inferens* Walker (Lepidoptera : Noctuidae) and screening for resistance in finger millet germplasm. Ph. D. Thesis. University of Agricultural Sciences. Bangalore (India).

Lingappa. S. 1987 Importance of physical properties of artificial diet for the pink stem borer. *Sesamia inferens* Walker. *Insect Sci. Applic.*, **8** (1) : 65-69.

Pant. N.C.. Gupta, P. and Nayar, J.K. 1960, Physiological studies on *Chilo zonellus* Swinh., a pest of maize crop. I. Growth on artificial diets. *Proc. natn. Inst. Sci. India*, (B) **26** (6) : 379-383.

Senthilkumar. P. and Siddiqui, K.H. 1993. Compounding artifical diets for rearing of the pink stem borer, *Sesamia inferens* Walker. *J. ent. Res.*, **17** (2): 81-90.

Seshu Reddy, K.V. and Davies, J.C. 1978. A medium for mass rearing of the sorghum stem borer, *Chilo partellus* Swinhoe (Lepidoptera: Pyralidae) and its use in resistance screeing. *Indian J. Pl. Prot.*, **6** (1): 48-55.

Sharma, V.K. and Sarup, P. 1978. Formulation of suitable artificial diets for rearing the maize stalk borer, *Chilo partellus* (Swinhoe) in the laboratory. *J. ent. Res.*, **2** (1): 43-48.

Siddiqui. K.H. 1977. Differential response of *Chilo zonellus* Swinhoe reared on artificial and natural diets in evaluating resistance in maize. Ph. D. Thesis. Aligarh Muslim University, Aligarh (India).

Siddiqui. K.H. and Chatterji, S. M. 1972. Laboratory rearing of the maize stem borer, *Chilo zonellus* Swinhoe (Crambidae: Lepidoptera) on a semi-synthetic diet using indigenous ingredients. *Indian J. Ent.*, **34** (2) : 183-185.

Siddiqui. K.H. and Sarup, P. 1980 Impact of vitamin E on egg production of the maize stalk borer, *Chilo partellus* (Swinhoe) reared in artificial diets. *J. ent. Res.*, **4** (1) : 45-50.

Siddiqui, K.H. and Sarup, P. 1982. Significance of ascorbic acid in artificial diets for mass rearing of the maize stalk borer, *Chilo partellus* (Swinhoe). *J. ent. Res.*, **5** (1): 91-93.

Siddiqui, K.H. and Sarup, P. 1982. Compounding of artificial diets with formaldehyde affect rearing of the maize stalk borer, *Chilo partellus* (Swinhoe). *J. ent. Res.*, **6** (2): 197-201.

Siddiqui, K.H., Sarup, P. and Marwaha, K.K. 1983. Formulation of artificial diets for mass rearing of the pind borer, *Sesamia inferens* Walker, in the laboratory. *J. ent. Res.*, **7**(2): 154-160.

Siddiqui, K.H., Sarup, P. and Marwaha, K.K. 1985. Effect of streptomycin in artificial diet on mass rearing of the maize stalk borer, *Chilo partellus* (Swinhoe). *J. ent. Res.*, **9** (1): 60-64.

Siddiqui, K.H., Sarup, P. and Marwaha, K.K. 1986. Deleterious effect of aureofungin in artificial diets formulated for the mass rearing of the maize stalk borer, *Chilo partellus* (Swinhoe) (Crambidae: Lepidoptera). *J. ent. Res.*, **10** (1): 74-78.

Siddiqui, K.H., Sarup, P. and Marwaha, K.K. 1988. Feasibility of compounding artificial diets with vegetable oils as source of natural vitamin E for increasing fecundity of the maize stalk borer, *Chilo partellus* (Swinhoe). *J. ent. Res.*, **12**(1): 61-68.

Siddiqui, K.H., Sarup, P., Panwar, V.P.S. and Marwaha, K.K. 1977. Evolution of base-ingredients to formulate artificial diets for the mass rearing of *Chilo partellus* (Swinhoe). *J. ent. Res.*, **1**(2): 117-131.

Singh, P., Unithan, G.C. and Delobel, A.G.L. 1983. An artificial diet for sorghum shoot fly larvae. *Entomologia exp. appl.*, **33**(1): 122-124.

Singh, R. and Sarup, P. 1986. Significane and optimal quanity of ascorbic acid in artificial diets for the mass rearing of maize stalk borer, *Chilo partellus* (Swinhoe) (Lepidoptera: Crambidae) *J. ent. Res.*, **10**(2): 173-178.

Singh, R. and Sarup. P. 1987. An improved artificial diet for mass rearing of the maize stalk borer, *Chilo partellus* (Swinhoe) (Lepidoptera: Crambidae) *Insect Sci. Applic.*, **8**(2): 181-185.

Sukhani, T.R. and Jotwani, M.G. 1979. A simple artificial diet for mass rearing of sorghum shoot fly, *Atherigona soccata* (Rondani). *Bull. Ent.*, **20**: 61-66.

Recent Advances in Indian Entomology
Editor: O.P. Lal
APC Publications Pvt. Ltd., New Delhi. 1996. pp. 335-342

Insect Pests of Groundnut and their Management

Y. MAHTO

Groundnut (*Arachis hypogea L.*) is believed to be a native of Brazil (S. America). It is a major oil seed crop in India, occupying about 8.6 million ha, of which 85 % is rain fed and only 15% is irrigated. The crop is cultivated in 45% of India's total oil seed area and accounts for 55% of the oil seeds produced in the country. Groundnut seeds contain approximately 55% edible oil and 25% protein. The haulms remaining after pod removal are used as nutritious fodder. Groundnut oil cake is an important cattle feed and a good soil amendment. Although India ranks first in area and production (7.2 million t) in the world, average yield is very low. The low yield in the rainy season is associated with low levels of input.

Average yield of semi spreading and spreading varieties is 1200-1400 kg of unshelled pods/ha, that of bunch type is 800-1000 kg/ha. Pods yield is 70-75% of kernals by weight. Yield of haulm (stalk) is usually 2-2.5 times that of pods.

Over 90 species of insects attack groundnut crop, but only a few are economically important over large areas. The various groups of insects attacking the crop may be categorised into the following heads :-

I. *Defoliators*

1. Tobacco caterpillar/army worm
2. Hairy caterpillars
3. Bihar hairy caterpillars
4. Gram pod borer
5. Groundnut leaf miner
6. Groundnut bud borer
7. Tussok caterpillar
8. Gray weevils

II *Sucking pests*

9. Aphids
10. Thrips
11. Jassids
12. Mites

III *Root and pod feeders*

13. White grubs
14. Termites
15. Jewel beetle
16. Ear wig
17. Wire worms and false wire worms
18. Sub-terranean ants

IV *Storage pests*

19. Groundnut bruchid 20 Red flour beetle and rice moth

21. Pod sucking bug.

I *Defoliators*

1. Tobacco caterpillar - *Spodoptera litura* (Fab.) (Lepi. : Noctuidae)

Groundnut fields in parts of Karnataka and coastal A. P. and Tamil Nadu are often attacked by this pest in the post rainy season.

The newly hatched larvae can be detected by the "scratch" marks on the leaf surface. The older larvae are night-feeders and are usually found in the soil around the base of the plants during the day time. The population density of the pest when found in large numbers, the field can be denuded and the larvae migrate to another field of groundnut in search of food. In areas where groundnut is grown in light soils the larvae (caterpillars) can bore into pods also.

In A. P. the insect has 12 generations a year, each lasting slightly more than a month in winter and less than a month in the hot season. In A. P. this pest was once limited to tobacco only but now it has adapted to groundnut because of continuous cropping.

2. Hairy caterpillars - (1) *Amsacta albistriga* Walk. and (ii) *A. moori* Butler. (Lepidoptera : Arctiidae). Several species of hairy caterpillars attack groundnut. Among them two are important named above. The former is found in S. India and the latter in N. India. Both have one generation in a year. Moth is brownish white, 41-50 mm in wing span. Female lays about 800-1000 eggs in cluster of 50-100 each on host. Both the species may be devastating, but are highly sporadic and unpredictably important in isolated pockets. Larvae are gregarious in nature, feed on leaves from undersurface, migrate from one field to another in search of food.

3. Bihar hairy caterpillar - *Spilosoma (Diacrisia) obliqua* (Walk. Lepi. : Arctiidae). This species is widely distributed in A. P., Assam, Bengal, Bihar, Haryana, M. P., Maharashtra, Punjab and U. P. Of late it has become an important groundnut pest in N. States. It has also been reported from Burma, China, Japan and Bangladesh.

4. Gram pod borer - *Helicoverpa armigera (Hubner) (Lepi. : Noctuidae).* This species is widely distributed in India, particularly in coastal A. P., Tamil Nadu and Karnataka in areas where cotton is extensively grown and insecticide application is heavy.

It completes seven or more generations in a year in S. Inida and 3-4 generations per year in N. India. Moth has a wing span of about 40 mm with dull brown forewings. Creamy eggs are laid singly on young leaves and flower buds. Damage caused by larvae to foliage is similar to that caused by tobacco and hairy caterpillars but the pod borer prefers flowers and buds.

Control : For *Spodoptera, Helicoverpa* and hairy caterpillars

(i) *Chemical* :During the first 50 days after emergence of seedlings, if defoliation exceeds 25% or if one or more larvae/plant is found application of (a) endosulfan @ 350 ml *a.i./ha* or (b) monocrotophos @ 300 ml *a.i.*/ha or (c) fenvalerate @ 100 ml *a.i.*/ha may be given.

Besides, there are number of predators e.g. spiders, birds and lady bird beetles, parasites and diseases - the natural enemies, which control the population densities of the pest.

5. Groundnut leaf miner - *Aproaerema modicella* Deventer *(Stermopteryx subscivella Zell.* (Lepi. : Gelechiidae).

This is a key pest of groundnut in many parts of India. It has a limited range of hosts, among which soyabean (*Glycine max)* is the most favoured alternative to groundnut. The adult moth is brownish grey, about 6mm long with 10 mm wing span. Shiny white eggs are laid singly on the underside of the leaflets close to midrib. The young larvae mine into the leaves.

The pest is of economic importance in Maharashtra and a potential pest in Gujrat. It is a major pest in A. P., Karnataka and Tamil Nadu. It is a minor pest of groundnut in Rajasthan and of lucerne in Punjab. Outside India it has been reported from Burma, China, Indonesia, Laos, Malaysia, Pakistan, Philippines, Sri Lanka and Thailand.

Losses due to this pest have been reported to vary from 24% to 92% by different workers. A severely attacked field looks "burnt" from a distance and epidemics can result in total crop loss.

Control : (i) Chemical insecticides like Dimethoate @ 200-250 ml *a.i.*/ha, or (ii) monocrotophos @ 150-200 ml *a.i.*/ha may be applied if more than five or more larvae per plant are found upto 30 days after seedling emergence.

6. Groundnut bud borer-*Anarsia ephippias* (Meyrick) (Lepi. : Gelechiidae).

This pest occurs in N. India and is considered to be a minor pest. Fully grown larvae are chocolate brown in colour and 10-15 mm long. Larvae prefer to bore into terminal buds and shoots.

7. Tussok caterpillar - *Euproctis* sp. (Lepi. : Lymantridae).

Tussok caterpillars, larvae of *Euproctis* sp. feed on groundnut leaves, buds and flowers. Larvae can be recognised by the presence of prominent, compact tuft of short hairs on the back, near the head. It is of minor importance.

8. Grey weevils - *Myllocerus* sp. (Coleop. : curculionidae).

Grey (ash) weevils eat groundnut foliage, particularly during the rainy season, notching or scalloping the edges of the leaves. They are of minor importance.

II. Sucking pests

9. Groundnut aphids - *Aphis craccivora* Koch. (Hemip. : Aphididae).

Aphids are found on groundnut and other leguminous crops throughout India.

Aphis craccivora reproduces without mating (in India). Individuals are capable of producing about 100 nymphs in their 5-30 days life span. These nymphs are dark brown in colour and turn into shiny black adults in about 10 days. when population density reaches a certain limit, winged individuals are found among the wingless forms. They fly away to form new colonies.

Aphid is a sporadic pest in India. Both adults and nymphs suck sap from tender growing shoots, flowers and pegs, causing stunting and distortion of the foliage and stems. They secrete a sticky fluid (honey dew) on the plant which is turned black by fungus. The blackened honey dew is sooty mould. Though it is a sporadic pest but some times there may be an outbreak during periods of drought stress. Heavey rain storms reduce aphid population densities and favour development of the fungal diseases that kill these insects. Aphids are particularly susceptible to predators like coccinellids, syrphids, lacewings and a number of parasites.

Although the feeding activity of aphid colonies can retard plant growth, this species is potentially more important as a serious vector. The most important aphid transmitted viruses are - (a) peanut stripe virus, and (b) the groundnut rosette virus complex. But this is not confirmed.

In case of severe infestation it can reduce yield by about 40%.

Control : Application of dimethoate @ 200-250 ml *a.i.*/ha under drought conditions when the crop is young, all terminal buds are infested and no lady birds, syrphids or lacewings are there, is recommended.

10. Thrips :- Thysanop. : Thripidae

Several species of thrips are there, the important ones are -

(i) *Scirtothrips dorsalis* Hood, (ii) *Caliothrips indicus* (Bangall),

(iii) *Frankliniella schultzi* (Trybom), and (iv) *Thrips palmi* Karny

Thrips of economic importance is *Frankliniella schultzi* (Trybom) which transmits tomato spotted wilt virus, the causal agent of bud necrosis disease of groundnut, leaf curl in green and black gram, necrosis in pea and tomato.

Thrips palmi transmits peanut bud necrosis virus which can cause wide spread plant death.

The most important thrips in groundnut are *Scirtothrips dorsalis, Thrips palmi* and *Frankliniella Schultzi.* It is virtually difficult to distinguish between species with naked eyes.

There are four instars in the life cycle of the insect - two nymphal, pre-pupal and pupal instars. Under optimal conditions, the immature stages last about 15 days. Adults live for 20 days and lay 40-50 eggs. They can be present at any time of the year but are most numerous in the post rainy season. They have a wide host range.

Nymphs and adults suck sap from the surface of the leaflets resulting, initially, in white patches on the upper and necrotic patches on the lower surface of the leaves. This consists of distortions of the young leaflets and patchy areas of necrotic tissues that puncture and split as the leaflets grow. Injury is normally seen in seedling.

Caliothrips indicus is of no economic significance.

Control : Application of dimethoate @ 200-250 ml *a.i.*/ha in the post rainy season is recommended. There after it is not necessary to use insecticides for thrips control.

11. Jassids :- (Leaf hoppers) (Hemip. : Jassidae).

Empoasca kerri Pruthi is the dominant species among various insects that infest groundnut. The common names of the species are

(i) *E. kerri* and *E. k.* var. *motti* (Pruthi, 1940)

(ii) *E. kerri* and *E. motti* (=*E. kerri* var. *motti*)

Recently it has become a serious pest in Saurashtra of Gujarat and Maharashtra. *Bachlucha* sp. is abundant in Western India, specially in Gujarat.

Initial injury appears on young leaves and veins become white. Affected leaves of young plants wilt. Prolonged exposure to jassids results in "V" shapped yellowings on the tip of the leaflets which may spread and the entire leaflet become yellow. This damage is also called "Hopper burn"

Both adults and nymphs suck the sap from below the young leaves. High density jassid populations are seen during August and September and February and March. High and well distributed rainfall and low temperature encourage jassid population.

Control : For hopper burn application of dimethoate @ 200-250 ml *a.i.*/ha is recommended.

12. Red spider mite :- *Tetranychus* sp. (Acarina : Tetranychidae).

It is a minor pest and sporadic. Infestation is generally localised. *T. telarius* is regarded as potential on groundnut in Pakistan. Probable species of the subcontinent may be *T. cinnabarinus*. Population build up may be rapid on multiple application of fungicides and insecticides.

Chemical control : Acarben, Azodrin, carzol, Galecron, Plictron, Omite and Trithion. may be applied.

III. Root and pod feeders

13. White grubs : (Coleop. : Melolonthinae). Several species of white grubs attack groundut but the most important one is *Holotrichia consanguinea. H. consanguinea* is very much common in North India and *H. serrata* in South India.

(i) *Distribution* : White grubs are serious in North, N. E. Gujarat, N. Rajasthan, Punjab, Haryana and W. Uttar Pradesh. In A. P. they are localised in parts of Anantpur district. In Indore (M. P.) white grub caused serious damage to groundnut during 1972-73. In Maharashtra it is serious in Digraj and Parbhani areas. In general white grubs are major pests in sandy loam or light red soils and not in clayey soils.

(ii) *Nature of damage* : Young grubs feed on rootlets and nodules. Older grubs devour the entire tap root and such plants wither and die. Maximum damage is caused during August to September. However, the damage is usually in patches.

(iii) *Control* : (a) Chemical - (i) Phorate 10% G @ 1.1-1.6 kg *a.i.*/ha, (ii) Carbofuran,

Isophenphos and fensulfothion @ 1 kg a.i./ha, and (iii) Diazinon G 5% @ 25 kg/ha.

(b) *Cultural* : Early sowing of the crop results in lower grub damage.

(c) *Mechanical* : Destroying adult bettles on host plants is necessary.

14. Termite : (Isoptera : Termitidae). Termites cause damage in several ways - (a) They enter the root system and burrow inside the root system. (b) They make holes in pods. (c) Remove soft corcky tissue from between the veins of pods. This scarification of pods - (called scarified pods) makes the pods susceptible to *Aspergillous* attack.

Generally *Microtermes* sp. and Odontotermes sp. (Isoptera : Termitidae) attack the crop and are serious on groundnut in W. Bengal, U. P., Rajasthan, M. P., Gujarat, Haryana and Punjab, but they are of minor importance in Bihar. As a matter of fact a total of 18 species of termites infest groundnut. The important ones are *Microtermes* sp. and *Odontotermis sp. O. obesus* causes this type of damage in M. P. It also feeds on pods'shell and scarify the pods. Occasionally it bores into pods and cause about 46% pod loss (10-45% in W. Rajasthan).

In M. P. 5-50% plant mortality has been reported due to termite. In Sudan 15% pre-emergence loss has been reported.

Control : (i) *Chemical* - seed dressing with suitable insecticide like aldrin or lindane is preferred.

(ii) *Cultural* : Harvesting of groundnut as soon as they are mature and early removal of the produce from fields will reduce chances of termite damage to pods.

15. Jewel beetle : *Sphenoptera indica* (Gory) (Coleop. : Buprestidae) and *S. perroteti* (Guer).

The adult is a shiny beetle, about 10 mm long and 3 mm wide. Eggs are laid singly on the main stem and the grubs, on hatching bore into the stem and tunnel into the root system, causing plants to die.

It is localised in some parts of S. India - Raichur (Karnataka), Anantpur and Kurnool in A. P. The affected plants can be pulled out. It causes damage upto 15% and is of minor importance.

16. Earwig : (Dermaptera : Forficulidae). *Anisolabis annulipes*(Luc.) (=*Euborellia stali*). It is mainly a pest of S. India and causes damage to pods ranging from 2.7 to 65%. In Israel it is a common pest.

It bores into tender pods and feeds on Kernels. The hole is plugged with excreta, sand particles and discoloured pulp. Spreading and semi spreading varieties are more susceptible than bunch varieties.

Control : (i) *Chemical* :- Application of phorate granules 5 G @ 1.2-1.6 kg/ha in furrow after irrigation and before earthing up operation has given good results. Besides (ii) Heptachlor 6%, endrin 5% and chlordane 5% dusts have also been found effective.

17. Wire worms and false wire worms :-

(i) *Wire worms* - coleop : Elateridae)

(ii) *False wire worms* - Coleop : Tenebrionidae)

Wire worms (click beetle larvae) and false wire worms (tenebrionid larvae) occasionally cause damage to groundnut. Adult wire worms (dusky brown beetles) occur on foliage and flowers. Some species apparently do not feed. Wire worms and false wire worms larvae are white or cream, shiny, hard-bodied and virutally indistinguishable. They feed on groundnut roots and pods.

18. Sub - terranean ants :-

(i) *Dorylus orientalis* Westwood

(ii) *D. labiatus* (Shuckerd) (Hymenoptera : Formicidae)

Groundnut pods suffer severe depredations by sub-terranean insects, particularly doryline, blind or red ants (*Dorylus* spp.). *D. orientalis* (an army ant) is a polyphagous pest and is distributed throughout the old and new world tropics. It causes damage to several agricultural crops like potato, cabbage, cauliflower and other vegetables, early sown groundnut and also kernels when lying in heaps after harvest, citrus, sugarcane, mustard and partly ripe mango fruits of dwarf variety. It causes heavy damage to groundnut pods in the field ranging upto 52%. It is rarely seen alone or above the soil surface. It makes several neat, round holes, 0.5-2.0 mm in diameter, and removes the seed completely, without leaving any soil inside the empty shell. Termites, in contrast, refill the pods, cavity with soil.

Control of soil insects : Soil insects are difficult and expensive to control with insecticides.

1. Applying carbofuran (3 G) granules in the furrow @ 1 kg *a.i.* per ha can be an effective prophylactic measure.

2 Application of 2.5 ml *a.i.* of chlorpyriphos/1 kg seed as a seed dressing can suppress white grub population.

3. Adult white grub can be controlled by spraying their feeding trees with carbaryl 50 WP per 2 gallon of water 3-4 times until mid July.

4. In endemic areas ploughing deeply after harvesting the crop can also reduce the population mainly through bird predation and the distruction of the pupae.

19. Pod sucking bugs :- *Elasmolomus sordidus*(Fab.) (=*Aphanus sordidus) (Hemiptera : Lygaeidae).*

In India it is considered as of minor importance but in Africa it is of considerable economic importance. In India it has been reported from Bihar, Maharashtra, Gujarat, T. Nadu and A. P.

Damage due to this pest is not easily visible. Both nymphs and adults suck oil from kernels from freshly harvested pods in fields and storage. Such pods have shell spotted with screta and kernels lose viability.

Control : Keeping rubbish heaps at various places can trap bugs. Such heaps could later be burnt.

IV Storage pests

Storing groundnut in the pods protects the seed against most post-harvest pests. The only exception is the groundnut bruchid (*Caryedon serrata*) reported from Gujarat.

The other insects that are commonly found feeding on shelled groundnut include

(i) Rust red flour beetle - *Tribolium casteneum* Herbst.

(ii) Saw toothed beetle - *Oryzaephilus surinamensis* Linn.

(iii) Almond moth - *Ephestia cautella* (Walker) and

(iv) Rice moth - *Corcyra cephalonica,* which are considered as secondary pests as they are not capable of infesting sound pods.

Control : (1) storing unshelled pods. (2) Completely dry pods should be stored. (3) Fumigation with celphos 3 gm tablet/sac of groundnut and then covering the sac with polythene sheet for 5 days.

Recent approach to augmentation of production

To achieve a sustainable production in groundnut the emphasis is being laid, of late, on

(i) Genetic improvement, which has been spectacular.

(ii) Integrating nutrient management (in conjunction with bio-fertilizers, organic manures, green manures).

(iii) Cropping system approach (crop geometry, intercropping,sequence cropping, relay cropping).

(iv) Integrated pest (disease, insect, weed) management (host plant resistance), cultural control, augmentation of bio-control agents, sex pheromones, bio- pesticides and judicious use of pesticides.

Recent Advances in Indian Entomology
Editor: O. P. Lal
APC Publications Pvt. Ltd., New Delhi, 1996, pp. 343-345

Management of Insect Pests of Rapeseed and Mustard Crop

S.K. PRASAD

Rapeseed and mustard crops are most important oilseed crops. In area and production they stand second to groundnut. But the yield of these group of oilseed crops are too low 906 kg/ha in India as compared to 3050 kg/ha obtained in Netherland. There are many factors for low yield. One of them is that the crop is severely infested by insect pests mainly mustard aphid, *Lipaphis erysimi* Kalt. The loss in yield caused by the aphid alone is about 30-70%. The aphid sucks the sap from leaves, buds, flowers and pods. It has also been reported to infest root of mustard crop (Singh & Singh, 1988). The infestation starts at seedling stage and continue upto pod stage. It leaves the mustard plant only when all the pods are mature and becomes hard which cause hinderance in penetration of aphid's proboscis.

Normally aphid appears on the crop in the month of November in alate form. Their population increases. The increase in aphids population is slow upto middle of January but the increase is very fast and reach to peak in the 2nd or 3rd week of February. Afterwards there is sudden decline in population. When the arrival of the aphid is at normal time late maturity varieties/species escaped from aphid attack. If the arrival of the aphid is delayed susceptible type like yellow sarson and brown sarson escaped from aphid's attack and late crops like *B. napus, B. Carinata* and *B. alba* are severely infested. From these observations it is concluded that time and intensity of aphid infestation are dependent on weather conditions prevailing in the mustard growing area. These factors are temperatures, relative humidity, sunshine hours, saturation deficiency & rainfall. Taking consideration the different weather parameter and apeterous aphids on the crop for four years a forecasting model has been formulated by Prasad *et. al.* (1984), through which we can estimate aphid population much earlier and prepare for its control. The model is Pt. = $a+x_1p$ (t-1) + x^2 mean temp. (t-1) + x_3 min. temperature (t-1).

If we study the presence of alate aphids in the surrounding area through adhesive traps and correlate with the aphid population on the crop, we can estimate the aphid's number on the crop without interfering with the plants. Similar studies have been carried out by Prasad & Phadke (1983) who have found that there was significant and positive correlation between them.

Before adopting any control measure we should also find out Economic Injury level (EIL) of the aphid so that any control measure is not uneconomical. This will help in adopting the control measure before aphid population reaching the EIL level. In case of mustard aphid EIL were worked out at different stages of plant growth by Prasad & Phadke (1983) and Singh & Singh (1987). According to Prasad & Phadke (1983) the EIL of mustard aphid at 70, 80, 90, 100 & 110 days after sowing were 3.3, 16.4, 15.7, 26.9 and 24.7/aphids' shoot respectively when phorate granules were applied and with aldicarb granules the EIL at above stages of plant growth were 3.7, 17.3, 15.9, 16.1 and 11.2 aphids/shoot respectively. Singh & Singh (1987) reported 9-19 aphids/shoot during 1st and 2nd week of January.

Mustard plants are also used as vegetables and it is not advised to use chemical at sowing time because there are chances of insecticidal residue in the leaves. Moreover, it is also not economical to use insecticide at sowing time because aphid arrive about 5-6 weeks after sowing. Uses of insecticide at sowing time may be recommended for the control of Painted bug, *Bagrada hilaris* Burm, mustard saw fly, *Athalia lugen* Klug and flea beetle, *Pyllotreta cruciferae* Goeze. Which damage the crop at early stages of plant growth. For the control of mustard aphid it was found that when crop was not sprayed upto 65 days it does not affect the yield significantly if the crop was exposed to aphid attack more than 75 days after sowing, there was economic loss (Prasad & Phadke, 1983). Therefore the crop should be sprayed between 65 & 75 days after sowing.

For controlling the aphid with chemicals contact and systemic insecticides belonging to organophosphate group like chlorpyriphos, oxydemeton methyl, manazon, phosphamidon, dimethoate etc. should be used. Among synthetic Pyrethroids, cyhalothrin and Fenvalarate were found more effective against the aphid (Prasad 1983).

There are many parasitoides and predators which reduce number of aphids in the field. These may be used to control aphid

The sowing of rapeseed and mustard crop should be completed upto 15th Oct. because delayed sowing attract more aphid infestation.

Another method of crop protection from aphid's attack is to sow resistant varieties. For obtaining resistant varieties large number of cultivars should be screened. It is difficult to count the number of aphids per shoot or per plant on large number of varieties. The plants should be categorised into 6 grades (0-5) depending upon aphid's population and its symptoms on the plant. For working out the mean aphid infestation index, the number of plants in each grade is multiplied by their respective grade and then the figure is divided by number of plants observed. The rapeseed and mustard were categorised into six grades in the following way:-

Grade No.	Description
0	Plant is completely free from aphid.
1	Plants having 1-15 aphids per inflorescence shoot. There is no symptom of aphid damage.
2	Plants having 16-100 aphids per inflorescence. Shoots and pods start curling due to aphids attack.

3	Plants having more than 100 aphids per shoot. Most of the branches are infested by aphids. Leaves start drying. Pods are curled.
4.	Each & every branch of the plant is fully covered with aphids and some of the branches starts drying.
5.	Plant is completely dry immaturely due to aphids infestation.

Example

Grade No.	No. of Plants in the grade	Weightage
0	2	0 × 2 = 0
1	1	1 × 1 = 1
2	4	2 × 4 = 8
3	2	3 × 2 = 6
4	1	4 × 1 = 4
5	0	5 × 0 = 0
Total	10	19

Mean Aphid Infestation Index (MAII = 19/10 = 1.9

References

Prasad, S.K. (1978). Chemical control of mustard aphid, Lipaphis erysimi Kalt. *Indian J. Ent.* **40**(3) : 328-332.

Prasad, S.K. (1978). Control of mustard aphid, *Lipap his erysimi* by granular systemic insecticides. *Indian J. Ent.* **41**(1) : 39-42.

Prasad, S.K. and Phadke, K.G. (1982). Yield infestation relationship and Economic injury level of mustard aphid. *Lipaphis erysimi* (Kaltenback). *Jour. Ent. Res.* **6**(2) : 117-122.

Prasad, S.K. and Phadke, K.G. (1983). Catches of alate *Lipaphis erysimi* (Kaltenback) on adhesive traps and relationship with aphids on yellow sarson (*B. compestris* Linn) and weather variables. *Jour. Ent. Res.* **7**(1) : 25-29.

Prasad, S.K., Phadke, K.G. and Mehrotra K.N. (1984). Forecasting of mustard aphid, Lipaphis erysimi (Kaltenback) on rapeseed crop. *Proc. Indian natn. Sci. Acad.* **B50**(2) : 146-153.

Singh, H. and Singh, Zile (1987). Economic Injury Levels of mustard aphid *Lipaphis erysimi* Kalt. in rapeseed (*Brassica campestris* var., brown sarson) (in) crop productivity. Published by *Hindustan Lever Research Foundation*, Bombay pp. 445-450.

Recent Advances in Indian Entomology
Editor: O. P. Lal
APC Publications Pvt. Ltd., New Delhi, 1996, pp. 346-352

Recent Advances in Apiculture

M. NAIM and P. B. SINGH

Apiculture is rearing and management of hive-bees for the benefit of mankind. Honeybees being predominant in this group, their rearing has become almost synonymous to the apiculture. This branch of insect science has gained tremendous signficance due to its role in increasing production of many agricultural crops besides providing valuable products like honey, beeswax, beevenom, pollen, propolis and royaljelly. As demand for the cereals, fruits, vegetables and edible oils are increasing day by day and population of pollinators are depleting due to the lack of sufficient forest cover and extensive application of pesticides, need for augmentation of apiculture in plain as well as hilly areas is being increasingly felt. Beekeeping has provided at least part-time employment to 2.28 lakh persons earning about 24.50 crore rupees in 1991-92. India produces about 10,000 tonnes of honey per year which is mainly utilized for medicinal purposes. There are at least four indigenous honeybees in our country viz., medium size bee, *Apis cerana* F., rock bee. *Apis dorsata* F., little bee, *Apis florea* F. and dammer bee, *Trigona irridipennis* apart from the introduced European honeybee, *Apis mellifera* L. which is being commercially exploited in Punjab, Himachal Pradesh, Haryana and some parts of Jammu and Kashmir, Bihar and Uttar Pradesh. Extensive studies have been made on European honeybee, *Apis mellifera* L. whereas little attention has been made to the indigenous species *A. cerana* F. Besides basic similarities at the generic level, there are significant differences in the morphology, anatomy, physiology, biochemistry and behaviour of these species. In the present review, mostly the works on Indian honeybee have been emphasised.

Physiology and biochemistry

Considerable progress has been made in physiology and biochemistry of the honeybee, *A. mellifera* L. Recently Robinson (1987) has demonstrated that quantity of juvenile hormone (JH) secretion in workers plays important role in determining the division of labour. He could induce premature foraging with the application of 50-250 ug JH analogue methoprene. Naik *et al.* (1988) analysed Nasonov gland pheromone of the Indian honeybee, *Apis cerana* F. and found to contain neval (Z)-citral - I as the main component. Whereas, Winston and Slessor (1993) succeeded in further identification and synthesis of one of the queen's mandibular gland pheromone complexes.

Digestive physiology

Banerjee and Saxena (1983) and Mishra (1983) reported presence of digestive enzymes, amylase, maltase, melizitase, sucrose trehalase, invertase and lactose in various parts of elementary canal, pharygea glands and post cerebral glands of *Apis cerana* F.

The digestive enzymes are released mostly as endoenzymes. These properties showed that Indian bees are well-equipped for digestion of starch, sugar, proteins and fats. However, the properties of various enzymes are still to be explored.

Circulatory system

The anatomy and histology of the circulatory system of *A. cerana* F. has been extensively dealt by Bhardwaj and Banerjee (1969). The composition of haemolymph, however, has attracted the attention of a number of workers. The protein concentration in the haemolymph was found to be 2.7 g per 100 ml in nurse bee, 2.34 g in drones and 1.3 to 1.8 g in foragers (Naim *et al.*, 1985, and 1988). The carbohydrate concentration in the worker was estimated as 2.1 g/100 ml, in drones 8.6 g/100 ml. The cations sodium concentration was 7.5 m eq/1 and that of potassium 46.94 m eq/1. The ratio of these cations was 0.16 (Naim, 1981). It has been shown that these characteristics change in a queenless colony and the protein contents of haemolymph of foragers increase from 1.8 g to 2.3 g/100 ml and that of drones reduced by half. The cation is also reduced but less in the queenless (1.34) colony. There is indication that there may be an androgenic low molecular weight protein present in the drones. Hagmeester (1973) immunologically compared the haemolymph protein of the four species of *Apis* and revealed interesting interspecific differences in the haemolymph protein spectra and antigenic properties. Immunologically *A. florea* F. and *A. dorsata* F. were found to be very different from *A. cerana* F. and *A. mellifera* L. Trehalose is known to be major circulating carbohydrate in haemolymph and supposed to be under harmonal control. This has been confirmed in the case of *A. mellifera* L.

Flight energetics

The Indian honeybee is an efficient forager. It has been estimated that it carries weight load ranging from 2.4 to 25 mg (Cherian *et al.*, 1946, Dhaliwal, 1970; Naim & Bhist, 1979). The load to body weight ratio is thus high in *A. cerana* F. when compared to *Apis mellifera* L. (Dhaliwal, 1970). The volume of the nectar carried solely depends upon the capacity of the stomach and is not related to the concentration of sugar in the nectar (Singh, 1971). On an average a worker bee carries 3 to 10 microlitres of nectar during one flight (Bisht, 1974). It is of interest to note that in Indian hive bee there is some sort of organization in the colony. The bees that go for pollen collection are comparatively of heavy body weight and light weight bees go for nectar collection. The flight ranges of the bees range from 900 m to 1000 m. An indirect evidence on the basis of glycogen reserve in the body of the honeybee and the rate of utilization of glycogen per minute of flight is suggested that the pollen forager had a flight time of about 25 min and flight range of 600 m to 1000 m (Bisht 1974). Although the Indian hive bee is known to utilize glycogen during flights the biochemical pathways of its break down have not been demonstrated. The carbohydrate as well as protein and fat metabolism awaits detail study in case of Indian hive bee.

Honeybee genetics and breeding

Attempts have been made to hybridize Indian and Italian bees without much success (Veeresh and Shylesha, 1990). For improvement of honeybee races or breeding of hybrids instrumental insemination of honeybees is necessary and with the help of this technique many hybrid bees have been developed in Italy and U.S.A. (Nazzi, 1992). In USA Starline and Midnite hybrids remain a standard by which many other stocks are often compared.

Foraging behaviour

The foraging behaviour of this species is dependent upon the prevailing ecological conditions. The foraging activity for pollen and nectar is significantly correlated with day temperature and humidity. The threshold temperature for foraging of *A. cerana* F. was found to be 14°C and that of *Apis florea* F. and *A. dorsata* around 17°C. The optimum temperature for foraging in the former was 20°C and in the latter two species it was 24°C during winter months (Bisht *et al.*, 1983). The defensive behaviour of Indian hive bee has three dominant characters : absconding, shimmering and aggressiveness. Absconding is a primitive trait having adaptive value as it saves the colony during dearth (Bisht and Pant, 1968). Shimmering is a trait by which the bees are able to ward off their predators. They produce some sound which ward off predators and no such sounds are produced by European bees.

The Indian bees can combat hornets and ectoparasite far more efficiently than that of *A. mellifera* L. Comparative studies on aggressiveness of the four species revealed that *A. mellifera* is the most aggressive followed by *A dorsata, Apis cerana* and *Apis florea* (Dhaliwal & Atwal, 1969).

Hive regulation

The honeybees are able to regulate the hive temperature. The Indian honeybee maintained the hive temperature between 37-38°C when the temperature varied from 2 to 14°C. *Apis cerana* F. neutralizes 70-80 per cent fluctuation in the environmental temperature. Compared to heating, the Indian hive bee is less efficient in cooling. It has been estimated that at 27-34°C temperature the honey bee spends the least amount of energy to regulate hive temperature. It is at this temperature, excess energy is available for food collection and brood rearing (Bisht *et al.*, 1979). That is why we get surplus honey in spring season when temperature ranges from 27°C to 34°C.

Queen pheromones

Presence of queen pheromone, 9-oxo-dec-trans-2-enoic from mandibular gland has been demonstrated in *Apis cerana* F., *Apis mellifera* L and *Apis dorsata* F. This substance attracts drones for mating. The other striking property of the queen pheromane is to inhibit ovary development in workers and formation of retinue around queen bee. This property of inhibiting ovary development has sufficient potential in insect pest management. Recent thinking on line has started and may prove a great success in future.

Honeybee diseases

Honeybees are afflicted to many diseases, these may be acarine, protozoan, fungal or viral. Acarine diseases are caused by ectoparasitic mites viz., *Varroa jacobsoni* Qudemans, *Pymotes herfsi, Tropilaelaps clareae* and endoparasitic mite *Acaraspis woodie.* Rennie. Among these, *V. jacobsoni* is a serious problem in many parts of the world but in India, its magnitude of damage is often limited. It has been reported from Delhi, Bangalore, Uttar Pradesh and Kashmir, *Tropilaelaps clareae* is another external mite which was observed to kill up to 50% late larval and pupal stages. It was found that fumigation with 5 ml of 85% formic acid per day for 13 days reduced *T. clareae* population without harming the brood or adult bees. Generally, sulphur dusting is resorted in non-honey flow period at monthly intervals to check external mites. The endoparasitic mite, *A. woodie.* has been reported from Himachal Pradesh and Uttar Pradesh and it can be controlled with 5-7 fumigations with Folbex strips at weekly intervals (Mishra and Sihag, 1987).

Among bacterial diseases, American foul brood, *Streptococcus pluton* White has been reported from Jeolikote. However, it is of negligible economic importance in India. Recently, Thai sac brood a viral disease caused major damage to honey bees in Kerala and Bihar which has no effective control measure. Hence more emphasis is needed to manage this disease. Presently the disease is managed to some extent by taking general hygienic measures and destruction/burial of infected colonies under the deep soil.

Pollination

Honeybees are efficient pollinators to many oilseed, vegetable, cereal and horticultural crops. Deodikar and Suryanarayana (1972) reviewed this aspect and mentioned that honey bees increase seed yield of fodder and grain legumes from 1.02 to 332.5, oilseeds from 1.01 to 125.05, orchard crops 1.06 to 103.5 and cotton from 1.02 to 1.49 times more. Honeybees increased seed yield of carrot, *Daucus carota* Linn. from 59.95 to 64.42 per cent in intra-varietal trials (Goyal *et al.*, 1989) whereas in the self pollinated mustard variety, *Brassica juncea* L. seed yield increased by 20 per cent (Naim *et al.*, 1994).

Considering substantial increase of crop yield through bee-pollination now hiring of bee colonies for this purpose is very common in many countries.

Toxicity of pesticides to honeybees

In present day agriculture, extensive use of pesticides is a common feature which results killing of honeybees. To save them, it is necessary to apply need based, less toxic pesticides and spray them in late hours to avoid early contacts, use selective formulations and repellents.

Sudden death of honeybees in large numbers near the hive is the first sign of bee poisoning. In case of BHC poisoning, bees became aggressive; DDT and other chlorinated hydrocarbons and organo-phosphorus compounds subdue the activity. Regargitation is a common symptom in organo-phosphorus compounds poisoning. Carbaryl and dieldrin slow down the bee activity as if they have been chilled. Queens may also be affected and show poor and abnormal egg laying and abdominal movements. Chlorinated hydrocarbons give a high initial kill but with a short residual action. BHC, dieldrin, and heptachlor are highly toxic whereas toxapene, endosulfan, DDT and endrin are comparatively safe. Organophosphorus compounds are highly toxic. Systemic insecticides are less toxic. Schradan has low toxicity to bees and can be applied safely at any time. Phorate, demeton and disyston are less hazardous with repellent action. Kasturi Bai and Reddy (1977) studied toxicity of 16 insecticides against *Apis cerana* and found dichlorvos as most toxic followed by methyl parathion, phosphamidon, quinalphos, carbaryl, fenitrothion, monocrotophos, oxydemeton methyl, dimethoate, carbofuran, carbofenothion, malathion, heptachlor, DDT, endrin and BHC. Whereas, Prakash and Kumaraswami (1984) observed phosalone and endosulfan as least toxic, monocrotophos, carbaryl moderately toxic and synthetic pyrethoids like cypermethrin, permethrin and decamethrin were the most toxic to the workers of *A. cernana*. Mishra and Verma (1982) also found endosulfan as least toxic as contact poison whereas fenvalerate as safest stomach poison.

Botanicals like nicotine, pyrethrum and acaricides are relatively safe to honeybees. Extracts of *Lantana* and sweet flag have repellent as well as insecticidal action. Herbicides like 2, 4-D have severe contact toxicity. Amitrol amine of 2, 4-D and TCA are less hazardous to bees. Fungicides like benomyl, bordeaux, copper compounds, dinocap, maneozeb, sulphur, thiram, ziram are non-toxic and to some extent organomercurials are toxic to bees. The use of repllents like R-874 (Hydroxy methyl octyl sulphide), Lantana leaf oil, citronella oil, ketones may reduce toxicity to bees or prevent bees picking poison (Veeresh and Shylesha, 1990). Melksham *et al.* (1988) described 54 chemical compounds as repellents. Out of these, he considered 34 as the most repellents. Some of the pesticides, like cypermethrin, 2-4 D, 2 & DB, deltamethrin demeton, DNBP, DNAP, flucythrinate, linuron, PCP, Permethrin, picloram, 2,4, 5-T were enlisted as pesticides repellent to bees. Rieth and Levin (1989) reported two phenylacetate-ester pyrethroids as repellent to the honeybees.

Use of honeybees in pest management

An unusual utilization of honeybee, *Apis mellifera* was studied in spreading the biological control agents by Corliss and Admams (1992). Bees proved effective in spreading beneficial bacteria, *Pseudomonas fluorescens* A 506 and *Erwinia herbicola* 318 to help outcompete bacteria which cause fire blight (*E. amylovora).* Bees also found to successfully spread nuclear polyhedrosis virus which affect *Heliothis virescens* and *Helicoverpa armigera.* The virus contained in a fine powder in a metal tray at the bottom of the hive was picked up by the bees and carried into the field.

Considering the multifarious role of honeybees in crop production and economic value of their products beekeeping is viewed as a rural industry for the integrated rural development of hilly areas in particular and plains in general

References

Banerjee, A. and Saxena, P. (1983) Digestive enzymes in the alimentary canal and associated glands of foragers of *apis cerana indica* F. Proc. Second Int. Conf. Apic. Trop. C1., 337-343. IARI, New Delhi.

Banerjee, S.K. and Bhardwaj, R.K. (1969) Nervous system and muscle innervations of *apis cerana indica. J. Nat. Hist.* **8**, 19-28.

Bisht, D.S. (1974). Flight energetics of Indian honeybee *A. C. indica*. Ph.D. thesis P.G. School, IARI, New Delhi.

Bisht, D.S., Naim, M. and Mehrotra, K.N. (1983). Studies on the role of honeybee in rapeseed production. *Proc. Second Int. Conf. Apic. Trop. C1*., 491-496.

Bisht, D.S. and Pant, N.C. (1968). Studies on brood rearing activity of Indian honeybee *Apis cerana indica* F. under Delhi conditions. *Indian J. Ent.,* **30** : 223-228.

Cherian, M.C., Ramachandran, S. and Mahadevan, V. (1946). Preliminary studies on the pollen carrying capacity of the Indian honeybee, *Apis indica* F. *Indian Bee J.* **8**, 109-111.

Corliss, J. and Admams, S. (1992). A bee line to biocontrol. *Agricultural Research* (Washington) **40** (7) : 10-13.

Deodikar, G.B. and Suryanarayana, M.C. (1972) Crop-yields and bee pollination. *Indian Bee J.,* **34** (3-4) : 53-64.

Dhaliwal, G.S. (1970). Observation on pollen carrying capacity of *Apis indica* F. and *Apis mellifera* L. *Indian J. Ent.,* **32** : 173-174.

Goyal, N.P., Singh, M. and Kandoria, J.L. (1989) role of insect pollination in seed production of carrot, *Daucus carota* L. *Indian Bee J.,* **51** (3) : 89-93.

Hagmeester, A. (1973). Immunologischer Vergleich der haemolymph proteins der 4 rozenten *Apis* Arterm, *mellifera, cerana, florea* and *dorsatae Proc. 7th Congr.* IUSSI, 151-156.

Kasturi Bai, A.R. and Reddy, C.C. 1977. Laboratory toxicity of some insecticides to *Apis cerana indica. J. apic Res.,* **16** (3) : 161-162.

Melksham, K.J., Jacobsen, N. and Rohdes, J. (1988). Compounds which affect the behaviour of the honeybee *Apis mellifera* L : A reveiw. *Bee World,* **69** (3): 104-124.

Mishra, R.C. and Verma, A.K. 1982. Relative toxicity of some insecticides to *Apis cerana indica* F. workers. *Indian Bee J.,* **44** (3) : 69-71.

Mishra, R.C. and Sihag, R.C. (1987). Apicultural research in India. 120 p. Haryana Agricultural University, Hissar

Naik, D.G., Gadre, R.V., Kapadi, A.H., Singh, M.K., Suryañarayana, M.C. and Kshirasagar, K.K. (1988). Nasonov gland pheromone of the Indian honeybee, *Apis cerana* F. *J. Apic. Res.,* **27** (4) : 205-206.

Naim, M. and Bisht, D.S. (1979). Pollen gathering capacity of Indian honeybee, *Apis cerana* F. *Indian J. Ent.,* **41** : 295-296.

Naim M. Rao, R.J. and Phadke K.G. (1985). Effect of queenless colony condition on some of haemolymph constituents of workers and drones of *Apis cerana* F *Indian J. Ent.*, **47**(1) : 127-31.

Naim, M., Bisht, D.S. and Rao, P.J. (1988). Haemolymph protein pattern of drones and queens of Indian honeybee, *Apis cerana. J. Ent. Res.,* **12** (2) : 98-99.

Naim, M., Rao, P.J. and Phadke, K.G. (1985). Effect of queenless colony condition of some of the haemolymph, constituents of workers and drones of *Apis cerana* F. *Indian J. Ent.,* **47** (1) : 27-31.

Naim, M., Prasad, S.K. and Singh, P.B. (1994). Role of Indian honeybee. *Apis cerana* F. in pollination of *Brassica napus* L. (Gobi sarson). *Shashpa* **1**(1): 89-90.

Nazzi, F. (1992). Fluctuation of forewing characters in hybrid honeybees from north eastern Italy. *J. apic. Res.*, **31** (2) : 27-31.

Prakash, R. and Kumaraswami, T. 1984. Toxicity of some insecticides to the Indian bee. *Apis cerana* F. *Indian Bee J.*, **46** : 15-17.

Rieth, J.P. and Levin, M.D. (1989). Repellency of two phenylacetate ester pyrethroids to the honeybee. *J. apic. Res.*, **28** (3) : 175-179.

Robinson, G.E. (1987). Regulation of honeybee age polyethism by jevenile hormone. *Behavioral Ecology and Sociobiology*, **120** : 329-338.

Singh, G. (1971). Sucrose solution inhibiting capacity of *Apis indica* F. in relation to various sucrose solution concentration and body weight. *Indian Bee J.*, **33** : 51-54.

Singh, Zile (1972). Propolis collection and its use. *Indian Bee J.*, **34** (1 & 2) : 11-19.

Veeresh, G.K. and Shylesha, A.N. (1990). Honeybees, an over view of Indian work In: Social insects an Indian perspective (Veeresh, G.K. Ed.) IUSSI. Bangalore.

Winston, M.L. and Slessor, K.N. (1993). Applications of queen honeybee mandibular pheromone for beekeeping and crop pollination. *Bee World*. **74** (3) : 11-128.

Recent Advances in Indian Entomology
Editor: O. P. Lal
APC Publications Pvt. Ltd., New Delhi, 1996, pp. 353-363

Recent Trends in Management of Problematic Non-Insects: Vertebrates

R.K. BHATNAGAR

Importance of Vertebrates in any natural ecosystem needs no introduction. These comprise important integrals or our biosphere. Most vertebrates have higher intelligence, varied degree of adaptability and movements. Their morphological and physiological features are advanced than lower groups in the animal kingdom. All these features coupled with behavioural features like communication provide them capability to subsist in alternate environment or habitat and even shift to other dietary items than taken in their natural environment/habitats. Therefore some vertebrates in certain situations create conflicts with our economic interests. Such vertebrates are referred as 'problematic' and necessitate management. Use of term 'pest' is avoided (excepting in case of rodents being 'notified' as pests) because of varied population status and multiple role of most vertebrates in ecosystem.

Management of problematic vertabrates is considerably different from conventional insect pest management. This is because of varied population status or decreased/ dwindling status due to various factors like overexploitation for various purposes, decrease in their natural habitats ('shrinkage of habitats') with deforestation and intensification of extensive monoculture of man's food plants, the agriculture, etc. Therefore to conserve vertebrate faunal heritage in our biosphere, numerous conventions, laws and even treaties have been formulated by most countries of the world. These comprise a subject in itself and are termed as 'legal aspects'. These necessitate use of management approaches that are in conformity with sound ecological and legal approaches etc. To ellucidate this statement perusal of activities of any mammal, say Blue Bull (*Boslaphus tragocamelus*) is suggested. Blue Bulls, despite their highly depredatory activities due to very high daily intake rate (approx. 25 kg/day/indivi.) and losses due to trampling in agricultural crop fields cannot be legally killed without local legal permission. Thus it is imperative to give a perusal to different activities of various vertebrate groups in problematic situations.

Objectives

To apprise agricultural entomologists, a brief introduction to vertebrates in problematic situations is being presented here. It is of obvious importance to them, as it is, they who have to frequently look into management of such faunal groups in problematic situations, viz., control of birds etc. Further, most agricultural scientists overlook these due to paucity of teaching courses in their study curricula. These comprise objectives of the present article. In the article conventional review of literature under each aspect is not being dealt in order to avoid lengthy coverage; instead selected references are being given at end.

Vertebrates in problematic situations

Unlike other faunal groups, most vertebrates have high economic importance in various perspectives ranging from roles in natural eco.-food chains to their utilisations in various forms for economic the scope of present article. All these points; besides population status and behaviour etc. warrant considerations in scientific management. These are also necessary as provide usages in development of management approaches in target situations. Locally, situations may considerably vary. So generalisation may not be always appropriate. For example, Common Dove, *Streptopelia decaocto*, has lately shown westward movements upto Holland in last ten years. It is now being reported spreading eastwards too. The species is best at sowing to sprouting stage in most crops in northern India; but is now being reported depredatory to Sunflower heads. The well-known insectivorous bird common Myna, *Acridotheres* spp. is now showing shift in dietary habits and is depredatory to some soft fruits. Therefore a perusal to depredatory activities is warranted on different groups of vertebrates.

Fishes

Many fishes are extensively exploited for food purposes as in inland fisheries development programmes in ponds etc. In these, some carnivorous fishes pose problems in particular in pond culture programmes on introduced herbivorous fishes which are preyed. Such carnivorous fishes warrant removal and are often referred as weed or as pest fishes. These deplinish food resources or prey of fish species under culture. Removal is done before introduction of species to be cultured. In large ponds removal, if not possible, it would be better to resort culture of large carnivorous species like Wallago and Bagarius spp. In other ponds, adults and finger lings of predatory fishes can be removed with netting or use of 50 ppm of phosphamidon. For this even Pongamia cake has been tried with advantage that immobilised fishes can also be used for other purposes. From fish nursery ponds sieving or netting and netting as exclusion devices over inlets are used.

With present day intensification of agriculture many chemicals are getting into water bodies. Pollutants from other sources like industrial effluents into water bodies are also affecting fish life. Vertebrate toxicologists and environmentalists are initiating extensive studies on various aspects. It is obviously important for agricultural scientists to monitor drainage of chemicals from fields to water bodies to reduce adverse effects on aquatic life. These offer interesting future lines of researches.

Amphibians

Amphibians like frogs and toads have high importance to farmers as natural biological control agents of surface dwelling insect pests; besides other usages like frog legs (Rana tigrina) and in various biological studies. Immature stages of frogs and toads prey on dietary items (phyto and zoo plankton) of fish fries adults of these amphibians often prey on fish fries and pose serious localised problems in fish nurseries. These can be controlled with sieving or netting and with use of nylon netting as exclusion devise.

Current researches are on culture of edible Ranids the common Bull Frog, Rana tigrina population of which has enormously decreased due to various factors connected with pesticide usages and pollutants. Exploitation of the species has also been severe for basic biological teaching programmes. Maintenance of natural population and its conservation present challenging problems as often clash with Insect Vector control for which water pools are liberally sprayed with insecticides in pools shared by both, mosquito larvae as well amphibians for breeding. Rana tigrina has very high export value-frog legs.

Reptilians

Most reptilians are predatory or carnivorous. However, some reptilians are herbivorous also and at times pose localised problems to plants. Therefore present account is being restricted to Rock Lizards (Agamids) of genus *Agama daudin*. These are distributed all along Himalayan foot hills and hills upto height of about 14000'. These have high adaptability to switch over to vegetarian diet when insect food is scarce. In the process these pose problems with nipping of apical and flower buds of cultured floriculture plants. At forest edges, *Agama himalayana*, at times pose serious problems to forest nursery, viz., to young seedling of Hazel Nut. These lizards are diurnal and live in rock crevices which can be closed with mud, when after sun set, these retrieve to crevices. During day time these can be sprayed with cold water which induces lizards to retrieve into crevices.

Current researches are on conservation of various reptilian groups, the Crocodiles, Chelonians, Snakes and many lacertilians. Most have high importance in natural eco-system. However, some species have drawn much attention and with concerted efforts have increased. It is still not known as to their eventual impact on prey population, viz., with increased introduced population of Crocodiles on riverain commercial fisheries. Thus studies on predator prey relationship provide interesting lines of researches, viz., Biotic theory of Pradhan on role of natural vertebrate predators on upsurges of Desert Locust etc. Similar to these are the results of extensive studies of Whitakers on role of Snakes on Natural rodent population-controlling effects. In fact general awareness is largely lacking in various perspectives to the extent that teaching courses on economic Zoology have been withdrawn from many teaching programmes. These need development as limit teaching and appreciation of ecology with exclusion of higher faunal components.

Birds

In animal kingdom, birds occupy important entity with their unique features, namely intelligence, adaptability flight, diet and various morpho. Physiological characters. These explain world-wide distribution of birds and their capacity to subsist on variety of dietary items. Economic importances thus are too numerous to enlist here. Therefore discussions are being limited to problematic relationships in agriculture-depredations in crops. The subject is now being paid high attention all over world for reducing crop losses due to birds. In general most birds that depredate crops have approx. 10% of their body weight as daily intake. This is high and in totality account high degree of losses.

Extent of losses in crops due to birds vary due to numerous factors like food and mode of intake (feeding pattern and behaviour), species composition, density stage of the crop and physiological state of the involved species etc.; besides movement or distribution of the involved species etc. In general with resident species like ploceids, phasianids and psittacids, depredations are localised and may vary from moderate to severe near roosting sites. With migratory species, depredations would be along their migratory passage. Since migratory species invariably remain in large congregations, extent of losses would be severe. Also, most migratory species have higher metabolic stress, therefore intake rates are higher. Some granivorous species at pre-breeding and to feed nestlings prefer insectivorous diets, losses to crops then would be reduced. Thus extent of losses considerably vary and (localised) generalisations are not appropriate. Studies on all these aspects are important, but are generally overlooked due to several limitations, in particular due to extensive field work which is lacking in most cases for want of 'bird tracking facilities'. For those having vehicles these aspects offer interesting research scopes.

Studies on relationships of birds as insect predator also offer interesting future lines of researches. These too involve extensive field studies as most species cannot be killed for analysis of gut contents to know dietary items. However indirect methods with visual observations and prey density with eventual assessment of impact of predator density on insect pest infestation (reducing effect) offer interesting lines of researches to agricultural entomologists. Unfortunately the aspect is much referred, but hardly any one takes up these studies. In these studies it is essential to know behaviour of bird and its status in the study area. In this connection example of Rosy Pastor *Sturnus roseus* is most appropriate. The species migrates into India from north west from July onwards. On passage before entering India it avidly predates hoppers of Desert Locust; but on further inward migratory passage it finds maturing millets and switches onto depredations of crops. Therefore it is imperative to give perusal to bird problems in agriculture.

Bird Problems in Agriculture are largely due to depredatory activities, i.e., feeding activities of birds that are herbivorous or are granivorous. In India, 181 sub-species, 154 species belonging to 72 genera referable to 27 families and 9 Orders are depredatory to agricultural crops. Discussions on all these are beyond the scope of present article. Therefore as an alternative, brief perusal to bird problem in different crops is being dealt with.

Cereals: Wheat, Barley and Rice are depredated from sowing to sprouting. In some areas young seedlings are also pulled out to feed upon soft shoots. Cutting of soft shoots and nipping of leaves are common in areas, migratory water birds or resident water birds are distributed. With grain setting to maturity several seed eating birds like ploceids, psittacids and buntings depredate. In some areas, congregations of migratory birds also depredate stacks in drying yards as is also seen in (drying yards) columbids etc.

Maize and Millets : In maize, losses are affected at sowing to sprouting and later from grain setting to hardening. Cob spathe is torn and soft or near hardened grain is taken. In some areas, this further promotes rodent damage on cobs. In millets, depredations commence from sowing to sprouting and later from grain setting to harvest. Extent of loss is generally high, if scaring is not done. Generally ploceids and psittacids affect loss with feeding. Psittacids cut away earheads from peduncle base and carry away to roost. World's most prolific bird pest, the African Weaver Bird and many ploceids of world have strong preference to millets.

To some extent resistance is afforded with tight spathes in maize. In millets large awans and persistent shed layer of anthers afford some degree of resistance. In sorghum varieties with drooping earheads or with compact grains, awns and with higher tanin content afford resistance to birds. Some of these factors, genetically induced or supplemented as mechanical measures have been used in control of damage by birds in these crops; besides of course with exploitation of bird behaviour-weak links. These provide interesting examples of multi-disciplinary nature of vertebrate control measures.

Pulses : Almost, all World, herbivorous bird species show strong preference to pulses due to nutritive values. Thus birds depredate almost all pulses from sowing to sprouting. Certain larger species like Cranes and Ducks also prefer soft shoots of Gram. Later various species, depredate from podding to drying stage in field, as well in drying yards.

Oil Seeds : Limited depredations are affected by various birds at sowing; but at sprouting and on to soft several species affect depredations. In crops having larger seeds as in Safflower and Sunflower moderate to severe losses are affected by Columbids, Phasiands, many water birds and at times by ploceids to sprouting seeds. Later losses are also affected to soft shoots by many species. At grain setting psittacids are most depredatory. In some areas like Karnataka now even Doves are reportedly depredatory to flower heads of Sunflower, Amongst Oil Seeds, Sunflower is most vulnerable to psittacids all over the world. In fact in India, cultivation of Sunflower has very limited area due to excessive damage by psittacids. However, with recent development by IARI of 'plastic bagging technique' bird damage is almost controlled. Use of this simple technique warrants to be brought to notice of cultivators of Sunflower.

Vegetables : Moderate losses are affected at sowing and in seedlings of certain crops with cutting of soft leave. At pod or at fruiting localised losses are affected, viz., in pea, tomato, ladies finger and other cole crops.

Fruits : Of almost all types are reportedly depredated by birds all over the world. Mostly seed eating and psittacids besides finches are involved in depredations. In India there are several species that damage fruits and also feed upon fruiting or apical buds which later create problems in fruit bearing capacities of plants. On soft fruits, like grapes and berries, even insectivorous birds like Sturnids are now damaging. This is 'shift in diet' and technically comprise mode of development of pestilent habits. There are several factors attributable to these parameters or aspects and offer interesting future lines of researches.

Management of Bird Damage in Crops : Comprise a multi disciplinary subject which is intricate due to varied population status of many of the depredatory species that are even protected by law. Further, many species have multiple roles in the eco-system. Therefore diverse types of approaches are used. Complexities are further aggravated due to flight and intelligence of birds. It is therefore imperative that all these factors are kept in mind for integrating approaches with innovative ideas to counter 'intelligence' of birds. Farmers of Rajasthan unknowingly and with experience have profitably found the use of old discarded audio tapes and strips of used discarded plastic bags in bird scaring around fields. The approach however is useful till 'behavioural habitation' is not brought about.

Now-a-days the subject of applied ornithology is being paid wide attention in many parts of the world and fair amount of literature is available. Therefore to avoid lengthy coverage, presently discussions are being restricted to gaps in our knowledge, current trends and future lines of researches with particular reference to India.

General Scaring : Technically scaring involves introduction of behavioural stress on bird receptors, viz., with sight, hearing and physiology (taste and body tempt. etc.). Thus these approaches are mechanical, as well chemical (to physiologically induce stress for a target response). Behaviourally some mechanical approaches also induce certain physiological responses in birds. Not much is so far known on degree of physio. Stress induced by such mechanical approaches, viz., fright; adverse effects on visual aquity etc.

Scaring for bird damage control has been todate popular all over the world from antiquity. There have been enormous innovative gadgetry for scaring even in developed countries. Now-a-days, development of each simple looking scaring approach has actually been after prolonged studies. An insight into these is therefore warranted.

So far numerous hypotheses have been putforth to explain the phenomena of scaring. Broadly, these encompass 'imprinting' of environment to which reflexes of birds are 'conditioned'. Any 'discrepancy' (introduced innovations) causes 'avoidance', 'withdrawal' and 'scaring' finally culminating in 'conditioning of reflexes' vary in different species and in various environment; besides depending upon 'cue' efficacy would always vary in different regions.

Thus technically, development of any vertebrate scaring method involve complex era of studies. Since use of these offer compliance of various technicalities and legalities for vertebrates, scaring approaches are continuously under study and development all over the world. Developed countries heavily bank upon these and comprise recent trends on applied ornithology for various purposes. For example enormous funds are being spent in developing scaring techniques for averting bird hits in aircrafts. In agriculture these are preferred in terms of economics.

Types of scaring approaches and scarers are numerous. Enlisting all these would involve wide coverage. Broadly these are, (1) Innovative or novelty based visual scarers (like balloon reflective ribbon and whrillers etc.); (2) Natural alarm stimuli inducing (like modles of predators, raptors and use of alarm calls); (3) Auditory (sound); (4) Optic : that induce visual stimuli. Reflective tapes encompass all these and with movement with wind habituation is delayed. Many of these approaches are mechanical in nature; some prefer to categorise these under mechanical measures.

Mechanical Measures : Also comprise approaches for trapping for various purposes and for exclusion devices. Studies on traps have been meagre, perhaps due to legality of trapped individuals. Lot of controversies exist on these aspects. Some opine that trapped individuals can be released elsewhere; while some opine that 'vacant spaces' would be filled with 'immigrants'. These in fact have been topic of debate between two developed nations in context to control of columbids. Still the approach is being practiced by several countries of the world. In India, despite with availability of 'automatic cage traps', aspect remains unutilised for want of legal cover Flood light traps have not yet been field tested for psittacids. Todate available studies are from IARI only in the country. These warrant studies in other regions.

There has been fair amount of development in nylon exclusion devices/measures-the nylon nets. However, programmes on cost reduction have not yet been thought of; besides development (making of nets) at village level. This development would provide cost reduction, as well as, offers enormous potential for employment generation at rural level.

Most recent development in mechanical measures is plastic bagging technique of IARI for controlling bird damage in Sunflower crop. It involves use of punched plastic bags for covering maturing flowerheads and works out economical than scaring with other approaches. This needs provision of coverage under extension programmes for promoting sunflower cultivation as todate is the only method of bird control for sunflower. Similar approach on maize using rubber rings cut from old bicycle tubes over cobs from IARI has also reported useful approach. In maize, wrapping of cobs with leaf is recently popularised technique from Punjab; but has actually been old method in many parts of the world. It was overlooked due to labour involved in coverage with leaf. However, the approach offers high usage for developing countries.

Habital Manipulation comprise another mechanical approach in bird control. It involves two basic features, discouraging from a habitat (with reduction in preferred habitats—food water, shelter etc.) and as an alternate, attracting (or distracting) to a 'target site' or an area. Later one also involves development of preferred habitats and even with 'feeding programmes'. In case of birds, reduction in roosting and nesting sites provide considerable reduction of depredation in adjoining fields. Studies on community structure and nesting preferences provide interesting clues to trees (nesting) not preferred by birds. These indicate utilisations of conventional studies on field ecology for applied usages.

'Shrinkage of habitats' as causes of reduced bird populations has been much talked off subject. However, factual development of habitats based on preferred food, nesting and roosting sites have todate been wanting. These are causatives for conflicts between conservationist and agriculturists in areas close to Sanctuaries, in particular in India. Feeding programmes for mammals on religious sentiments are unknowingly well developed (as for Black Buck by Vishnois); but are unheard in India on Birds to reduce problems of migratory bird congregations to which India is one of the 'treaty signatory'. These comprise important lacunae.

Cultural Practices involve utilisation of cultural techniques for avoidance of depredatory aves at crop vulnerable stage(s). Thus agronomical practices like extra irrigation immediately after sowing and covering seed furrows with raked over earth reduce bird damage. Cultivation of lesser susceptible varieties (maize, sorghum, pearl millets, wheat) have been profitably utilised in many parts of the world. However, these warrant further development and there is a need for studies on varietal susceptibility in different regions. Besides these, most lacking part on the aspect has been avoidance of influx of large migratory bird congregations. In USA, even migratory bird alerts are used. In India, as yet no efforts have been made to avert loss due to such congregations with manipulations of crop maturity or sowing date of crops. Even intercropping techniques have been wanting for averting bird damage. Utility of these parameters are evident from inter-cropping of indigenous lentil variety, 'Bhat' in Himalayas which is quite resistant to bird damage in comparison to soybean, 'Bragg'. Similarly studies on 'trap crops' or 'lure crops' warrant studies, viz., winter maize for reducing bird damage in sunflower etc. These aspects comprise recent trends of studies in applied agricultural ornithology and its present lacunae.

Chemical Approaches : These have been priority study areas all over the world as provide informations on various aspects; besides ease in application and economics. Numerous chemicals as seed dressings and as sprays on grain bearing panicles/ earheads or flower heads have been tried all over the world. Almost all the insecticides have been tested; but none has been reported providing consistent effective results against any species of the world. Many botanicals have been tried with poor results. In fact use of chemicals and botanicals for reducing bird damage in crops reported to be nearly 500 year old effort. Limited success led or drew attention to physiological basis for altering feeding behaviour or related behavioural features for affecting reduction of bird damage; with induction of fright, immobilisation, stress and gustatory repellents etc. So far numerous studies have been made from different parts of the world and some of the approaches are being used with varied success as an integral of integrated approaches or with combination of methods.

Repellents : So far studied are gustatory and tectile. Gustatory repellents are chemicals that on ingestion cause at pre or post-ingestinal disturbance by affecting cerbro cortical region of brain. This leads to development of feeding aversion and which further leads to development or 'conditioning' of 'feeding aversion'. This according to some hypotheses turns into 'imprinting of conditioned aversion'. In other words affected birds learn to avoid so treated material. Many chemicals induce aversions in different degree; but certain carbamates (Methiocarb and Landrin) and others like Lithium Chlorides show prolonged effects. Thus Carbamates-Methiocarb is popular all over the world against many species. Responses, however, have been poorly studied in different species. It is still not well studied as to induction of target response with varied concentration (dose effect). It has been shown in Landrin that dose effect warrants to be studied for all gustatory repellents.

Tactile repellents so far used are waste by products which are jelly like in consistency or have adhesive property. These on touch are 'detested' by pest species as give them a feeling of 'insecurity' due to difficulties in 'take off'. These have been popular for 'urban bird control', but in agricultural aspects have been of limited use.

Immobilising Compounds effect cerebro cortical region and nervous system in various way which culminate or appear as varied behaviour. This has scaring effect to non-affected. Certain chemicals like 4-Amino Pyridine N-Oxide at 3% conc. induce fright and erratic behaviour but is toxic in higher conc. With induction of fright scaring effect is more. The compound is widely used in USA while in Europe Alpha Chloralose in 2% conc. as baits is preferred for immobilising and removal of affected birds. Affected birds not retrieved generally recover after 20-40 minutes. In most of such compounds todate dose effect is precisely not known due to various difficulties in field conditions and often upto 20% mortality takes place. To date safer chemicals for Vertebrate Mgmt are not available. This is despite the fact such chemicals are also widely used in 'immobilising darts' for trapping and movement of large mammals and also for therapeutic usages in human beings. Further cheaper chemicals are still not available.

Chemical reproduction inhibitors have been tried with limited success in bird or other vertebrate management. Numerous chemicals have been reported variously affecting reproduction, viz., with thinning of egg shells, reducing yolk production, nestling mortality and shrinkage of tests etc. Almost all the insecticides and various pollutants show various effects that can be related to effect reproduction. So far generally used chemical in bird management is Azacholesterol or Ornitrol. Major lacunae in use of such chemicals is wanting need of 'selective baiting technique' and to 'bait larger population simultaneously', besides these, studies period of 'bait exposure' etc. Most of such chemicals are carcinogenic too. Thus warrant utmost care. However, studies on effects of various chemicals on birds and other vertebrates offer interesting lines of researches. These aspects are currently of major interests with problems related to environmental pollution etc., and studies can be useful with birds as 'biological indicators'. These need impetus and development.

Lethal Baits have been tried with limited success with use of numerous toxicants of low LD 50 values. Thus any insecticide can be used, subject to availability of 'legal cover' Generally used toxic baits are in conc. of 1 to 3% baits in preferred carrier (feed), viz., with Carbofuron, Methyl or Ethyl Parathion. Use of toxic baits too necessitate 'obtaining of selectivity', other safety precautions; besides other parameters cited in above para. In India such studies on effects of different insecticides on vertebrates have been wanting.

Test procedures on effects of chemicals are quite different for vertebrates and are cumbersome. While results can be useful for various usages, even for human beings no attention is being paid on these. Even teaching courses are not yet available in most Universities and consolidated literature is hard to find in India. (Two chapters by Bhatnagar in Proceedings of Summer Instt., Entomology Divi. IARI-1991-Edtd by Prakash Swaroop - provide complete information).

Mammals

In relation to agriculture only herbivorous mammals like rodents, hare or rabbits, deers and large mammals like Blue Bulls and Elephants at times pose serious problems. On these (with exception to notified rodents) and also on bats squirrels, Bears Wildlife Act provides complete protection. Thus presently permissible approaches are non-lethal, i.e., broadly scaring and exclusion. Commonly employed approaches for scaring are 'watch and ward', 'firing of crackers' or use of firing of 'blank cartridges' (without lead pellets). To limited degree, use of reflective ribbon provides scaring of small deers. Many have suggested 'translocation of large mammals'. This however requires cumbersome process with wildlife permission under the act. Essential for this approach is development of habitat with sustainable food resource for wild mammals in the habitat, these are planned to be 'released'. Attracting to a desired areas with 'feeding points/programmes' have not yet been tried in India. It is done with cultivation of cheaper 'lure crops'. Bishnoi tribe in Rajasthan conserve Black Bucks in fair large herds and 'feeding programmes' are profitably used. Exclusion measures for mammals are also expensive and cumbersome, like fencing, 'stockade fencing with wooden stakes', electric fencing and trenching (for wild Elephants). All these are expensive. However, electric fencing for 'conditioning of reflexes with imprinting' has not yet been studied and remains a major lacunae in the approach. Utilisation of this behavioural feature is very well seen in management of even 'man eating' tigers in Sunderbans. Therefore obvious needs are subjecting every approach to deep scientific study. In most depredatory mammals, behavioural studies in particular to foraging have remained unstudied. Study of these offer many clues (weak links) for exploitation in management. Studies have revealed that a Blue Bull, needs minimum two metres of space for a 'stride' to jump over any type of fence. Fencing if so erected would check intrusion of Blue Bulls into field. Thus two rows of Cacti fence at inter distance of 1.5 metres would check or effect 'exclusion' of Blue Bulls. This is profitably used with high success in many parts of Gujarat. Even cultivation of non-preferred crops can be an appropriate approach, viz., cultivation of Cummin ('Zeera') in infestation area of Blue Bulls.

For agriculturists, certain Carnivore Mammals, like Jackals, Wolf, Wild Dogs ('Dhole'), Tigers and Panthers, Cats Leopards are at times serious problem to animal husbandry. So far only feasible approaches are scaring and watch ward using trained dogs. This is very effectively used in most developed countries in large ranches. In handling wild carnivores, it is most essential to take care against rabies commonly transmitted by Jackals even to pet dogs and dairy animals and finally to human beings.

Flying Mammals like Bats and Squirrels (in hills) are at times depredatory to horticultural crops. For these netting as 'exclusion' is the only management approach. However, for Bats, use of Electric Bulbs lighted over fruit plants also provides considerable relief from damage.

Rodents are notified pests in most countries of World and extensive literature is available. These are therefore not being dealt with in this article. However, on these too, there are several aspects that todate remain unstudied. Behavioural studies and preferences, population, movement, durable baits, enviromentally safe approaches, traps not causing shyness in 'self operating mode' for trapping more individuals etc. are some of the aspects on which not much is known. Even use of vertebrate predators has remained poorly studied. For urban rodent control, safe and disposable traps like Japanese Adhesive Traps have still not been developed in India for want of simple formula to increase viscosity of indigenously available by product (like Petrovis). Utilisation of rodents for various biochemical studies is universal. However, todate in India, pathogen free breeding stock is not available for pharmacological studies. This is however a million dollar industry in USA.

Recent Advances in Indian Entomology
Editor: O. P. Lal
APC Publications Pvt. Ltd., New Delhi, 1996, pp. 364-367

Vermiculture

R.K. BHATNAGAR and CHITRA SRIVASTAVA

Earthworms are segmented worms and hence from latin word, 'Vermi' is generally used. Zoologically, there are 4500 species presently reported from whole world. However, many species remain to be identified and many areas have not yet been intensively surveyed. Generally, there are two types of earthworms, (1) Aquatic (Microdili) comprising 280 species and terrestrial (Megadrili) comprising 3320 species in whole world. In totality, these are reported to constitute nearly 80% of the invertebrate biomass. This is after evaluation of density of invertebrate (terrestrial) biomass in many agro-eco-systems of the world. Currently this productivity is exploited with mass culture for poultry, animal feed and for vermi composting etc.

General activities

Life activities of earthworms has intrigued man from antiquity. There are many old records indicating high productivity of plants where 'casts' (faeces) of earthworms were found. Charles Darwin was, however, first biologist to initiate studies in 1881. He gave an estimate that 18 tons of soil per year is upturned by a small population in a natural habitat. He also gave an estimate that earthworms would provide a layer of 2 over entire globe every year. Darwin termed earthworms as "best natural ploughmen". Gradually, earthworm drew wide attention and in last 50 years only, studies have been from all parts of the world. In India, however, studies are of quite recent period.

Importance in soil

Earthworm activities on soil comprise fragmentation, turn over, aeration, dispersion, supplementing nutrition, grazing of microflora, conversion of ammonia and nitrates, etc. etc. with 'casting' (faeces excretion) and continuously feeding on soil surface, numerous processes are brought about that have highly beneficial roles to soil fertility and 'soil health'. Earthworms are therefore of wide usages as biofertilising agents, etc.

Harmful activities

Like all living forms in the biosphere, some of the other activity of earthworm clashes with our economic interests or with maintenance of ecological interests. Earthworms too in some situations present certain conflicts. Noteworthy are with accumulation (within their body) of pesticides and other pollutants in soil. This causes secondary hazards to predators. It is still not well known, if besides foot and mouth disease (most dreaded viral disease of mammals) other diseases are spread by earthworms. However, it is also reported that Salmonella infestation is reduced by 30% in soils inhabited by earthworms. These aspects provide interesting future lines of researches

Economic importances of earthworms are too numerous. Worm culture is therefore being variously exploited ranging from use of earthworms as biological indicators of pollution to production of vermi-compost and management of organic waste etc. Present discussions are therefore being restricted to vermiculture for vermi-composting these necessitate an insight into feeding activities of earthworms.

Feeding and casting

On average, earthworms feed 100-300 mg per gm of body weight and pass out ingested material within 24 hrs. While passing out, digestive processes (enzymatic action, microbe action and mucus) convert available ammonia, nitrates, phosphorus, potassium and magnesium etc. into assimiable forms for plants. These with "casts"are deposited on soil surface. Thus "casting" and feeding activities comprise important functions to soil and therefore to plants. In these processes leaf litter is taken in (as feed) and conversion to humus is done. Thus earthworm feeding activities can also be utilised for management of organic wastes for profitable utilizations.

It is estimated that in nature 1.4 to 77.8 tons per hectare annually is converted by earthworms (leaf litter). Therefore with culture of suitable species (vermi-culture), fertilizer can be produced economically. In nature, therefore, earthworms have high importance in maintenance of soil health.

Vermi-technology

It involves "wormic way" conversion of assimiable organic wastes into useful products. Technically it is biodegradation of waste biomass for sustainability of soil. The process comprises, vermi-culture, vermi-conservation, vermi-composting, management and conservation.

Currently these are large industries (agro-based) in many developed countries of the world. Japan annually imports from USA around 3000 tons of live earthworms for waste management and about 20 times quantity of live worms for Eel culture.

Thus worm or vermi-culture can be a profitable rural industry or a biological fertilizer factory for every farmer in India. According to an estimate 30 to 60% fertilizer needs can be met up with vermiculture waste conversion in India.

Uses of vermiculture

Vermiculture is used for many purposes, some of these are : (1) organic waste mgmt; (2) pollution control, (3) waste land development, and (4) vermi-compost production, besides use of earthworms for animal and poultry feed production etc. For biologist, earthworms can be fully used as biological indicators of various pollutants. For these various species like Eisenia foetida is widely used. For these, it is imperative to give a perusal to characteristics of species used in vermiculture.

Characteristics of species for vermiculture

Vermiculture involves mass rearing of earthworms in quite an unnatural environment with readily available feed, the organic waste. Obviously, it necessitates that species to be cultured have tolerance to stress, survive in high organic media should have high fecundity and should have high intake as well conversion rate.

Above characteristics are seen in species of Eudriliidae, Megascolecidae, Otochaetidae, Oenerodriliidae and Maniligastriidae. Generally *Eisenia foetida* and *Eudrilus eugeniae* are used as these show all characteristics.

Most of earthworm species used in vermiculture mature or attain reproductive stage in 6 weeks. Cocoons numbering approx. 100/pair are laid in 3-6 months, and from each cocoon 3-7 individuals are produced. This reproduction rate is very high and is to be maintained for effective vermiculture. In fact any species can be used for vermiculture, if reproduction rate and some connected features can be managed.

Vermiculture

Vermiculture is simple process of raising earthworms and is generally done for production of vermicompost.

Vermiculture can be done in any container with a hole in bottom for drainage of water. This is filled with feed mixture to provide a 'habitat substrate' which is in three layers each of 3 to 4 cms. high. First layer is washed sand, second layer is washed and dried saw dust and top layer is pesticide free garden soil. Over this water is sprinkled to provide approx. 40% moisture. Over this on a corner a ball of standard feed is put.

Standard feed is made of animal dung (10%), rice polish (1.0%), wheat bran (1.0%), gram bran (1.0%), vegetable waste or peels (1.0%), and egg shell powder (0.1%). These are mixed moistened and left in shade for 2-3 weeks for maturing. Then balls are made and put over substrate in the container.

In place of standard feed, in large scale vermi-composting pits, leaf litter or assimiable organic waste, sewage etc. can be used with 'allowed maturity' of 3-3 weeks for rapid conversion. The leaf litter is collected and broken. It is then moistened and left for 7-15 days to mature in shade.

Cast production and conversion

Immediately commences with introduction of worms. This is done in relation to size of culture trough or pit. Casts are shed over top. Production rate of cast is about 1.5 kg/150-200 gm biomass of earthworms. Conversion of litter into vermicompost depends upon assimiability by worms, etc. Vermicompost production rate is approx. 1 ton/month from 150 kg worm biomass. This is highly efficient and economical process of composting with least management expenditure.

Collection of cast and vermicompost

Vermi cast is hand picked and filled in bags. If possible, cocoons on surface are separated. Unfortunately, 90% zoologists have not seen earthworm cocoons. However, with little care these can be differentiated and separated for further culture. Cocoons in most cases are similar to 'Coriander Seeds'. Vermicast is also a nutritive material for plants.

For separation of vermicompost, converted material is upturned in shade from small containers. From large pits it is digged out and left under shade on clean surface. Generally adult worms move down and compost is separated. Worms are collected and reused in composting or can be sold to others.

Nutrients of vermi cast and compost

Worm converted leaf litter is comparable to any good quality mature leaf litter manures. Nutrients are more or less same, if not supplemented through standard feed. Some opine rather one school opines that this a way even "tailor made" vermicompost can be produced with supplementing or enriching any of the nutrients. This, however, depends upon worm feeding and physiology, etc.

Analysis of casts show all the macro nutrients, N (7.37%), P_2O5 (19.58%), K_2O (15.39%) with pH (3.39).

Conclusion

Vermiculture and vermi-composting is a simple process for production of rich organic manure at any level : for kitchen garden to field level. It offers immense potentials in self employment at rural level and comprises finest example of useful/ utility of Economic Zoology.

Recent Advances in Indian Entomology
Editor: O.P. Lal
APC Publications Pvt. Ltd., New Delhi, 1996, pp. 368-369

Role of Solvents and Emulsifiers in Designing the Toxicity of Insecticide Emulsifiable Concentrate

P. SIRCAR

Majority of the modern synthetic insecticides at a low field dosage cannot be applied in its technical form due to problems encountered in their uniform distribution over a large area. Consequently, it is thus necessary to formulate them. Insecticide formulation could thus be defined as "the act of bringing it into a form to provide an end-use product capable of producing a desired biological response on a target organism". It is now well appreciated that the toxicity of insecticide varies with the type of formulation, its method of preparation, various components and their ratios in relation to the species, development stages of insect and its environment.

Formulation and emulsion

Among the three commonly used formulations, *viz.*, dust, wettable powder and emulsifiable concentrate (E.C.), E.C. are relatively more effective and one of the most widely used among conventional agricultural formulation. An insecticide emulsion is a liquid formulation, which on dilution with water would yield a stable emulsion. It primarily consists of a toxicant, solvent, emulsifier, stabiliser and other adjuvants, if necessary.

Effect on toxicity

Out of these, solvent and emulsifier affect the toxicity of insectides to a great extent. The speed and degree of penetration through the insect cuticle depends mainly on the solvent used. The solvents used are of commercial grade and differ widely in their chemical composition from batch to batch. Evidently, a slight change in their chemical composition and structure results in formulated products differing widely in their physico-chemical and biological properties. Surfactants/emulsifiers are amphiphillic moleules with a polar hydrophillic head group and a hydrophobic tail. These are classified into three major categories depending upon their nature of ionisation : anionic, cationic and non-ionic. The role of solvents and emulsifiers in assisting penetration through insect and plant cuticle, thereby enhancing the apparant bioactivity of insecticides, is of great significance. Hence, it is imperative to evaluate the role of different solvents and emulsifiers in designing the toxicity of insecticide emulsion.

Role of solvent

Toxicity of DDT has been observed to more pronounced in an oil solution than as dry film. Malathion diluted in oil has been found to be more effective against desert locust as compared to concentrated malathion. Effect of five solvents on the toxicity of chlordane emulsion differed significantly against cotton aphid. Radiobiological assay technique recorded different rate of penetration with different solvents. The solvents ethnol, acetone, glycerol used with Y-BHC resulted in relatively quicker knockdown of housefly as compared to other solvents. Solution of DDT in mineral oil were more toxic to DDT resistant flies than solution in acetone. For normal susceptible fly strains, acetone solutions were about four times as toxic as the oil solution. Effect of fifty-eight solvents with Y-BHC was studied against crickets, which varied significantly against. males and females. It is reported that there is an optimum concentration of each solvent, above and below which the toxicity is affected adversely. Against *S. litura,* aromex has been found to be most suitable solvent for malathion and xylene for endosulfan. Solvent mixture has been found to have marked effect on the toxicity of methyl parathion to *S. litura.* Significant role of solvents has been reported for synthetic pyrethroid emulsion.

Role of emulsifier

Emulsifier level has been reported to have significant role in stabilising the toxicity of an emulsion. In agricultural sprays emulsifier should help in quick breaking of emulsion, while in animal dips it should be quite stable. The emulsifier Triton X-100 has been reported to increase the toxicity of OP insecticides against red-spider mites. The emulsifier Tween-80 has been found to be most suitable for lindane and aldrin emulsion against desert locust. Out of eighty seven surfactants, nine in combination with zectran caused greater mortality of cotton boll weevil. Emulsifier mixture caused great variation in the toxicity of methyl parathion emulsion against *S. litura.* Emulsifiers have also been found to play significant role in designing the toxicity of emulsifiable concentrate formulation of synthetic pyrethroid insecticides against important insect pests of agriculture.

Scope

Since development of insecticide is expensive and self-limiting, the development of newer formulation is the most pragmatic approach to improve the bioefficacy of existing ompounds. All the more, it is indispensable and relevant to the economy of a developing country.

Recent Advances in Indian Entomology
Editor: O.P. Lal
APC Publications Pvt. Ltd., New Delhi, 1996, pp. 370-373

Role of Synergists in Insect Pest Management

SWARAN DHINGRA

Pest managament has been one of the most important developments in pest control since the advent of modern organic pesticides. Pest management continues to be an important step in our ability to deal with pests in a manner compatible both with the environment and with production economics. Compatability with the environment can be achieved by avoiding three 'R's. i.e., population resistance, resurgence and replacement.

Resistance of populations to pest management *tactics* is the most important of the three 'R's. It is a genetic-evolutionary phenomenon most frequently associated with the use of pesticides. To avoid pesticide hazards in environment the emphasis has started to move towards, the use of minimum quantities of pesticides, selective and biodegradable as well as insecticide mixtures. The idea of using mixtures is to slow the development of resistance. The insects developing resistance to one compound would succumb to another in the mixture. Success with this approach is enhanced when the insecticides mixed are synergistic and have similar decay rates and short environmental stability. Also, mixtures should be applied early before resistance occurs to any one fo the components. It has been shown that chemicals with contrasting modes of action or detoxification pathways show some delay in the onset of resistance. Instead of applying mixture of insecticides, synergists can also be used to get desired results. Synergist enables a fixed amount of certain toxicants to be more poisonous than they would be if applied alone under otherwise identical conditions. Synergists act at the biochemical level, often by inhibiting the microsomal oxidases that inactivate many, but not all, toxic substances. If the enzymes are inactivated, the toxic substances persist longer at the site of action and can therefore do greater damage. So synergists, the protector chemicals or more appropriately the detoxification inhibitors of toxicants in the biological system helping in slowing the development of resistance have attracted global attention during the past four decades. Build-up of resistance to carbaryl in houseflies was largely prevented by using a 1 : 5 mixture of carbaryl and MDP synergist PBO. Both piperonyl butoxide and S, S, S- tributyl phosphorotrithioate synergised the action of fenvalerate in resistant strains of houseflies. The presence of the synergist had little effect on the LC_{50} of the pyrethroid to susceptible flies (Faruham, 1973). Thus, when resistance is exclusively caused by metabolic factors, synergists such as piperonyl butoxide, sesamex and tributyl phosphorotrithioate have in some instances greatly decreased the rate at which resistance develops in selection pressure experiments carried out in the laboratory. It should not, however, be assumed that the presence of a synergist always delays to a significant degree the development of resistance. Chen and Sun (1986) in the larvae of the diamond back moth, *Plutella xylostella*, observed a high level of resistance within five generations to fenvalerate applied alone but the rate of development of resistance was somewhat less to a mixture of fenvalerate and piperonyl butoxide. Strong cross-resistance was observed to three other pyrethroids (Permethrin, cypermethrin and deltamethrin) but with only a slight increase of resistance to some OP insecticide. Mehrotra (1989) while suggesting the different strategies to contain pesticide resistance as well as to protect the environment from the adverse effects of pesticides recommended the use of synergist which will enhance the toxicity of a given pesticide by inhbiting the detoxification mehanism. Dhingra *et al.* (1979) reported that even against naturally tolerant adults of the sweet potato weevil *Cylas formicarius* Fab. to DDT, piperonyl butoxide among methylene dioxy-phenyl compounds, synergised pyrethrum. Further, the synergistic effect was evident when dillapiole Karanjin, HYOXYD-45 were separately mixed with p. p. DDT; piperonyl butoxide with lindane; Karanj oil with carbaryl; and karanjin with dichlorvos (Dhingra & Sarup 1979). Out of 48 combinations tested against lindane resistant strain of *Tribolium castaneum,* the well-known pyrethrum synergist piperonyl butoxide exhibited synergism with pyrethrum, high-lighting the possibility of using piperonyl butoxide with pyrethrin in preventing the development of resistance to lindane in *T. castaneum*. Although the same combination showed additive action against the normal strain of *T. castaneum* (Dhingra & Sarup 1981).

Dillapiole, a waste material in the essential oil of *Anethum* Sowa Roxb. and dihydrodillapiole were the first successful synergists reported in India (Gulati and Parmar, 1969 a b). These products revealed either additive or synergism with different groups of insecticides except malathion at various proportions against different test insects and were comparable in activity with piperonyl butoxide, an internationally recognised standard synergist. Another material of considerable interest is the non-edible oilseed plant Karanja (*Pongamia glabra* Vent.). Its constituents alongwith their uses have been reviewed by Parmar *et al.* (1976). They in 1975 reported that this oil considerably synergised the activity of lindane, heptachlor and endosulfan at both 1 : 1 and 1 : 5 (insecticide : oil) levels against *Tribolium castaneum* DDT, BHC, methoxy chlor, aldrin, dieldrin and toxaphene were either antaganised or poorly synergised. Dhingra and Sarup (1981) observed additive effect of Karanj oil with pyrethrum, DDT, lindane, diazinon, dichlorovos and phoxim against *Tribolium castaneum* (lindane resistant strain), whereas, lindane and diazinon were antagonised. However, it matched with PBO in enhancing the rate of knockdown of *Musca domestica* Linn. and both knockdown and mortality of *Periplaneta americana* Linn. (Parmar 1977).

Karanjin, one of the major furanoflavonoids present in the oil was evaluated by Parmar and Gulati (1969) as a pyrethrum synergist and was found to be a significantly inferior to PBO against *Musca domestica* Linn. Dhingra *et al.* (1979 a & b) observed that Karanjin antagonised pyrethrum and lindane against *Cylas formicarius* but it considerably enhanced the activity of DDT, dichlorvos and carbaryl. Attri *et al* (1973) also concluded that both Karanj oil and Karanjin antagonised pyrethrin against *Tribolium castaneum* Herbst. but Pongapin. a methylene dioxy analogue of Karanjin. was mildly synergistic.

Singh et al (1976) screened about 40 extractives from 23 plants as pyrethrum synergists and reported that besides *Anethum sowa*. extractives from *commiphora mukul*. *cymbopogon martini* Var. Motia. *Quisqualis indica* and *Thevetia nerifolia* exhibited synergism against *Tribolium casteneum* Herbst. Enssential oils from *Blumea lacera* lacera D.C. and *Blumea malcomii* Hook have also been reported to possess synergistic property with pyrethrins against female houseflies (Gupta *et al*.. 1977). Datta et al. (1978) reported that 'reetha' shell power synergised DDT against *Anopheles stephensi, Aedes aegypti, culex fatigans, Periplaneta americana, Tribolium castaneum* and *Rhizopertha dominica,* maximum and minimum activity being against *A. aegypti and T. castaneum,* respectively.

The potential advantages of using non-toxic synergists thus, include, lower cost, decreased environmental impact, and improved control of resistant insects. But, what concerns most is the biological effect of these synergised pesticides, which, at various stages of environmental alteration, can come in contact with many forms of biological systems. In literature not much information is available for the estimation of residues of mixture of non toxic chemicals with insecticides. Handa (1979) developed a rapid colorimetric method for estimating dillapiole residues in wheat grain. The method involved reacting dillapiole with sulphuric acid to liberate formaldehyde which in turn reacts with Chromotropic acid to yield a pink coloured complex having an absorption maxima at 575 nm. The method can detect 1-15 g dillapiole per ml. Excessive residues of dillapiole and dihydrodillapide in combination with DDVP were detected on wheat grain but simple treatments like washing, chapati making etc. considerably reduced the same (Awasthi, 1974). Therefore residues tolerance of stable synergists require careful evaluation.

A real benefit of using synergists can only be derived if the synergised toxicants can be used on agricultural crops. Of particular significance is the toxicological information on the synergised toxicants particularly in relation to the behaviour of such formulations in the mammalian bio-system. It is suggested that trials are justified as mixtures seem more promising than other management strategies so far proposed. Mixtures seem unlikely to give appreciably faster build-up of resistance than the sequential use of single compound and at best, they could greatly delay it.

References

Attri, B.S., Gupta, S.C., Mukerjee, S.K., Parmar, B.S. and Singh R.P. 1973. Evaluation of Pangapin as a pyrethrum synergist. *Pyrethrum Post* **12**: 87.

Awasthi, M.D. 1974. Fate of DDVP and Synergised DDVP in/on stored grain wheat Ph.d. thesis. Post Graduate school, I.A.R.I. New Delhi.

Datta, M.K., Jat, N.R., Deshmukh, S.N. and Bhatnagar, A.K. 1978. Reetha Shell powder (RSP) in pesticide formulations. I. RSP in DDT water dispersible powder. *Indian J. Ent.* **40** : 142.

Dhingra, S., Prakash Sarup and Agarwal, K.N. 1979. Synergistic activity of some non-toxic chemicals in mixed formulations with pyrethrum against the adults of *Cylas formicarius*. Fab. *J. Ent. Res.*, **3**(1) : 96-103.

Dhingra, S. and Prakash Sarup. 1979. Evaluation of some non-toxic chemicals as synergists for different insecticides in mixed formulation against *Cylas formicarius* Fab. *J. Ent. Res.*, **3** (2) : 131-141.

Dhingra and Prakash sarup 1981. Effect of some non-toxic chemicals in mixed formulations with different insecticides against the lindane-resistant strain of *Tribolium castaneum* (Herbst) (Coleoptera : Tenebrionidae) *J. Ent. Res.*, **5** (1) : 1-11.

Gulati, K.C. and Parmar, B.S. 1969a. Improvements in or relating to methylene dioxyphenyl derivatives. *Indian Patent No.* 119536

Gulati, K.C. and Parmar B.S. 1969b. Improments in or relating to dillapiole and dillapiole bearing components for use as synergists for insecticides. *Indian Patent No.* 128129.

Gupta, S.C., Khanolkar, U.M., Koul, O. and Saxena, B.P. 1977. Pyrethrin Synergistic activity by the essential oils of a few *Blumea* species. *Curr. Sc;* **46** : 304.

Handa, S.K. 1979. Colorimetric determination of dillapiole in wheat. *J. Assoc. off. Anal. chem.* **62** : 203.

Mehrotra, K.N. 1989. Pesticide resistance in Insect-Pests Indian scenario. Pesticide Research Journal **1** (2) : 95-103.

Parmar, B.S. 1977. Karanja (*Pongamia glabra*) seed oil as a synergist of pyrethrins. *pyrethrum Post* **14** : 22.

Parmar, B.S. and Gulati, K.C. 1969. Synergists for pyrethrins (ii)..... Karanjin *Indian J. Ent.* **31** : 239.

Parmar, B.S., Attri, B.S., Singh, R.P. and Mukerjee, S.K. 1975. Karanj oil as a synergist for chlorinated insecticides. *Pesticides* **1** : 29.

Parmar, B.S., Sahrawat, K.L. and Mukerjee, S.K. 1976. *Pongamia glabra* constituents and uses. *J. Scient. Ind. Res.*, **35** : 608.

Singh, R.P., Tomar, S.S., Attri, B.S., Parmar, B.S., Maheswari, M.L. and Mukerjee, S.K. 1976. Serarch for pyrethrum synergists from some too tanicals. *Pyrethrum Post* **13** : 91.

Recent Advances in Indian Entomology
Editor: O. P. Lal
APC Publications Pvt. Ltd., New Delhi, 1996, pp. 374-383

Advances in the Study of Insect Pathogens : Bacteria, Fungi, Protozoa and Nematodes

K. L. SRIVASTAVA

Insect pathogens include those microorganisms that cause disease, debility, morbidity and/or mortality in insects. These pathogens belong to various groups of microorganisms such as virus, bacteria, fungi, protozoa and also nematodes. Way back in 1835, Lodi observed one disease caused by fungus in silkworm. Pasteur (1870) isolated microsporidian causing pebrine disease in this insect. Metchnikoff (1879) used fungus, *Metarrhizium anisopliae* to control an insect pest, *Anisoplia austriaca.* de'Herelle observed mortality in locust by a bacterium, *Coccobacillus acridiorum.* Information on microorganisms isolated from insects has been given by various authors (Steinhaus, 1946; Steinhaus, 1949; Steinhaus, 1963; Burges and Hussey, 1971; Poinar and Thomas, 1978; Burges, 1981; Srivastava and Ramakrishnan, 1982; Ramakrishnan and Kumar, 1977; Jayaraj, 1987). A list of important genera recorded as pathogenic to insects is given below :

Bacteria : *Bacillus* sp. *B. cereus, B. popillae, B. subtilis, B. thuringiensis, Pseudomonas aeruginosa, Serratia marcescens* and *Streptococcus faecalis.*

Fungi : *Aspergillus flavus, Aschersonia* sp. *Beauveria bassiana, Cephalosporium aphidicola, C.lecanii, Coelomomyces indcus, Entomophthora aphidis, E.coronata, E.exitalis, E.freseni, E.grylli, E.muscae, Fusarium equiseti, Hypocrella olivacae, Macrosporium* sp. *Metarrhizium anisopliae, Paecilomyces farinosus, Spicaria relei.*

Virus : Nuclear polyhedrosis (NPV), Cytoplasmic polyhedrosis virus (CPV), Granulosis virus (GV), and Pox virus.

Protozoa : *Adelina triboli, Bacillidium* sp. *Farinocystis tribolii, Mattesia dispora, Microsporidium* sp. *Tetrahymena* sp.

Nematodes : *Agamermis* sp. *Binema* sp. *Blatticola blatticola, Cameronia biovata, Chitwoodiella chitwoodiella, Grycola grycolla, Hexamermis* sp. *Isobinema flangeiicara, Leidynema appendiculata, Merimis* sp. Mirzaiella asciatica, and *Thelastoma indiana.*

Pathogenic microorganisms, under favourable conditions cause diseases, and bring about changes in the behaviour and body structure of insects. They produce some general symptoms as, loss of appetite, cessation of feeding, diarrhoea, gut paralysis, regurgitation, convulsion, sluggishness, general paralysis and death. Insects attacked by fungi exhibit characterstic symptoms i.e. mummified growth of mycelium and spores on the outer surface of the body of host. The mummified growth may be of different colours. For example, Powdery white - a characterstic of *Beauveria bassiana;* Green, *Metarrhizium anisopliae;* Greenish yellow, *Aspergillus flavus;* and yellowish colour - a characterstic of *Paecilomyces farinosus.* Insect attacked by fungi show following symptoms : Cessation of feeding, weakness, disorientation, moribund and death. Viruses too cause some changes in the body of insects. Infected larvae turn white or pale in colour and become flaccid. Infected individual moves towards tops of plant and depict "hanging posture". Pathogenic protozoans show general symptoms i.e. small size, morphological deformities, lethargy, difficulty in moulting, reduction in feeding, loss of balance, white opaque cuticle and black spot on integument. Nemade infection show such symptoms as loss of appetite, flaccidness, less active, mottled appearance, spotty colouration, and change in colour, as dirty white, light brown, dark brown or ocherous brown and finally death.

Acquireing knowledge of symptomology act as prelude to sorting out infected individual and subjecting them for isolation and identification of pathogens. The name of the important pathogens and their identification characters are mentioned below :

Bacteria

For identification of bacteria, body of insects was cut into pieces, haemolymph was pooled and then with the help of bacterial loop it was transfered in Nutrient Agar petriplates. Colonies of bacteria are picked up, on the basis of colour, size, shape, elevation etc. These are then subjected to various tests such as : Gram's, spore and crystal staining and physiological, biochemical and serological tests.

The characters of different pathogenic bactaria are given as follows :-

(i) Bacillus thuringiensis : Gram positive, aerobic, motile, rod shaped, sporeformer, crystalliferous and cause toxemia in insects. This bacterium forms colonies that are large, irregularily round and white in colour. This is facultative and grows in artificial media such as Nutrient Agar and Nutrient Broth.

(ii) Bacillus popillae : Gram positive, anaerobic, mesophylic, rod shaped, sporeformer and obligate pathogen. It grows in haemolymph of scarabeids. It does not grow in Nutrient Agar or in Nutrient broth medium.

(iii) Serratia marscescens : Gram negative, rod shaped, motile, aerobic, and facultative. It grows on Nutrient Agar. The colony of this bacteria sometimes appear red.

(iv) Pseudomonas aeruginosa : Gram negative, rod shaped, motile, aerobic, egg yolk reaction positive, and the colony of this bacterium appear blue green in colour.

(v) Streptococcus faecalis : Gram positive, round or coccus in shape, non motile and aerobic. It grows on Nutrient Agar. Colonies of this bacterium appear round, convex, translucent and colourless.

Fungi

Soon after collection of infected hosts these are cut into pieces and each piece is placed in seperate Potato Dextrose Agar petriplate for growth of hyphae and spores of fungi. With the help of lacteriological loop this growth is picked up and put for staining in bactophenol blue. Fungi are identified on the basis of structure of conidia, and also with the help of other tests such as physiological, biochemical and so on..

(i) Aschersonia taitensis : Conidia produced by phialids in slimy heads, non-septate, fusiform, conidiophores arranged in sporodochia on scale insects or white flies.

(ii) Aspergillus flavus : Conidia in dry, long chains, conidiophores consisting of an unbranched stripe terminating in a vesicle bearing conidiogenous cells and/or nictulae, arising singly from insect body.

(iii) Beauveria bassiana : Conidia produced by sympodial conidiogenous cells, conidiogenous cells in clustures or singly, conidiophores without swollen vesicles, conidiogenous cells with a swoolen basal part terminating in rachis.

(iv) Entomophthora virulenta : Primary conidia discharged, formed on host's exterior or with external access through a ventrally open cavity. Primary conidia variously shaped, uni- to multinucleate but not in a hymenium lining a ventrally open cavity. Primary conidia variously shaped, and number of nuclei, primary conidiophores simple or branched, zygospores (or azygospores) with hyaline to coloured to very dark outer wall layer, never formed within the layer of two gametangia, villose spores and microdonidia absent, most species not cultured on Potato Dextrose Agar.

(v) Fusarium roseum : Conidia produced by phialides in slimy heads, one or more-septate, usually curved.

(vi) Hirsutella entomophila : Conidia produced in slimy heads, non-septate, conidiophores not arranged in sporodochia, phialides mostly single with a swoolen basal part abruptly tapering to a thin long neck, conidia single or few and covered by a slime sheath.

(vii) Isaria purchasi : Conidia in dry long chains, conidiophores without vesicle. Phialides differently arranged. Conidiophores loosely arranged in synnamata, conidia in long divergent chains. Phialides with distinct neck, conidiophores are irregular or vertically branched elements.

(viii) Metarrhizium anisopliae : Conidia in dry, long chains, conidiophores without vesicle. Phialides differently arranged, Conidiophores closely packed in sporodochial structure, conidia in colums.

(ix) Nomuraea releyi : Conidia in dry, long chains, conidiophores without vesicles. Phialides differently arranged conidiophores loosely arranged in synnamata or single, conidia in long divergent chains. Phialides very short nacked, conidiophores bearing dense whorls of branches and phialides.

(x) Paecilomyces farinosus : Conidia in dry, long chains, conidiophores without vesicle, phialides differently arranged, condidiophores loosely arranged in synnamata

or single, conidia in long divergent chains, Phialides with distinct neck, irregular or vertically branched.

(xi) Verticillium lecanii : Conidia produced by phialides in slimy heads, non-septate, conidiophores not arranged in sporodochia, phialides awl-shaped and in whorls or vertically branched conidiophores.

Protozoa

Diagnosis of protozoan disease on the basis of symptomology is rather difficult affair, as insects dying of this disease show very general symptoms. For identification of protozoan pathogens the diseased host is crushed, body fluid of mascerated tissues are stained and the infected stage of protozoa are observed and subjected to histopathologicae, biochemical and serological tests.

Not many protozoa hve shown remarkable success in the control of, insect pests. However, two protozoa have shown some degree of success. Characters of these are given below :

Nosema bombysis : This belongs to class microsporidia. It produces spore singly and posses sporont diporous with nuclei in diplokaryotic arrangement.

Adelina tribolii : This belongs to sub-class coccidia. This is parasite in fat body and gut wall of insect. Its spore is spherical to sub-oval, macro- and microgamete present, cysts with 10 or more spores, each with two spdronts.

Virus

Insects suffering from virus disease show characterstic symptoms. These are : change in body colour, body rupture and oozing out of infected haemolymph and hanging-on posture on porlegs of dead larve. For identification, infected tissues are smeared, stained and observed under microscope. Other tests such as histological, histochemical, biochemical and serological are also carried out.

Nuclear polyhedrosis virus (NPV) : It has shown much promise in insect control. Identifying characters are : Virions are rod-shaped, comprising a lipoprotein envelop around a protein capsid containing DNA-protein core, inclusion bodies contain major polypeptides mol. wt. 30,000 dalton, virions contain 11-25 polypeptides, viral genome is a single molecule of double stranded circular supercoiled DNA with mol. wt. 50-100 $\times 10^6$ dalton. Replication of virus occur in nuclei of host cells.

Nematode

These are comparatively well developed organisms. Two nematodes have shown much promise. Characters of these are given below :

Mermis subnigrescens : This is long, thin, thread like and think nemas. More than 100 times longer than thick. Musculature degenerated in whole oesophagous, only a thin oesophageal duct. Adult in coiled state as well as larvae of nematode occur in insect body.

Steinernema carpocapsae : This is popularly known as DD-136 nematode. Nemas less than 100 times longer than thick, both ends tapering, oral opening with three lips, short stroma, no stylet, oesophagus with corpus, isthmus and bulb well musculatured, stroma variable, shallow, bulb with reduced valve, giant female in first generation, bacteria kill host, mature nemas or larvae found in insect.

Production

Microbes produce some metabolites in the growth medium that on testing prove lethal to population of insect hosts. These substances are known as "Toxins" The toxins produced by *Bacillus thuringiensis* and entomogenous fungi are given as follow :-

(A) Bacillus thuringiensis : It produces (i) Endotoxin, also known as crystal, or parasporal body, (ii) Thermostable exotoxin (iii) Thermolabile exotoxin (iv) Enzyme lecithinase (v) Enzyme protease and (vi) Spore. (B) ***Fungi*** : (i) *Aspergillus :* It produces, Aflatoxin, Kojic acid, Aspochriasin, and Nigragillin (ii) *Penicillium* : produces, Citinin, Rubratoxin, A, Patulin, and Ochrasin (iii) *Fusarium* : produces, T-2 toxin, Fusarubin, Fursaric acid, Javanicin, Fusario toxin, and Trichothecenes. and (iv) *Metarrhzium* : produces, Dextruxin-A and Dextruxin-B (v) *Verticillium* : produces, Bassianolide.

Table 1. Commercial products of insect pathogen.

Product	Pathogen	Host
BACTERIA		
Agritol	*Bacillus thuringiensis*	Lepidopterous larvae
Agritol		
Bactobacterium		
Bacilase		
Bactospeine		
Baktukal		
Bakthane		
Bathurin		
Bioasp		
Biolep		
Biospore		
Biotrol		
Dendrobacillin		
Dipel		
Parasporeine		
Plantiab		
Sporeine		
Thuricide		
Doom	*Bacillus popillae*	*Popillia japonica*
Japademic		
Milky spore		
VIRUS		
Gypchek	Nuclear polyhedrosis	Gypsy moth

Lecont virus	virus	Pine sawlly
minosarnuo virus	NPV	-do-
Neocheck-S		Sawlly
SAN-404		Allalla looper
TM Biocontrol		Tussack moth
Virion		Gypsy moth
VPN-80		*Autographa calliornica*
VPN-82		*Spodoptera exigua*
FUNGI		
Boverin	*Beauveria bassiana*	Leptinotarsa
Collego	*Colletrotrichum gleosporioides*	*Aeschynomena*
Devine	*Phytophthora palmivora*	*Morrenia odorata*
Mycolal	*Verticillium lecanii*	Aphids
Mycar	*hirsutella thompsoni*	*Phyllocopterate* sp.
Vetalec	*Verticillium lecanii*	Aphid

These toxins have been exploited for mass-production on commercial scale.

Most of the pathogens have been produced on commercial scale and these are available in the market as microbial pesticides with various trade names. These are given in Table 1.

Microbial pesticides are available in various formulations. These can be used as per the need and according to circumstances. For example, against lepidopterous posts, acquous suspension, liquid concentrate and wettable powder can be used. To check the population of borers, like top shoot borer of sugarcane, granules may be of good use. Wettable powder may be a good choice for house hold pests.

Name of formulations of microbial pesticide, name of pathogens and commercial trade names are given in table 2.

Unlike chemical pesticides, resistance of insects to microbial pesticides is not a common occurence. It was thought for quite long that insects can not develop resistance to biopesticides, as these are the biological products. However, some reports on pest developing resistance have now appeared. These are mentioned below with a note of caution that such important factors must remain in the minds of the user who can do better planning so as to achieve maximum insect control.

Insect pathogens, now a days, are being mass produced on commercial scale. They are available in the market as microbial pesticides under field conditions. Results of some of these are given in table 4.

With the advent of biotechnology considerable attention has been focussed on construction of genetically engineered microorganisms having superior qualities of attributes of insect pest control. Development of transgenic plants and production of insecticides based on hormones and nucleopeptides have become possible now.

Table 2. Formulations of biopesticides.

Formulation	Product	Pathogen
Aqueous suspension	Dipel 2 ×	*B.t.*var. *kurstaki*
	Thuricide(HPC)	
Emulsifiable suspension	Dipel 4L	
	Dipel 8L	
Flowable concentrate	Foray Cobit	*B.t.*var *tenebrionis*
	Novodor	*B.tk.*
Granules	Dipel 10 G	
Liquid concentrate	Sok bt	*B.t.k.*
Wettable powder	Baldimos	*B.t. israelensis*
	Elcar memestrin	NPV
	Dipel 2 ×	*B.t.k.*
	Thuricide HP	*B.t.k.*
Ready to use Suspension	Doom	*Bacillus popillae*
	Agrovir	
	Madex	GV
	Granupone	

Table 3. Pest resistance to microbial pesticide.

Pest	Pathogen	Resistance
Plodia interpunctella	*B.t.k.*	27 fold resistance
Culex sp.	*B.t.isrealensis*	70 fold
Heliothis virescens	*B.t.k.*	30 fold
Plutella xylostella	*B.t.k.*	36 fold
Pieris brassicae	GV	Showing resistance
Lymentria dispar	NPV	-do-

The notable examples are :

Baculoviruses are potent killer of insects. They are host specific and also possess some good qualities that help make it best tool of insect control. One major disadvantage in it is that there exists a considerable time lag between the time of application and the cessation of feeding. One approach, to overcome this, has been to insert a gene into the genome of *Autographa californica* NPV (AcNPN) that encodes juvenile hormone enzyme (JHE). JHE initiates metamorphosis and causes

Table 4. Field trials with microbial pesticides.

Formulations	Crop	Insects
BACTERIA		
B. thuringiensis	Castor	*Achoea*
B. t.	Cauliflower	*Pieris, Plusia, Adisura, Spodoptera, Euproctis, Anomis,*
Bactospeine	Lablab	*Plutella*
Biotrol	Cabbage	*Spodoptera*
Dipel	Bengal gram	*Heliothis*
	Bhindi	*Earias*
	G'nut	*Amsacta*
	Crucifers	*Plutella*
Thuricide	Rice	*Scirpophaga, Sesamia*
	S'cane	*Bissetia, Sesamia*
	Knolkhol	*Plutella*
B.popillae		
Doom	Bajra	White grub
FUNGI		
Beauveria	Banana	Banana pests
		Leptinotarsa
Entomophthora	Brassica	*Myzus, Lipaphis Threoaphis Rophalosiphum*
Fusarium	Coffee	*Coccus viridis*
	Brinjal	Mealy bug
Metarrhizium	Sugarcane	*Pyrilla*
Verticillium	Coffee	*Coccus viridis*
NEMATODE		
Steinernema	Maize	*Spodoptera*
	Tobacco	*Heliothis*
	Turnip	*Agrotis*
	Potato	*Leptinotarsa*
	Rice	*Popillia japonica*
	Forest tree	*Dondroctenus Curculio*
Neoplectana	Lemon	*Oemona hirta*
carpocapsae	Corn	*Heliothis zea*
	Cabbage	*Pieris brassicae*
PROTOZOA		
Nosema melolonthae		*Melolontha melolontha*
Nosema locustae		*Melanoplus glasnoti*
Thelohania hyphantriae		*Hyphantrea cunea*
Adelina triboli	Stored product	Stored grain pest
Vairimorpha nectatrix		*Trichoplusia ni*

cessation of feeding in insects. Other approach, for example, in the AcNPV is to delete the "Egt" gene that encodes the enzyme ecdysteroid UDP-glycosyl transferase. Expression of "egt" prevents moulting and causes prolonged feeding in insect. Once it is deleted, virus becomes more potent. Insertion of *B.t.* toxin gene is another possibility. A polyhedrin-positive construct increaes the LD_{50} value and when complete insecticidal crystal protein of *Bt.* is engineered the altered virus inhibited the feeding of *Pieris brassicae.*

Quality of the fungal pathogen can be improved by protoplast fusion technique. In one example, autotrophic mutants of *Beauveria bassiana, B. brongniatrii* and *B. amorpha* were treated with xymolase or with novozyme to achieve fusion.

Quality of *B.thuringiensis* has been improved by several ways. In japan one variety of *B.thuringinsis* has been engineered that is effective against the cutworm but harmless to silkworm.

Genetic engineering methods have been applied in production of "Transgenic plants". *B. thuringiensis* toxin gene or genes have been inserted into the plants so that plants themselves can manufacture such toxins. A gene, which is responsible for the production of crystal toxin, a glycoprotein, has been isolated and partially characterised. This gene can be engineered into plants through *Agrobacterium* mediated Ti-plasmid system. In this way transgenic plants of tobacco has been engineered, which cause 75-100% mortality in *Manduca sexta* and also cause reduction of feeding and indreased mortality in Insect species such as *Heliothis zea, H. virescens* and *Spodoptera exigua.* Many more transgenic plant are in the process of engineering, they are : tomato, cotton, rapeseed, rice, potato, apple vegetables and walnut.

Insect pathology has made tremendous development in the recent past. Judging from the historical records, one finds that insect pathology started with the stage of isolation of microorganisms from soil, water, phyllophane and insects and testing their pathogenicity against insects to development of technology for mass production and finally bringing industrially produced insect pathogens to a stage of commercialisation. On the literature front, first phse started when all available literatures on microorganisms related to insects, found widely scattered, were twined in the form of a book entiled "Microbiology of Insect" (Steinhaus, 1949). The real impetus in the subject occured when definition, history, and other important aspects of insect pathogens were wisely defined and elaborately explained. Other books, of course appeared later (see references). On the research front, advances have been made on identification of pathogens, possible uses of bacteria, virus, fungi, protozoa and nematodes, in control of insects, study of toxins of pathogens, determination of host sprectum of pathogens, possibilities of resistance of insects to pathogens, safety testing, development of technology of mass production of pathogens, standerdisation, formulation of product, improvement of strains of pathogens through biotechnology, development of genetically engineered transgenic plants and economics of microbial control. Mention of recent development in all aspects of insect pathology is not possible in the manuscript bound by limitation. Therefore, advance made only in the selected areas that has direct bearing in control of insects such as knowledge on possible potent pathogens, symptoms caused by pathogens, characters of identification of pathogens, availability of commercial product of microbial pesticides, formulations, information on improved strains of pathogens and on transgenic plants have been incorporated which it seems prove its worth not only to insect pathologists with entomological background but also to other readers interested in the subject.

References

Burges, H.D. 1981. Microbial control of Pests and Plant diseases 1970-80, Academic Press Inc. (London) pp. 860.

Burges, H.D. and Hussey, N.W. 1971. Microbial control of insects and mites. Academic Press London. pp. 860.

Jayaraj, S. 1985. Microbial control and Pest Management. Center for plant protection studies. Tamil Nadu Agricultural University, Coimbatore, India. pp. 278.

Poinar, G.O. Jr. and Thomas, G.M. 1978. Diagnostic manual for the identification of insect pathogens. Plenum Press, New York. pp. 218.

Ramakrishnan, N. and Kumar, S. 1977. Biological control of insects by pathogens and nematodes. *Pesticide Annual.* pp. 32-47.

Steinhaus, E.A. 1949. Insect Microbiology. Comstock Publ. Co. Ithaca, New York. pp. 763.

Steinhaus, E.A. 1949. Principles of Insect pathology. MeGraw Hill Book Co. Inc. pp. 757.

Steinhaus, E.A. 1963. Insect Pathology - an advanced treatise. Vol.2 Academic Press London. pp. 689.

Srivastava, K.L., Chaudhari, S. and Gitanjali Jayachandran. 1993. Genetic engineering of microbial agents for pest control. In: Integrated Pest Management. Edited by Dwivedi, B.K. pp. 202-206.

Srivastava, K.L. and Ramakrishnan, N. 1982. Microbial control of crop pests. In: Agricultural Entomology. Vol. I. Edited by All India Scientific Writer's Society. New Delhi. pp. 117-138.

Recent Advances in Indian Entomology
Editor: O.P. Lal
APC Publications Pvt. Ltd., New Delhi, 1996, pp. 384-387

A New Approach in Insecticidal Formulations

V.S. SAXENA

Chemical insecticides provided a reasonable and effective answer for insect control until resistance to these compounds began to develop. The short term remedy was to increase dosage rates, often to dangerous limits and when this failed the only resource was to employ new formulations or new compounds.

The potential advantage of synergism to encounter resistance was recognised with the characterisation of enzymes responsible to insecticide resistance (Sternburg 1954, Morello 1964). Discovery of compounds like 1,1, bis (p-chlorophenyl) methyl carbinol (DMC) and N, N, dibutyl-p-chlorobenzene sulfonamide, known as WARF antiresistant, as an effective inhibitors of DDT resistance mechanism (Moorefield and Kearns 1955; Fales and Bodestine 1961) fascinated researchers with a hope to avert worldwide DDT resistance, further aggrevated the problem more dangerously as the insects when selected over their combinations with DDT developed resistance greater than DDT alone Wilkinson, 1968).

India is the fourth biggest consumer of pesticides for its various public health and plant protection programmes. Among the technical grade pesticides manufactured, hexachlorocyclohexane or benzene hexachloride (HCH or BHC) always occupied the maximum share which contains only 13% lindane or α-BHC responsible for insecticidal properties. Therefore in a programme for alternate uses of lindane by products it was observed that α-isomer which constitutes 60-70% of HCH technical grade when suitably blended with DDT augmented its bioefficacy manyfold against various resistant pests of agricultural and medical importance and discounted the earlier observations of Moorfield and Kearns (1955), and Wilkinson (1968) with potent synergists like DMC and WARF anti-resistant. Further detailed study led to the discovery of first viable Antiresistant insecticide formulation 'Power Pack' DDT 25 WP from this laboratory and finally patented Indian Patent No. 153987 (Saxena and Tomar 1982).

Antiresistant Power Pack DDT 25 WP, thus, prepared incorporating α-HCH as formulation adjuvant proved to be significantly effective against *Poecilocerus pictus* (Fab.), *Schistocerca gregaria* (Forsk), *Galerucilla birmanica* (Jacoby), *Tribolium castaneum* (Herbst), grubs of *Trogoderma granarium* (Evert), *Culex quinquefasciatus* Gilas; *Aedes aegypti* (Linn), *Anopheles stephensi* (Liston), *Xenopsylla cheopis* (Roths), and *Musca domestica nibalo vicina*, all highly resistant to DDT. (Saxena and Tomar 1983; Sexena et.al. 1984, 1986).

Apart of encountering DDT resistance, counteraction of cross resistance with improved DDT formulation was also evident. Dieldrin and Lindane resistant strains of *T. castaneum* were 5.2 and 8.47 times as susceptible as DDT resistant strains to antiresistant Power pack DDT 25 WP (Saxena 1983, 1989). Control of *P. pictus* and *S. gregaria with* antiresistant Power pack DDT 1% dust @ 20 kg/ha is of utmost significance being naturally tolerant to DDT. A deposit of 33.3 mg/insect *Anopheles aulicifacies* gave 66.68% mortality comparable to BHC 10% dust containing 1.3% lindane.

Similarly deposit of 100 mg (a.i.)/sq.mt. gave effective protection from houseflies and mosquitoes for minimum ten weeks. Houseflies insensitive to 6000 mg p'p'DDT/ sq.mt. were equally susciptible to improved DDT formulation. Only 2.5% healthy adults could emerge when selected on antiresistant Power Pack DDT 25 WP as compared to 24.19% in the case of DDT selection pressure. Likewise α-HCH exhibited potentiation in the case of pyrethrins and carbaryl (Tomar and Saxena 1986, 1990).

Metabolic studies *in vivo* confirmed DDE as main metabolite in DDT resistant *T. Castaneum* when treated with p'p'-DDT. Initially 0.0838 µg DDE (0.951%) recovered from whole body homogenate reduced to 0.121% in DDT + PBO treatment. Percentage of DDE reduced to maximum i.e. 0.0109 µg (0.09%) in whole body homogenate of DDT resistant *T. Castaneum* treated with the combination of DDT + α-HCH (Saxena and Mukerjee 1989). Such synergistic effect was not evident towards mammals. Augmented toxicity of DDT in antiresistant Power Pack DDT 25 WP formulation against different resistant insect and counteraction of dieldrin and lindane resistance may be due to inhibition of wider group of gluatathion? dependent enzymes (Goodchild 1970) by α-HCH which penitrates faster than other HCH isomers (Kurihara 1974).

Acute toxicity data for antiresistant Power Pack DDT 25 WP (LD 50 : oral; 1223.2 mg/kg (male) and 1122.0 mg/kg(female) ; dermal, 10000 mg/kg body weight (male) and (female) on *Rattus rattus* profounded its safe use in public health programmes and locust control operations (Saxena and Tomar, 1982; Saxena et al., 1984).

During the course of study in the programme of recycling of lindane by-product, α-HCH was also recorded as potent insect growth regulator (IGR) to different nymphal instars of *S. gregaria* when administered topically or through food or injected into the haemolymph. Individuals of different instars fed on cabbage leaves treated with 0.1 and 0.3% α-HCH ingested 124 to 450 µg α-HCH in the case of 3rd and 4th instars whereas the 5th instars hoppers consumed 375 to 1125 µg - HCH leading to 52% deformed adults. Similarly 1 µg α-HCH/5 µl alcohol when injected into the haemolymph of 4th and 5th instar nymphs caused 20 and 33.3% morphogenetic effects in respective instars with a poor percentage of normal healthy adult emergence. Interestingly the substitution of cholesterol to enhance the intake of plant sterols resulted in dose dependant reduction in deformities and suggested interference by α-HCH in the biosynthesis of moulting hormone responsible for ecdysis. The abnormalities noticed were (i) incomplete ecdysis, (ii) damaged or underdeveloped wings and (iii) crippled legs. Apart from developmental abnormalities, prolongation of nymphal period and reduced food intake was also evident in different treatments (Saxena and Rao 1990). This was the first report on α-HCH as insect growth regulator. Similar observations were also recorded on *Spodoptera litura, Diacrisia obliqua*, and *Dysdercus koenigii* (Saxena 1993). Among the survived female adults antigoindotropism resounded the post ecdysal disturbance in ecdysteroid titre of ovaries contemplating vitelogenesis (Hadegorn *et.al.* 1975 and Gande, *et.al.* (1979) in their successive delayed progenies (Saxena 1994).

At last antiresistant DDT formulations are helpful to encounter resistance in various pest control operations and restricts the residual population to multiply.

References

Fales H. and Bodenstein O.F. (1961). Promosing synergists for DDT. *Soap & Chem. Spec.* **37** : 77.

Gande A.R., Morgan E.D. and Wilson I.D. (1979). Ecdysteroid levels throughout the life cycle of desert locust. *J. Insect Physiol.* **25** : 669.

Goodchild B and Smith J.N. (1970). The separation of multiple forms of housefly. 111-Trichloro 2,2, bis (p-chlorophenyl) ethane (DDT) dehydrochlorinase from Glutathion S-aryl transferase by electrofocusing and electrophoresis. *Biochem. J.* **117** : 1005.

Hagedorn H. H; O'conoor J.D. Fuchs M.S., Sage B., Schlaoger D.A. and Bohan N.K. (1975). The ovary a source of β-ecdysone in adult mosquito *Proc. Nat. Acad. Sci.* USA *72* : 3255.

Kurihara N., Uchida M., Fugita T and Nakajima E. (1974). Studies on BHC isomers an related compounds. VI. Penetration and translocation of BHC isomers in cockroaches and their co-relation with physiological properties. *J. Pesti. Biocem. Physiol.* **4** : 12.

Moorfield H.H. and Kearns C.W. (1955). Mechanism of action of certain synergists for DDT against DDT resistant houseflies. *J. econ. Entomol.* **48** : 403.

Morello A. (1964). Role of DDT hydroxilation in the resistance. *Nature* (London) **203** : 705.

Saxena V.S. and Tomar S.S. (1982). Indian Patent No. 153987 Fel. 18. 1982 (To the Secretary, Indian Council of Agricultural Research, New Delhi).

Saxena V.S. (1983). A new DDT formulation to counteract lindane resistance in *Tribolium castaneum* Herbst. *J. ent. Res.* **7** : 60.

Saxena V.S. and Tomar S.S. (1943). A novel DDT formulation I. Control of insects resistant to DDT. *Indian J. Ent.* **5** : 49.

Saxena V.S., Pillai M.K.K., Sharma V.P., Mukerjee S.K. and Tomar S.S. (1984). A novel DDT formulation II. Control of insect vectors resistant to DDT *Indian J. Ent.* **46** : 89.

Saxena V.S., Bhanotar R.K. and Srivastava Y.N. (1986). Improved DDT formulation for the control of desert locut and grasshoppers. *Pesticides* **8** : 22.

Saxena V.S. (1989). Counteraction of dieldrin resistance in *Tribolium castaneum* Herbst. with new DDT formulation. *J. Insect. Sci.* **2** : 53.

Saxena V.S. and Mukerjee S.K. (1989). Effect of α-hexachlorocyclohexane on the metabolism of DDT in DDT resistant Tribolium castaneum Herbst. *Pesticide Research Journal* **1** : 39.

Saxena V.S. and Rao P.J. (1990). α-hexachlorocyclohexane : a potent insect growth regulator. *Curr. Sci.* **59** : 1250.

Saxena V.S. (1993). Antagonistic effect of α-hexachlorocyclohexane on endocrine control metamorphosis in insects. *Pesticide Research Journal* **5** : 54.

Saxena V.S. (1994). α-hexachlorocyclohexane induced female susceptibility in *Schistocerca gregaria Pesticide Research Journal* **6** : 92.

Sterhburg J., Kearns C.W. and Moorfield H.H. (1954). DDT-ase an enzyme found in DDT resistant houseflies. *J. agric. Fd. Chem.* **2** : 1125.

Summerfood W.T. (1954). Review of synergism among halogen containing synergists. *J. agric. Fd. Chem.* **2** : 310.

Tomar S.S. and Saxena V.S. (1986). Synergistic effect of α-hexachlorocyclohexane with pyrethrins. *J. agric. Biol. Chem.* **50** : 2115.

Tomar S.S. and Saxena V.S. (1990) Synergism of α-hexachlorocyclohexane with N-methyl carbamate insecticides. *Indian J. Ent.* **52** : 618.

Wilkinson C.F. (1968). The role of insecticide synergists in resistance problem. *World. Rev. Pest Control* **7** : 155.

Recent Advances in Indian Entomology
Editor: O.P. Lal
APC Publications Pvt. Ltd., New Delhi, 1996, pp. 388-392

Principles of Bioassay Techniques

D.S. SINGH

Bioassay which is considered a newer development in the field of insecticides, has been used for more than 70 years in the field of pharmacology where this technique was used for assaying the potency of different drugs. It is now a recognised method for the quantitative studies of certain reaction of various substance on living organism. Bioassay in wider sense means that measurement of potency of any stimulus which may be physical, chemical physiological, biological or psychological by means of the reaction that it produces in the living matter.

A large number of literature is available on technique of bioassay and various factors affecting the result of bioassay (Busvine, J.R. 1957, Finney, 1964. There is a high degree of correlation between the estimation made by bioassay and chemical assay techniques. (Kavadia and Rattan Lal, 1967, 1968, Krishanaiah 1970; Hameed and Rattan Lal 1971). As compared to other conventional chemical techniques, excluding modern G.L.C., the sensitivity of the bioassay technique has been often reported not only to compare favourably but sometime even more sensitive than chemical method.

The results of bioassay technique is used for a variety of purposes and catogorised as follows :

A. Screening of potential insecticides.
B. Relative toxicity of pesticides.
C. Fundamental studies on the acion of insecticides.
D. Quality testing of formulated insecticides.
E. Insecticide resistance in insect.
F. Determination of PT, RT, LT, LC and LD value.
G. Estimation of insecticide residues.

A. Screening of potential insecticides

A number of laboratories and private concerns are engaged in the development of new chemicals which are to be biologically tested to determine their insecticidal or repellent properties. These tests should be done on more than one insect, because of insect specificity to insecticides and different modes of insecticidal entry in insects. Chemicals showing some toxicity are to be tested in greater details.

B. Relative toxicity of pesticides

With the discovery of DDT a large number of synthetic insecticide has been introduced. It is necessary to test these products, before recommending to carry out large scale field trails. Bioassay technique has been successfully adopted to test the relative toxicity of pesticides and their proprietery product for the proper selection of toxicant to be tried against a particular pest in field. Thus, the values of relative toxicity is a ready reckoner for the proper selection of pesticide and their formulations for pest control, and also provide the basic data for the detection of resistance in insects.

C. Fundamental studies on the action of insecticide

It is imperative to carry out bioassay test against suitable test insect for fundamental studies on the action of insecticides to correlate the chemical constitution with the toxicity of chemicals and to determine the mode of action.

D. Quality testing of formulated insecticides

For checking the quality of marketed insecticides, the samples from shops and consumers level are tested for their biological performance in laboratory.

E. Insecticide resistance in insect

One of the most important use of bioassay has been the detection of resistance in insects and determinations of its intensity. The changes in LC50/LD50 of an insecticide for an insect pest with the passage of time indicates variation in its susceptibility. Changes in resistance after exposure causes, cross resistance to insecticides, effect of synergists in over coming resistance, the number of inheritable factors concerned etc. are all best studied by means of a reliable bioassay method.

F. Determination of PT, RT, LT, LC and LD values

with the help of bioassay technique, PT, RT, LT, LC and LD values of different pesticide can be calculated.

G. Estimation of insecticide residues

The synthetic insecticides are toxic to insects as well as higher animals and human beings. Therefore, it is necessory to see that no dangerous residues are left on food and fodder which may prove hazardous to the health of man and animals. Estimation of insecticides residues on crop plants by bioassay is an established method. The bioassay is equally and often more sensetive than chemical assay. The main advantages of the bioassay technique are that there is no need of any clean-up procedure for extractives of insecticide treated plant parts and it does not require sophisticated costly equipment, which is usually needed for chemical estimation.

Standardization of bioassay technique

The primary requirement of bioassay test is the standardization of method to an extent to get consistent and reproducible results.

For this we have to first consider the source of variations. It has been shown by various workers that there are various sources which affect the result of bioassay. These variations affecting bioassay may be broadly classified under following headings. Various factors affecting the bioassay results are :

(i) Equipment for application of insecticides and their formulations.

(ii) Formulation of insecticides.

(iii) Test insect.

(iv) Test conditions.

(v) Choice of concentration and time of observations.

Equipment for application of insecticides and their formulation

This aspect of bioassay has been comprehensively discussed in the book entitled "A critical review of the technique for the testing insecticide" Busvine J.R. 1957, Commonwealth Institute, London. Selection of method depends on the mode of entry and type of pesticide formulation. It is also varied whether insecticide is applied in the form of dust or spray or gas. A critical review of literature indicates a gradual evolution in the laboratory equipment for application of insecticides.

Dusting apparatus : The precision dusting appliances are based on the same principle *viz.*, blowing of dust in close space and then allowing the dust to settle on the surface required to be treated.

Spraying apparatus

The most common and older type of sprayer was Peet and Grady chamber used for testing the flying insect. An older type of popular sprayer is spray tower. This type of sprayer was devised as early as 1942 and undergone numerous modification and improvements brought by various workers. The Thrid type of sprayer is that in which spray material goes first up in the tower and then allowed to settle on the object to be sprayed.

Other type of applicator

A. 1. Injecting individual insects with various type of graduated capillary tubes, hypodermic syringes etc.

2. Topical application on individual insects for carrying out the fundamental studies.

B. *Immersion and dipping technique* : This method was not so popular. It is used for aquatic insect.

Formulation of insecticides : Emulsifier, diluents, amount and quality of solvent, particle size of the carrier etc. affect the toxicity of formulation.

Test insect : The test insect or organism is most important variable because a number of internal and external factors affect the susceptibility of the test organism. Some of the important internal factors are specificity, stage and age, sex and weight and the important external factors are rearing condition viz., temperature, humidity and nutrition.

Specificity : The response of different species to various insecticides is variable, for example, DDT the leader of modern insecticide is known to be practically ineffective against desert locust and this insecticide would not have been discovered if tested against this pest. Differences in susceplibility may exist in insect belonging to same group. A marked specificity was observed in their susceptibility to insecticides against *Cylas formicarious* and *Myllocerus undecimpustulatus* belonging to same family. A similar specificity has been observed in two insects belonging to same genus *viz;* *Aphis craccivora* and *Aphis gossypii.*

Stage and Sex : The response of toxicity of insecticides varies considerably with age and stage of insect. It is common experience that eggs are most resistant to most of the common insecticides. The larvae of *Trogoderma granarium* is a serious pest and very difficult to kill during its larval stage than adult. Similar results are also obtained with successive instar of the locust against BHC. Several factors are responsible for this variation but only three are main factors *viz;* (1) Surface area per unit body wt. (2) percentage of wax in the exuviae and (3) The probable thickness of the exuviae wax. The first two being negative correlation with LC50 and thrid showed positive correction.

Sex : Generally male insects are reported to be more susceptible to insecticides than female. But in certain cases females are reported to be more susceptible. In the case of *Schistocerca gregaria* and *locusta migratoria,* male and females are more or less equal susceptible to dinitro-orthocresl.

Size and weight : It has been reported that resistance of *Euproctis lunata* DDT as stomach poison increases with the increase in body weight. It has been also suggested that more or less constant amount of DDT per gram body wt when ingested begins to have an inhibitory effect on further feeding. Therefore the test insect which is used in bioassay test should be of the same size and weight.

External factors

Rearing conditon A. Temperature : Development of any insect, its metabolism activity and consequently its susceptibility to insecticides are greatly affected by temperature and humidity under which it is reared.

B. Humidity : Humidity appears to have lesser effect on the susceptibility of insect than temperature. A gradual decrease is found in the mortality of *Tribolium confusum* with increase in humidity. In case of desert locust LC_{50} which is obtained at higher humidity is slightly lower than what is obtained at lower humidity however, these values are insignificant.

Nutrition

The host plant or food on which the insects are reared also influence the susceptibility. *Spodoptera litura* has different LC50 value for a toxicant when reared on different food. Susceptibility also changes with starvation. *T. confusum* susceptibility increases with starvation to Ccl_4 while it decreases in the case of *Euproctis lunata* as stomach poison.

Conclusion

The affect of insecticides & their formulations on biological performance of insects can be precisely studied by standardised technique of bioassay, which is cheap and simple.

References

1. Busvine. J.R. 1957. A critical review of the techniques for testing insecticides commonwealth Institute of Entomology. London 208 pp.

2. Finney. D.J. 1964. Probit analysis University Press Cambridge pp. 1-318. Bentley House 200 Euston Road, London - N.W. 1.

3. Hameed. S.F. and Rattan Lal 1971. Residues of malathian on cabbage cauliflower and knol-khol eximated by bioassay and chemical - assay methods. *Indian J. Ent*; **33** : 326-37.

4. Kavadia. V. and Rattan Lal 1967. Dissipation of insecticide residue from crop of Indian rape-seed 1. Treatment with BHC WP and lindane E.C. *Indian J. agric. sci.* **37** (5) : 419-27.

5. Kavadia. V. and Rattan Lal 1968. Residues of malathian on Indian rape-seed, *Brasica Campestris* var. *dicotoma* watt. *Indian J. agric.* **38** (4) : 708-13.

6. Krishnaiah, K. 1970. Persistance of dimethoate in some Cruciferous crop (Cauliflower, Cabbage and 'Sarson'). Thesis for the degree of Ph.d. in Entomology of IARI, New Delh-12. (unpublished).

7. Pradhan, S. and Jotwani, M.G. 1965. Principle of Bioassay of insecticides. "In advances in Agric. Sciences and their application". Golden Jublee Commemoration Volume of Madras Agric. J. Coimbatore pp. 636-647.